The Great Sperm Whale

THE
GREAT
SPERM
WHALE

RICHARD ELLIS

*A Natural History
of the Ocean's Most
Magnificent and
Mysterious Creature*

UNIVERSITY PRESS OF KANSAS

Published by the
University Press of Kansas
(Lawrence, Kansas 66045),
which was organized by the
Kansas Board of Regents and
is operated and funded by
Emporia State University,
Fort Hays State University,
Kansas State University,
Pittsburg State University,
the University of Kansas, and
Wichita State University

Library of Congress
Cataloging-in-Publication Data
Ellis, Richard, 1938–
  The great sperm whale : a natural history of the
ocean's most magnificent and mysterious creature /
Richard Ellis.
    p. cm.
  Includes bibliographical references and index.
  ISBN 978-0-7006-1772-2 (cloth : alk. paper)
  1. Sperm whale. I. Title.
  QL737.C435E437 2011
  599.5'47—dc22

                                    2010048266

British Library Cataloguing-in-Publication Data
is available.

Printed in the United States of America
10 9 8 7 6 5 4 3 2 1

The paper used in this publication is recycled and
contains 30 percent postconsumer waste. It is acid
free and meets the minimum requirements of the
American National Standard for Permanence of
Paper for Printed Library Materials Z39.48-1992.

This book is for my grandchildren

Lochlan Davies-Guest, Stella Juniper Adams,

Felix Leo Ellis Adams, and Michael Hope

Rockwell Kent's 1930 interpretation of Moby Dick for Random House edition of Melville's classic.
(Rockwell Kent Gallery and Collection, Plattsburgh State Art Museum)

But still another inquiry remains; one often agitated by the more recondite Nantucketers. Whether owing to the almost omniscient look-outs at the mast-heads of the whale-ships, now penetrating even through Behring's straits, and into the remotest secret drawers and lockers of the world; and the thousand harpoons and lances darted along all continental coasts; the moot point is, whether Leviathan can long endure so wide a chase, and so remorseless a havoc; whether he must not at last be exterminated from the waters, and the last whale, like the last man, smoke his last pipe, and then himself evaporate in the final puff.

Comparing the humped herds of whales with the humped herds of buffalo, which, not forty years ago, overspread by tens of thousands the prairies of Illinois and Missouri, and shook their iron manes and scowled with their thunder-clotted brows upon the sites of populous river-capitals, where now the polite broker sells you land at a dollar an inch; in such a comparison an irresistible argument would seem furnished, to show that the hunted whale cannot now escape speedy extinction. . . .

Wherefore, for all these things, we account the whale immortal in his species, however perishable in his individuality. He swam the seas before the continents broke water; he once swam over the site of the Tuileries, and Windsor Castle, and the Kremlin. In Noah's flood he despised Noah's Ark; and if ever the world is to be again flooded, like the Netherlands, to kill off its rats, then the eternal whale will still survive, and rearing upon the topmost crest of the equatorial flood, spout his frothed defiance to the skies.

—Moby-Dick

Contents

PREFACE / IX

In Which the Author Explains Why and How He Wrote This Book,
Who Helped Him, and Who Didn't

I. PHYSTY / 1

We Open with the True Story of the Fortuitous Rescue of a Baby
Sperm Whale from the Beach, and His Successful Return to the Sea

II. MR. MELVILLE'S WHALE / 8

The Most Famous Whale in the Greatest American Novel, Brought
to Life Before Your Eyes by H. Melville, R. Ellis, and Others Too
Numerous to Mention; Slur (in a Manner of Speaking) of Three
Movies; In Addition to Countless Television Documentaries, Spin-
offs, Rock Bands, Operas, and Comic Books

III. WHENCE THE SPERM WHALE? / 55

From the Shadows of Deep Time Comes the Great Cachalot, Today's
Miraculous Paradigm of Adaptive Evolution

IV. THE (UN)NATURAL HISTORY OF THE SPERM WHALE / 93

In Which We Meet the Great and Wondrous Sperm Whale, Deep-
Diving Champion, Vocalist, Teuthophage, and Source of Scrimshaw
and Ambergris

V. THE SOCIAL LIVES OF SPERM WHALES / 143

Does the Animal with the Largest Brain of Any Animal That Has
Ever Lived Have a Complex Social Life? (What Do You Think?)

VI. BATTLE OF THE GIANTS / 174

Wherein the Greatest Foes of the Depths Battle to the Death
(Or Do They?)

VII. "I'll Have the Calamari . . ." / 191

The Sperm Whale's Menu Consists Almost Exclusively of Squid, but Not Necessarily Giants

VIII. Making Contact / 224

The Beginning (and the End) of the Sperm Whale Fishery, from the Depths of the Seven Seas to the Poles, the Antipodes, and Elsewhere

IX. How to Catch a Whale / 240

A Veritable Instruction Manual for the Capture and Processing of the Sperm Whale, as Scribed by Herman Melville and Contemporaries (Not to Mention the Japanese and the Soviets)

X. The War on Whales / 265

Around the World in Pursuit of Sperm Whales, Used in the Manufacture of Oil, Meat, Margarine, Nitroglycerin, and Other Valuable Products

XI. "Can Leviathan Long Endure So Wide a Chase?" / 305

The Catastrophic Intermingling of the Lives of Men and Whales, with the Unfortunately Inevitable Outcome for the Whales

Appendix. The Adventures of a Whale Painter / 315

In Which We Follow the Career of R. Ellis, Whale Painter Extraordinaire, from Denver (Denver?) to Buffalo (Buffalo?) and (Finally!) New Bedford

References / 335

Index / 359

Preface

THIS BOOK IS the natural successor to other books I have written, including several that included extensive discussions of the sperm whale. In *The Book of Whales*, first published in 1980, the chapter on *Physeter macrocephalus* is by far the longest in the book, and later, in *Men and Whales* (1991), I spoke at length about the nature of this species and the convoluted history of the sperm whale fishery. The juvenile sperm whale that we named Physty stranded on Fire Island, New York, in 1981, and while I wrote a couple of magazine articles at the time, it wasn't until 1993 that I wrote the children's book about this whale's rescue. There are very few whales—sperm or otherwise—in *Imagining Atlantis* (1998), *Tiger Bone and Rhino Horn* (2005), *Tuna: A Love Story* (2008), and *On Thin Ice: The Changing World of the Polar Bear* (2009), but every other book I've written has had sperm whales plowing powerfully through the pages. How could I ignore the sperm whale in *Monsters of the Sea* (1994)? It is the attacker of ships and the scourge of vindictive whaling captains (however ficticious)—in short, the quintessential threatening sea monster. In *Deep Atlantic* (1996), *Physeter* appears as the paradigm of deepwater cetaceans, conducting its business far from sunlight and farther still from prying human eyes. In the 1998 *Search for the Giant Squid*, the whale appears as the predator in chief of the greatest of all cephalopods, generating those (largely apocryphal) stories of the battle of the titans in the depths. And when the cachalot appears in *The Empty Ocean* (2003), it is because the great whale, hunted intensively for centuries, is so greatly reduced in numbers that innumerable whale-sized cavities have been left in its beleaguered habitat—which is as threatened as the whale itself.

This is not simply a cut-and-paste compilation of my previous writings on the sperm whale; it is an attempt to pull together all those disparate discussions and add substantial new material that I hope will flesh out the story of one of the largest, most mysterious, and historically most important animals that ever lived. (Herein you will find the

Dancer Mikhail Baryshnikov is shown here wearing a T-shirt decorated with blue whales drawn by Richard Ellis in 1974, the year Baryshnikov defected from the Soviet Union. (Animal Welfare Institute)

chronicle of the sperm whale's evolution, which, as far as I know, has heretofore not appeared in the popular literature.) Because of my fascination with the sperm whale, I have also painted its portrait many times, as posters, prints, book illustrations, easel paintings, and murals for the walls of museums. Because my exposure to sperm whales reaches back to the first mural, painted in 1978, I can say that my "sperm whale period" spans some four decades. During that time, I have perched on the shoulders of countless experts, seeking enlightenment, information, and guidance on the subject of this book, the great enigma that we know as the sperm whale.

When I began as a newly minted exhibit designer at the American Museum of Natural History in 1967, among my first assignments was the design of the blue whale model that now hangs in the Hall of Ocean Life. I worked with mammalogists Dick van Gelder and Karl Koopman, and although the model we hung was obviously not a sperm whale, it was the first whale species I ever researched, and thus can be legitimately listed as the origin of my whale-related career. This career intensified when I painted the portraits of ten great whales for the January 1975 issue of *Audubon* magazine, to accompany the article "Vanishing Giants," written by David Hill. It was authorized by Les Line, *Audubon*'s editor, who also elected to put my painting of a sperm whale on the cover of the magazine, the first painting used as a cover illustration in three decades. This painting also appeared on the cover of my *Book of Whales* in 1980. (A sperm whale painting was also featured on the jacket of *Men and Whales,* although it was not painted by me.) Bill Schevill of Harvard University and Wood's Hole looked at the "Vanishing Giants" paintings before I submitted them to the magazine, and much to my delight, he approved my interpretations—although he said that the humpbacks were "too gaily spangled."

In 1981, I was nominated to serve on the American delegation to the International Whaling Commission, and during the decade that I attended meetings in Washington, England, Argentina, Sweden, and Scotland, I encountered dozens—perhaps hundreds—of cetologists, conservationists, and even whalers, who filled in the yawning gaps in my knowledge of whales and whaling. If I were to name them all—or even if I remembered them all—the list would take up half this book. Around this time, the little sperm whale we called Physty showed up on the beach at Fire Island, not far from my home in Manhattan, and during the course of the "rescue," I worked with veterinarian Jay Hyman and with whale conservationists Bill Rossiter, Michael Sandlofer, and Sam Sadove. To determine what to do with a sperm whale in "captivity," I called my friend Ken Norris at UC Santa Cruz, and he's the one who suggested I get in the water with Physty and try to determine where the whale's popping noises were coming from. Ken Norris died in 1998, and because he encouraged me and so many others to study the miracle of the cetaceans, I will be forever grateful to him. David and Annie Doubilet took many of the photographs I eventually used in the book I wrote about this adventure in 1991.

In 1978, Dick Wehle of Buffalo, New York, commissioned a sperm whale mural for his private museum (which he began because of his name), and because I really didn't want to spend three months in Buffalo (that was the winter they got 18 feet of snow), I painted the 20-foot-long mural in my dining room. I'd like to thank my (then) wife, T.A., for tolerating a mural-painting studio in the middle of our apartment. (Our kids, Elizabeth and Timothy, thought it was very cool.) For the New Bedford Whaling Museum, director Dick Kugler commissioned the 100-foot-long Moby Dick mural that now hangs in the museum's Lagoda Room, and I painted it in a loft on Duane Street in lower Manhattan that belonged to (and was occupied by) my friend Bobby Leacock and his wife, Robin.

While working on *Men and Whales,* I visited whaling stations and historical sites in such far-flung places as South Georgia, Norway, Australia, New Zealand, Iceland, South Africa, the Azores, and Japan, and also the more accessible New Bedford, Nantucket, Vancouver, Newfoundland, California, and Mexico. In all those places, personnel and historians graciously assisted in my studies, and again, naming them all would be far too cumbersome. Besides, in *The Book of Whales* and *Men and Whales,* I listed all the people who helped me at museums, libraries, whaling stations, whaling ships, and whale-watching cruises, and many of those individuals therefore contributed to my ongoing concentration on my eponymous subject.

Disclaimers notwithstanding, a lot of material was researched and written specifically for this book, and for expertise on Melville, I'd like to thank John Bryant, Tom Inge, and Elizabeth Schultz; for recent material on the biology and history of the cachalot, I thank Trish Lavery, Hal Whitehead, Jan Straley, Phil Gingerich, Hans Thewissen, and Mark Uhen. As with *Men and Whales,* much of the historical material came from the collection of the New Bedford Whaling Museum, and I am grateful for their support and cooperation. Photographs of living sperm whales underwater were provided by Tony Wu, Doug Siefert, and Eric Chang, and I think it is the inclusion of their spectacular images that differentiates this book from any other sperm whale book—of which there are more than a few.

Mike Briggs of the University Press of Kansas once again demonstrated the requisite faith and enthusiasm to publish a book on such an enigmatic subject. (In 2003, he published my *Sea Dragons: Predators of Prehistoric Oceans;* long-extinct 60-foot-long plesiosaurs and giant dolphinlike ichthyosaurs are the very definition of arcane natural history.) After almost 40 years of writing about such esoterica—giant squid, extinct marine reptiles, the narwhal as unicorn, Steller's sea cow, the destruction of marine wildlife, and so on—I have begun to realize that books written in the interest of conservation may not have saved many animals. But because writing books is what I do (well, OK, I paint pictures, too), I am more than satisfied with the efforts I've made, if not to save animals in trouble, then at least to identify their problems—and even, in some cases, propose a solution, no matter how naive and idealistic. Will this book "save" the sperm whales? I don't know. Will it enhance our understanding of the great bluff-headed whale? I think it will. And that, of course, is the point: to provide a window onto the life and times of what I truly believe is the most fascinating animal on the planet.

From my first book, *The Book of Sharks,* where I used it as an epigraph, I have relied on Henry Beston's timeless quote to demonstrate the need for a completely different way of looking at animals, aquatic or terrestrial. (In *The Outermost House,* Beston was actually reflecting on the magical way shorebirds unified their movements as they flew over his Cape Cod beaches.) Here it is again:

> We need another and a wiser and perhaps a more mystical concept of animals. Remote from universal nature, and living by complicated artifice, man in civilization surveys the creature through

the glass of his knowledge and sees thereby a feather magnified and the whole image in distortion. We patronize them for their incompleteness, for their tragic fate of having taken form so far below ourselves. And therein we err, we greatly err. For the animal shall not be measured by man. In a world older and more complex than ours they move finished and complete, gifted with extensions of the senses we have lost or never attained, living by voices we shall never hear. They are not brethren, they are not underlings; they are other nations, caught with ourselves in the net of life and time, fellow prisoners of the splendour and travail of the earth.

The 20 years Stephanie and I have been together have been the richest and most productive in my professional and personal life. Of all the literary figures in history, Sancho Panza has to be the one who least resembles Stephanie, but I want to acknowledge her support of my lifelong quest for windmills to charge at—made more complicated because my targets were often underwater—and her unwavering support of my campaign to make the mysteries of the natural world more accessible, and my quixotic attempt to save the oceans, one species at a time.

I

Physty

THE LITTLE WHALE first appeared off the beach at Coney Island in Brooklyn, New York City. The whale was reported thrashing in the surf at about five o'clock on the afternoon of April 15, 1981. The coast guard was notified. They responded by driving the whale back offshore with a 41-foot cutter. The whale beached itself again the next morning, but not until it had managed to move some 40 miles from Coney Island to Oak Beach on Fire Island. To have moved 40 miles in 10 hours is no great accomplishment, until you take into account the actual appearance of the whale on April 16; it was so debilitated that it could not maintain an upright attitude in the water but remained floating listlessly on its side.

By an amazing stroke of luck, the whale had beached itself the second time within a mile of an empty boat basin, so instead of being allowed to die on the beach or being directed back out to sea, the whale was towed—backward—to the boat basin, and thus became the first sperm whale ever held in captivity. In the past, at least three newborn sperm whales have been kept alive briefly in various facilities: one in Bermuda in 1932, one in Boston in 1976, and one in Seattle in 1979. These neonates had become separated from their mothers whose protection and nursing they needed, so despite all the efforts made to save them, they were doomed, and none of them lived more than a few days.

During its first night in the boat basin, the whale seemed to have tested the net that had been stretched across the opening, but by the next morning—when I first saw the animal—it had become quiescent again. It was lying on its right side with just its left eye, left flipper, and left lobe of the tail visible above the surface. My first thought when I

arrived at the marina at Robert Moses State Park was, "I'm too late. The whale has died." I arrived when it was between breaths—but when the time between breaths is more than five minutes, one can easily assume that the whale is not breathing at all. Then it raised its blowhole out of the water, exhaled noisily, took in another breath, and submerged its nose again. Its tail stock had been scraped raw where the tow rope had been applied, and blood oozed into the water from its flukes, where it had cut itself while pounding its tail on the beach the night before. We saw no other injuries, and the gashes that had been reported in the newspapers were no more than the normal white ventral markings, visible on the flanks in an irregular pattern.

Although it was evident that the whale was somehow stressed—it probably would not have stranded otherwise—no one really knew what was wrong with it. There was the usual speculation about faulty navigation, parasites, and even a "suicide wish" (I was even asked if I thought the whale had beached itself in a protest against the whaling industry), but the fact remained that we had no idea why this animal had come ashore. If the whale had died and a necropsy had been performed, we would have been able to find out what was wrong with it, but the whale did not die—at least not during its visit to Fire Island—so we couldn't determine the cause or nature of its illness. Later, tests would be conducted that would indicate that the whale had pneumonia, but we could not know whether there were any other problems. All we knew was that we had a very large, very sick animal on our hands.

We measured the animal at 25 feet and estimated its weight at five tons. (The measuring was easier.) Because sperm whales' teeth do not erupt until the whales are about 10 years old, and because this whale would often open its mouth underwater and we could see that it had no teeth, we concluded that it was about five years old. It was tentatively decided that the animal was male, and he acquired the name Physty, a nickname for *Physeter*, the generic name of the sperm whale. The name was also a cetological rendering of the word *feisty*, pronounced the same way, meaning "spirited and frisky" as well as "irritable and touchy."

Divers and a marine mammal veterinarian (white wet suit) using swabs to take samples from Physty's blowhole. The little whale had pneumonia. (Anne Doubilet)

For the first few days, the animal barely moved at all under its own power. Rather, it drifted with the wind and the currents, occasionally coming into contact with a piling, at which time it would thrash feebly. This suggested to us that sperm whales are unaccustomed to things coming up on them from behind, and the thrashing reaction expressed surprise. It was usually the tail that hit the pilings. The animal mostly kept its head away from these obstacles. It was clicking regularly, although the sounds were unlike any I had ever heard before. They were more like pops than clicks, and they were usually emitted in a short, collective series. I heard the sounds compared to the turning over of a small outboard motor, and even the cracking of a great knuckle. When we eventually got into the water with the whale, we could hear—and feel—these clicks directed toward us. They were strongest immediately in front of the animal, which was lying on its side.

Ken Norris of the University of California, Santa Cruz, was at that time the foremost authority on sperm whales in America. I called him in California and told him we had a juvenile whale in captivity, and I asked him whether there was anything he wanted me to do. He asked me if I could get in the water with it, and when I said yes, he said, "See if you can figure out where the sounds come from." The next day, I went to the boat basin with my wet suit, mask, snorkel, and flippers, prepared to swim alongside Physty. The water in the boat basin was only about eight feet deep, so it wouldn't be a particularly dangerous dive, but none of us had the faintest idea of how the whale would respond to people next to it. I entered the water and swam slowly toward the whale, who was still making those cracking noises. I felt the noises more than I heard them. As I got closer, the sounds got louder, until I was directly in front of him. I reached out and put my hand on the flat of his nose. He made a "pop" so loud that it knocked my hand off his nose. I moved my hand up, so that it was higher on his nose, nearer the blowhole. The sound was weaker. I returned to the middle of the nose, and the whale popped my hand off again. It appeared that the sounds were coming from the middle of his nose, exactly where a sperm whale has a mysterious organ inside, known as the *museau du singe,* "monkey's muzzle." Inside the whale's nose, the *museau* consists of a pair of horny "lips" that may be popped apart to make a sound. People had known of the existence of the *museau du singe,* but they didn't know its function. They still don't, but if the place where the sounds originated is so close, it strongly suggests a connection between the *museau du singe* and the creation of the whale's noises.

The next day, Physty lay listlessly on his side. One of the veterinarians announced, "We've got a dying whale here. We've got to do something." We tried to get a garden hose into his mouth to serve as a stomach tube for the oral administration of more antibiotics, but we couldn't pry his jaw open. He hardly moved, even when we attempted to force a two-by-four into his tightly clenched jaws. Rubber fenders were hung down from the dock to keep him from further scraping himself on the barnacle-encrusted pilings. When I left Physty that evening, there were divers in the water with him, laying their hands on him gently, wanting to be with him during his last moments. It was Thursday, April 23, and I was sure I would not see the whale alive again.

During the next day, divers were able to jam some squid packed with chloramphenicol, an antibiotic, into the corner of the whale's mouth. Michael Sandlofer, one of the divers, later told me that his entire arm up to the elbow had been in the whale's mouth, and surely that was the first time a human has had such an experience with a living sperm whale. Surprisingly, the whale seemed to recuperate. No one really believed that the medication could work that quickly, even if we had known the correct dosage—or, for that matter, what was wrong with the whale. For whatever reason, however, Physty rallied, and by Satur-

April 1981. The author being interviewed on *The Today Show* by Tom Brokaw on the subject of Physty. In the background is the only illustration they could find of a sperm whale. (Richard Ellis)

day, April 25, he appeared to be swimming strongly. Those of us who had been discussing a necropsy changed our focus to setting him free.

When the wind dropped enough to allow Zodiacs in the water, the nets that had closed off the marina were cut, and Physty was driven out of the enclosure where he had spent the last eight days, through the Fire Island Inlet, and out into the Atlantic. The Zodiac and several coast guard vessels herded the whale for about five miles, and then they lost sight of him. We do not know whether Physty survived his return to the sea. We do know that he did not reappear on our local beaches, so we can hope that he was somehow able to find and rejoin the school he came from. I have tried not to oversentimentalize what happened, but it is difficult to avoid some emotion about the nature of the event, and about the whale itself. We had spent eight days in the company of one of nature's great mysteries. How many people had ever seen a sperm whale blow its forward-angled spout? How many had been able to see its dorsal hump, its big square head, its off-center blowhole? In the past, only those engaged in killing these mighty animals could describe their unusual features firsthand. We watched a young sperm whale— the first one most of us had ever seen—as it breathed and swam, and we heard it as it clicked and popped. Of course we were happy to see the whale released, and happier still that it did not beach itself again.

After eight days of "captivity," Physty appeared to be healthy enough to be herded out of the enclosure in the boat basin. We never saw him again. (Richard Ellis)

Perhaps a child waving good-bye to the whale said it best: "Good-bye and thank you, Physty. I hope we never see you again!"

Misinformation and misconceptions abounded. All that really happened was that the whale was rescued from a beaching, kept in captivity for eight days, medicated, fed a couple of squid, and then set free. It is unlikely that an animal so stressed that it beached itself twice in two days, then so debilitated that it could not swim upright, could be completely cured in a week. It is also true that no one involved had ever had any experience with a live-stranded sperm whale; many of the decisions were thus spontaneous and not based on precedent or experience. Although all of us wanted to see the whale survive, we tried not to have our vision clouded by sentiment. The event was an important one because we could actually observe a living sperm whale, something that had not happened before under controlled conditions. As it was, numerous opportunities were missed, including one to put a radio beacon on the whale when it was released. A tag was brought from the University of Rhode Island at Narragansett, but in the tactical and legal confusion of releasing the whale, it was never applied.*

*In a newspaper interview conducted approximately two years after Physty's release, biologist Sam Sadove, one of the participants in Physty's rescue, said he was flying over the Atlantic when he saw a group of sperm whales. One of them had the telltale marks left on Physty's tail by the nylon rope used to tow him into the boat basin. "Physty was seen alive," Sadove said. "Exactly when and where I won't answer. It was a significant amount of time after the stranding for me to be sure he recovered.

Several years later, I wrote and illustrated a children's book: *Physty: The True Story of a Young Whale's Rescue.* It was published by Running Press of Philadelphia, and its creation was a joy. It took me a day to write the text and five months to do the illustrations. I painted little Physty and his mother hunting squid; Physty threatened by a big nasty shark; Physty washing ashore at Fire Island (where—in the book, anyway— the first people to see him were my children); and Physty being driven from the boat basin after he perked up. To show that all these events actually happened, we used photographs of various aspects of the story on the back of the jacket, taken by me and by David and Annie Dou-bilet. The Physty paintings were exhibited at various museums around the country; they finally came to rest at the New York Aquarium at Coney Island, close to the spot that the little whale first came ashore.

I saw the scar on the tail flukes—and I knew it was Physty. He'll have those scars for as long as he lives. And a sperm whale can live for more than 70 years." In 1997, with Timothy Scott, Sadove wrote an article for *Marine Mammal Science* entitled "Sperm Whale Sightings in the Shallow Shelf Waters off Long Island, New York." The authors recorded several surface sightings of sperm whales southeast of Montauk Point. They hypothesized that they might be feeding on schools of the local squid species, *Loligo pealei* and *Illex illecebrosus*, in water that was no more than 180 feet deep. Scott and Sadove do not mention Physty, although he could have been one of the whales off Montauk, before or after his stranding.

II

Mr. Melville's Whale

HERMAN MELVILLE was born in New York City in 1819. At the age of 18, he shipped aboard the New Bedford whaler *Acushnet* in 1842, jumped ship in the Marquesas Islands after 18 months at sea, and, after finding temporary refuge among cannibal natives, escaped on the Australian whaler *Lucy Ann* headed for Tahiti. From there he enlisted as a seaman on the frigate *United States* and returned to Boston in 1844. He began writing novels based on his experiences, the first of which was *Typee* (1846), then *Omoo* (1847), followed by *Mardi* (1849), *Redburn* (1849), and *White-Jacket* (1850). He moved to Pittsfield, Massachusetts, in 1850, where he began the project for which he will always be remembered.

Moby-Dick, published in 1851, is a massive, mysterious novel that is generally considered to be the consummate achievement of American literature—the Great American Novel. Told by the narrator, Ishmael, it is superficially the story of Ahab, a mad whaling captain who obsessively pursues the white sperm whale that took off his leg, but even though the novel contains the most detailed and important descriptions of 19th-century Yankee whaling, it is far more than a whaling yarn. As the whaleship *Pequod* voyages around the world, Ishmael describes the mates Starbuck, Flask, and Stubb; the harpooners Queequeg, Daggoo, and Tashtego; and the rest of the motley crew drawn into Ahab's ultimately fatal obsession. It has been called an elegy to democracy, a tract on the nature of religion, an investigation of man's relationship to the natural world, and a conflict between the eternal forces of good and evil. Unappreciated when it was first published, *Moby-Dick* has come to be recognized as one of the greatest novels ever written in English. It has been translated into countless languages and has been the subject of any number of movies, including *The Sea Beast* (1926), *Moby Dick* (1930), John Huston's 1956 version, and a 1998 made-for-television movie, with Patrick Stewart as Ahab, that is four hours long.

Of his nominal subject, Melville wrote, "The sperm whale, scientific or poetic, lives not complete in any literature. Far above all other hunted whales, his is an unwritten life." But after *Moby-Dick*, it would be hard to argue that the sperm whale's life was unwritten. Melville himself devoted a good proportion of the book's 135 chapters to various aspects of the life of the sperm whale, including lengthy discussions of its food preferences, its blubber, its brain, its head, its tail, its oil, and its habits—not to mention a detailed description of a certain white whale, whose history, character, and disposition occupy the core of the book.* (The other human beings, including Ishmael, Tashtego, Daggoo, Queequeg, Starbuck, Stubb, and Flask, are primarily actors in the drama of the whale.) Indeed, Melville did a bang-up job of writing the life of the sperm whale and the sperm whaler. He had shipped out on the New Bedford whaler *Acushnet* in 1841, and after a hard 18 months at sea, he was able to chronicle the details of whaling operations from firsthand observations. Although he has given us the best description of Yankee whaling we will ever have, his disquisitions on blocks and tackle, sails and rigging, and ships and boats have made the novel insufferably long and boring for many modern high school students. I was not one of them. I loved the details of whales and whaling then; I delight in them now; and they will, as you will see, form the backbone of this book. You may, if you like, call me Ishmael. (After all, somebody has to survive to tell this story. . . .)

In the century and a half that has passed since the publication of *Moby-Dick*, more has been revealed to researchers and scientists than Herman Melville could ever have foreseen in his wildest cetological imaginings. The New England open-boat sperm-whale harpoonery, at its peak when he was writing *Moby-Dick*, ended within two decades and was replaced by a mechanized version, where whalemen aboard

---

*The title of Melville's novel is *Moby-Dick*, with a hyphen; the name of the whale is Moby Dick, without. As Howard Vincent wrote in *The Trying-out of "Moby-Dick,"* a study about the writing of the novel, "The use of the hyphen in all references to *Moby-Dick* the book, and the omission of the hyphen in all references to the whale, are in accordance with Melville's own example."

Barry Moser's portrait of Herman Melville, which first appeared as the frontispiece in Joseph Epstein, ed., *Literary Genius* (Philadelphia: Paul Dry Books, 2007), and is used here courtesy of the artist.

smoke-belching diesel catcher boats shot the whales with cannons and flensed them aboard 500-foot-long floating factories. As did the Yankees before them, 20th-century industrial whalers killed the whales for oil, but now the oil was used for products Melville never heard of: nitroglycerin and margarine. (In 1851, there probably wasn't an industry dedicated to the production of pet food either, but by the 1970s, many whales were killed to feed household dogs and cats.) Although the killing of sperm whales continued long after the last of the square-rigged whalers was scrapped, the 20th century saw a concentration on those whales that had once been considered too powerful and too fast for men in rowboats. Here is what Melville said of the blue whale, the largest of all whales, which would be hunted to the brink of extinction by the oil-hungry British, Dutch, Japanese, and Norwegians of the 20th century:

Sulphur Bottom: Another retiring gentleman, with a brimstone belly, doubtless got by scraping along the Tartarian tiles in some of his profounder divings. He is seldom seen; at least I have never seen him except in remote southern seas, and then always at too great a distance to study his countenance. He is never chased; he would run away with rope-walks of line. Prodigies are told of him. Adieu, Sulphur Bottom! I can say nothing more that is true of ye, nor can the oldest Nantucketer.

Early whalers didn't hunt the blue whale because they couldn't catch it, and couldn't kill it if they did. They thus knew virtually nothing about it and can be excused for having such limited knowledge of its attributes. It is, in fact, the largest animal that has ever lived on earth, reaching a length of 100 feet and a weight of more than as many tons, almost twice as heavy as the largest sperm whale. Melville summarily dismissed this "retiring gentleman," as he did the bowhead (which he regularly conflated with the right whale), the fin whale, and all the smaller whales, dolphins, and porpoises, so he could concentrate on the sperm whale, "without doubt, the largest inhabitant of the globe; the most formidable of all whales to encounter; the most majestic in aspect; and lastly, by far the most valuable in commerce, from which that valuable substance, spermaceti is obtained." As of the publication of *Moby-Dick*, it could be argued that Melville knew more about sperm whales than anyone else, not only because he had seen them killed and processed, but also because he had read almost everything ever written about the fish and the fishery up to that time. (Although he certainly knew better, Melville amused himself by referring to the whale as a "spouting fish with a horizontal tail.") In chapter 32, "Cetology," he lists the "men, small and great, old and new, landsmen and seamen, who have at large or in little, written of the whale: The Authors of the Bible; Aristotle; Pliny; Aldrovandi; Sir Thomas Browne; Gesner; Ray; Linnaeus; Rondeletius; Willoughby; Green; Artedi; Sibbald; Brisson; Marten; Lacépède; Bonterre; Desmarest; Baron Cuvier; Frederick Cuvier; John Hunter; Owen; Scoresby; Beale; Bennett; J. Ross Browne; The Author of Miriam Coffin;* Olmstead; and the Rev. T. Cheever."

In the decade after Melville deserted the *Acushnet* in 1841 in the Marquesas Islands, he was probably thinking about a book that could somehow incorporate his whaling observations and experiences. *Typee*, his first novel, was about the adventures of a castaway on a South Seas cannibal island, and it was followed two years later by *Omoo*, a sequel, where the main character is a whaleman who fetches up in Tahiti. Neither of these books was particularly successful, although Melville acquired a reputation as "the man who lived among the cannibals." In 1846, for the journal *Literary World*, he read and reviewed J. Ross Browne's *Etchings of a Whaling Cruise*, about which he wrote, "Enough has been said to convince the uninitiated what sort of a vocation whaling is. If further information is desired, Mr. Browne's book is purchas-

---

*The author of *Miriam Coffin* was Joseph Coleman Hart, whose 1834 novel about a Nantucket woman who makes a fortune in the whaling business and then loses it all was a best seller when first published. *Miriam Coffin, or The Whale-Fishermen* is a Nantucket classic, based on the life of the Tory merchant Kezia Coffin (1723–1798), which presents a detailed picture of the island when it was the whaling capital of the world. In his introduction to the 1995 reprint, Nathaniel Philbrick wrote that the book "was not only America's first whaling novel, it is also the first book to provide an in-depth account of Nantucket's history."

able in which they will find the whole matter described in all its interesting details." Melville published *Mardi* in 1849 and *White-Jacket* in 1850, and by May of that year, he wrote to Richard Henry Dana (who had published *Two Years Before the Mast* in 1840) that the "whaling voyage" will be "a strange sort of book I fear; blubber is blubber you know, tho' you may get oil out of it, the poetry runs as hard as sap from a frozen maple tree—& cook the thing up, one must needs throw in a little fancy, which from the nature of the thing, must be ungainly as the gambols of the whales themselves. Yet I mean to give the truth of the thing, spite of this." For Melville, the facts often came from other authors' accounts.

Thomas Beale (1807–1849) and Frederick Debell Bennett (1806–1859) were probably Melville's sources for information on sperm whales and whaling in the 19th century. Obviously limited by the state of biological knowledge during their time, Beale and Bennett still stand as the foundation for much of what was known—or suspected—about the natural history of this leviathan. In *The Trying-Out of "Moby-Dick,"* a book about the writing of Melville's masterpiece, Howard Vincent points out "that the primary source book for Melville in composing the cetological section of *Moby-Dick* was Thomas Beale's *Natural History of the Sperm Whale*." Melville himself referred to this book and to Bennett's *Narrative of a Whaling Voyage around the Globe from the Year 1833 to 1836* as "exact and reliable," and he used them liberally in his preparation of *Moby-Dick*. Melville's other source, especially for those sections not directly concerned with the sperm whale, was Scoresby's 1820 *Account of the Arctic Regions with a History and Description of the Northern Whale Fishery*. Those who have followed Beale and Bennett (among others, Melville, Scammon, Bullen, Ashley, Scheffer, and me) have been more than willing to acknowledge their debt to these two intrepid British surgeons who, fortunately for science and literature, found themselves aboard whaleships in the heyday of the early whaling industry.*

Henry T. Cheever (1814–1897) was born in Hallowell, Maine, and

*Although Melville acknowledges Beale as one of his sources, it would not have been necessary for him to have a copy of Beale's book. Between 1833 and 1843, the *Penny Cyclopedia* was published in London by the Society for the Diffusion of Useful Knowledge. Volume 27 contains a detailed description of the sperm whale, which is introduced thus: "To render the following abridgment of the description by Mr. Beale (who, in his excellent work on the 'Natural History of the Sperm Whale,' has done

graduated from Bowdoin College in 1834. After being ordained as a minister in 1840, he began a world-spanning evangelical career and served as associate editor of the *New York Evangelist*. Among his books were *The Island World of the Pacific; Life in the Sandwich Islands* (that is, Hawaii); and his best-known work, *The Whale and His Captors, or The Whaleman's Adventures and the Whale's Biography, as Gathered on the Homeward Cruise of the "Commodore Preble,"* published in 1850. This homeward cruise, which began in Tahiti and concluded in Boston, took 236 days. Because mid-19th-century whaleships were not ordinarily equipped with libraries, one must assume that Cheever did the research for this book once he came ashore, for his summaries of whaling history and the industry are prodigious, and were probably not the subjects of sailors' scuttlebutt. For example, here is his description of the extent of the industry in New England:

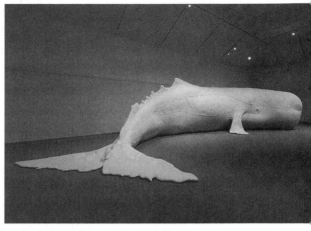

In 2009, Philadelphia artist Tristin Lowe created *Mocha Dick*, a 52-foot-long ghostly white sperm whale made of industrial wool felt stretched over an armature and an inflatable device that makes it appear properly solid and muscular. (Tristin Lowe)

> By the Whaleman's Shipping List at the commencement of 1844 or thereabouts, there were employed in the whale fishery, from the ports of the United States, six hundred and seventy-four vessels, five hundred and ninety-three of them then at sea, chiefly from New Bedford, Nantucket, Sag Harbor, New London, Stonington, and Newport. Allowing for the average of thirty souls to a ship, which is a moderate computation, there were then more than twenty thousand persons prosecuting this trade.

Cheever's book, which was published too close to Melville's own whaling novel to have had much of an influence on *Moby-Dick*, covers much of the same itinerary (but in reverse—South Seas to Massachusetts—and with a happier ending) and includes the good reverend's observations aboard the *Commodore Preble* under Captain Lafayette Ludlow, as the ship worked its way from New Zealand to Hawaii, eastward

―――――

more to elucidate its habits and form than any other writer), we prefix, as he himself does, his cut [drawing] which is by far the most accurate published figure extant of the Spermaceti Whale."

around the Horn, into the South Atlantic, the Gulf Stream, and then to Boston, whaling as she goes, with the Reverend Cheever preaching all the way home. He was a particular advocate of suspending whaling activities on the Sabbath, but his views did not always prevail. Of the apostate whalemen, he wrote, "Some are of vicious, low stock, vicious education, and an incurable addiction to vice. Others are of good families, from religious communities, and have been taught to fear God and keep his commandments. A few of them profess godliness. All of them are alike in this, that they are rational, accountable men, under obligation to keep God's law, and having man's natural right to a need of the Sabbath for rest and religious worship."

Toward the end of his account, he wrote, "When, lately, we were in the midst of a herd of sperm whales, it was my inward earnest prayer that God would give [the captain] good success in their capture, in order that he might yet realize a profitable voyage for his owners at its close, and enter port with a full ship, after all its first losses and misfortunes." Alas, it was not to be. Cheever's prayers and attempts at enforced Sabbath keeping notwithstanding, the *Commodore Preble* came home half empty—but at least it made it, unlike the *Pequod.* Cheever follows the tradition, as did Melville, of acknowledging those writers who preceded him and who provided information that he could not obtain from his own observations. "Other interesting matter of a miscellaneous character pertaining to the whale fishery," he wrote, "is to be found in the appendix to a work of J. R. Browne, called *Etchings of a Whaling Cruise,* and in a volume entitled *Incidents of a Whaling Voyage* by F. A. Olmstead."

Cheever's narrative of the voyage of the *Commodore Preble* is almost as comprehensive as Melville's account of the *Pequod. The Whale and His Captors,* however, although providing extensive details of sperm whales and sperm whaling from firsthand observations, as well as filling in the gaps with yarns about whalemen being lost at sea, whales bashing ships with their heads, the uses of spermaceti oil, and other topics that Cheever heard about while sailing homeward, was no *Moby-Dick.* In addition to its almost endless descriptions of the methodology of whaling, Melville told a story. Cheever's book is a prosaic chronicle of 19th-century sperm whaling, but as Stuart Frank of the New Bedford Whaling Museum wrote in a 1989 article about illustrations of *Moby-Dick,*

Moby-Dick is also a ripping good "industrial novel about the oil business," providing a satisfactorily exotic peep into the gurry-encrusted shipboard life of the American whaleman—which perhaps accounts for at least part of the reason why the book continues to be read by anyone other than college professors and their undergraduate wards. In point of fact, it is its poetic texture, towering symbolism, and compelling metaphors that set the book apart from more pedestrian accounts of the whale hunt. Yet, while the philosophy may be abstruse and the many digressions seemingly detract from "the plot," the story itself is straightforward and easily summarized, and the prose is rich in texture, sometimes filled with humor, sometimes gently sentimental, and often cunning in its irony and forceful in its invective. Perhaps most tellingly, the book is authoritative. There can be no doubt that the author knows something firsthand about the whaling business, that he has ruminated at length on its practical dimensions as well as its transcendental meanings, and that the business-end of a harpoon is as familiar to him as the business-end of a pen.

As for Melville's eponymous whale, there was no shortage of reference material either. In 1820, the Nantucket whaleship *Essex*, captained by George Pollard Jr., was cruising on the line, far to the west of the Galápagos Islands in the Pacific. When sperm whales were sighted, three boats were lowered, and Owen Chase harpooned a whale, which gave his boat a "severe blow with his tail," punching a hole in the hull. Chase stuffed some jackets into the hole and headed back to the *Essex*. Once back aboard, he saw a gigantic whale ("as well as I could judge about eighty-five feet in length") lying quietly about a hundred yards away. The whale spouted two or three times, then disappeared. When Chase saw him again, he was charging toward the ship, and crashed into it with his head. Chase realized that the whale had made a hole in the *Essex*, and that the *Essex* was filling up with water and beginning to sink. Then the whale returned and smashed into the *Essex* again, completely caving in the bows. As the captain returned, he hailed the mate and shouted, "My God, Mr. Chase, what is the matter?" "We have been stove by a whale," answered Chase.

Melville knew the true story of the *Essex* because he had actually spoken to William Chase, the son of the mate of the ship that was stove

The hapless whalemen have every reason to be terrified. They are confronting a kind of sperm whale that nobody has ever seen before: one with an extraordinary number of teeth in its upper *and* lower jaws. (New Bedford Whaling Museum)

by a whale in the Pacific. Melville acknowledged this debt when he wrote, "I have seen Owen Chace [*sic*], who was chief mate of the *Essex* at the time of the tragedy; I have read his plain and faithful narrative; I have conversed with his son; and all within a few miles of the scene of the tragedy." Owen Chase kept a journal of the events of that tragedy and the ensuing voyage that he titled *Narrative of the Most Extraordinary and Distressing Shipwreck of the Whale-ship Essex, of Nantucket; which was Attacked and Finally Destroyed by a Large Spermaceti-whale, in the Pacific Ocean*. The description and the quotes that follow are taken from that journal:

> The whale charged the ship, smashing into the bow: "The ship brought up as suddenly and violently as if she had struck a rock and trembled for a few seconds like a leaf." The *Essex* began to take in water, and as Chase signaled for the other boats to return, he saw the whale, "apparently in convulsions, on the top of the water about a hundred rods [a rod is 16.5 feet] to leeward. He was enveloped in the foam of the sea that his continual and violent thrashing about in the water had created around him, and I could distinctly see him

smite his jaws together, as if distracted with rage and fury." As the ship settled into the water, the whale charged again.

I turned around and saw him, about one hundred rods ahead of us, coming down apparently with twice his ordinary speed and, it appeared to me at the moment, tenfold fury and vengeance in his aspect. The surf flew in all directions about him, and his course towards us was marked by white foam a rod in width, which he made with the continual violent thrashing of his tail. His head was about half out of the water, and in that way he came upon us again and struck the ship.

Later in his narrative, Chase reflected on what had transpired, and he concluded that the whale's attack was intentional:

Every fact seemed to warrant me in concluding that it was anything but chance which directed his operations. He made two separate attacks upon the ship within a short interval, both of which, according to their direction, were calculated to do us the most injury. By being made ahead, they thereby combined the speed of the two objects for the shock. To effect this impact, the exact maneuvers which he made were necessary. . . . His aspect was most horrible and such as indicated resentment and fury. He came directly from the shoal which we had just before entered—and in which we had struck three of his companions—as if he were fired with revenge for their sufferings.

This was the first recorded instance of a sperm whale's attacking a ship, and because we know that young Melville had read Chase's 1821 account, it is not difficult to imagine his desire to incorporate this incredible tale into his whaling novel. In fact, he made it the climax.

The crew was able to salvage some bread, water, a musket, and a few tools, and they spent the first night with their boats lashed to the *Essex* before it sank. For the next 97 days, the crew floated around the South Pacific in three little whaleboats, at the mercy of broiling sun and terrible storms. They were unable to figure out where they were, where they wanted to go, or how to get there. (They knew South America lay to the east, but they also knew it was 2,000 miles away.) They used up the little food and water they had, then began to suffer the pangs of gut-wrenching hunger and dehydrating thirst. As their shipmates began to die, they dumped the bodies overboard, but when they realized that they were disposing of the only food they were likely to see, they began eating their dead. They knew that they would starve to death if they waited for their shipmates to die, so they drew straws to see who would die so that the others might live. Of the 20 men who left the *Essex*, five survived the ordeal. They had covered 4,500 miles and spent three months in open whaleboats until they were rescued by the British frigate *Indian*.*

A real ship-sinking sperm whale named Mocha Dick was one of the important inspirations for Melville's white whale. Around 1810, this whale began a malicious rampage, sinking everything from lumber carriers to whaleships. In the *Knickerbocker Magazine* of May 1839, Jeremiah N. Reynolds, an officer in the United States Navy, published a piece entitled "Mocha Dick: The White Whale of the Pacific." Reynolds was aboard the whaler *Penguin* as it headed for Santa Maria for repairs, and he fell into a shipboard conversation with the mate, who believed that "whaling was the most dignified and manly of all sublunary pursuits," and who, "in order to prove that he was not afraid of a whale," had run his boat up against the side of an old bull, leapt to the back of the fish, sheeted his lance home, and returned to the safety of the ship. This same mate, who remains nameless throughout Reynolds's narrative, relates the story of Mocha Dick.

As described by the mate, Mocha Dick was "as white as wool . . . from the effect of age, or more probably from a freak of nature. . . . On the spermaceti whale, barnacles are rarely discovered; but on this *lusus naturae* [freak of nature] they had clustered, until it became absolutely rugged with the shells. In short, regard him as you would, he was a most extraordinary fish; or in the vernacular of Nantucket, 'a genuine old sog' of the first water." The crew utters "in a suppressed tone,

---

*Without the cannibalism, the story of the wreck of the *Essex* was incorporated into *Moby-Dick*, but in the year 2000 was retold by Nathaniel Philbrick in his National Book Award–winning *In the Heart of the Sea*. Like Melville before him, Philbrick had access to Owen Chase's narrative of the wreck of the *Essex*, but he also found the diary of cabin boy (and survivor) Thomas Nickerson, which added more flavor to the story. Melville's epic tale ends with the sinking of the whaleship and the death of all but Ishmael; *In the Heart of the Sea* begins where *Moby-Dick* left off. The story is fascinating enough, but Philbrick fleshes it out with descriptions of Yankee whaling, Nantucket history, survival at sea (and its converse), cannibalism, and something the other accounts ignore: what happened to the survivors after they returned home.

the terrible name of Mocha Dick!" and lowers the boats after the great white whale, his back studded with irons. The mate harpoons him "deep into his thick white side," and they are towed "onward in the wake of the tethered monster" until the whale lessens his impetuous speed. Another boat closes in—"Good heavens," shouts the mate, "hadn't they sense enough to keep out of the red water!"—and is promptly upended by the whale's flukes. The mate is about to cut the whale loose to save the floundering whalemen, but when he sees the captain approaching in another boat, he says, "The captain will pick them up, and Mocha Dick will be ours after all!"

The white whale dives: "By this time two hundred fathoms of line had been carried spinning through the chocks, with an impetus that gave back in steam the water cast upon it. Still the gigantic creature bored his way downward, with undiminished speed. Coil after coil went over, and was swallowed up." Just as they are about to cut the line to keep the boat from being dragged under, the tension lessens, and the whale rises. He tows the boat again, and this time the mate is close enough to plunge a boat spade into his back, fatally wounding him. "The dying animal was struggling in a whirlpool of bloody foam, and the ocean far around was tinted with crimson. 'Stern all!' I shouted, as he commenced running impetuously in a circle, beating the water alternately with his head and flukes, and smiting his teeth furiously in their sockets, with a crashing sound. A stream of black, clotted gore rose in a thick spout above the expiring brute, and fell in a shower around, bedewing, or rather drenching us, with a spray of blood." When they tried out the carcass "seventy feet from his noddle to the tips of his flukes"—they found no fewer than 20 harpoons in him, "the rusted mementos of many a desperate encounter." Despite his final flurry, this was not the end of Mocha Dick: "It is of course impossible that Mocha Dick was killed," wrote Howard Vincent, "for he is deathless. Every reader of *Moby-Dick* knows this."

Reynolds acknowledges that "the particulars of the tale were in some degrees highly colored," but he goes on to note that "the facts presented may be a fair specimen of the adventures which constitute so great a portion of the romance of a whaler's life. White or not, Mocha Dick was a real whale; named not for his color, but for the island of Mocha, off the coast of Chile." Beginning his vendetta sometime around 1810, this whale continued to attack ships and boats—not only those associated with whale killing—throughout the Pacific. In 1840, some 200 miles off Valparaiso, the British whaler *Desmond* lowered after a huge whale, only to have him turn on the boats and destroy two of them. Two of the whalemen did not make it back to the ship. The whale fit the description of Mocha Dick, a 70-foot behemoth with an eight-foot white scar across his head. Two months later, the Russian whaler *Serepta* was working some 500 miles to the south when Mocha Dick breached spectacularly between the ship and two boats that were towing a dead whale. He smashed one of the boats, and the other quickly cut the carcass loose and headed for the presumed safety of the ship. The great whale lingered in the vicinity as if standing guard.

The following year, the British whaleship *John Day* was working off the Falkland Islands when the lookout sighted a huge whale. They lowered three boats, two of which were smashed to kindling by the vindictive whale. The migrations of sperm whales were as poorly understood in 1840 as they are today, so when Mocha Dick appeared off the coast of Japan in 1842, no one thought it unusual. He smashed a lumber schooner, but its buoyant cargo kept it afloat long enough for three whalers to come to the scene. The *Yankee, Dudley,* and *Crieff* lowered a total of six boats, and in a mad frenzy, Mocha Dick slashed and smashed his way through two of them, swallowed two of the men who were thrown into the water, and destroyed the bowsprit and jib boom of the Scottish whaler *Crieff.* Throughout the literature of whaling, the death of Mocha Dick was claimed by many whalers, but his resurrection was the work of Herman Melville. Howard Vincent believes that Mocha Dick was the inspiration for Moby. He says, "Mocha Dick—why Melville changed the name to Moby Dick has never been satisfactorily explained—was a bone fide whale, the 'terror whale of the Pacific, of awesome size and with a reputation for malice justifying his selection as the villain of a novel.'"*

A year after *Moby-Dick* was published, not far from where the *Essex* had been stove, the whaleship *Ann Alexander* out of New Bedford was struck by a wounded sperm whale. On November 5, 1851, the *New York*

*Mocha Dick lives on. In 2009, Philadelphia artist Tristin Lowe, "whose practice delves into the crude and rude, absurd and abject, pushing low-brow, low-tech methods and materials toward unexpected ends," created *Mocha Dick,* a 52-foot-long ghostly white sperm whale made of industrial wool felt stretched over an armature and an inflatable device that makes it appear solid and muscular. The whale was exhibited at the Philadelphia Fabric Workshop in 2009 and at the Williams College Museum of Art in 2010.

Times ran this headline: "Thrilling Account of the Destruction of a Whale Ship by a Sperm Whale—Sinking the Ship—Loss of the Boats and Miraculous Escape of the Crew." On August 20, as Captain John Deblois recounted the story, two boats had been lowered for whales on the offshore grounds near the Galápagos. When a bull whale was struck, he charged the whaleboat *"and lifted open his enormous jaws and taking the boat in, actually crushed it into fragments as small as a common-sized chair!"* In the other boat, Captain Deblois picked up the crew members who had been dumped into the sea and returned to the ship, but as he watched from the rail, the whale charged the ship, hit just abreast of the foremast, and with its head bored a great hole in the hull, allowing the Pacific Ocean to rush in. As the *Ann Alexander* began to sink, the crew put to sea in two boats, and within a fortnight, they were picked up by the whaler *Nantucket*. Five months later, the *Rebecca Simms* harpooned and captured a bull sperm whale, said to be carrying two irons from the *Ann Alexander* and splinters from the ship's timbers embedded in its battering-ram head. How Melville would have loved this ending!

Herman Melville died in 1891, so he missed what was probably the last attack by a sperm whale on a whaling ship by 11 years. In 1902, on the so-called 12–40 ground in the South Atlantic, a thousand miles off the coast of Brazil, all three of the *Kathleen*'s boats were fastened to whales when Captain Thomas Jenkins spotted a whale coming directly toward the ship:

> Instead of that whale going down or going to windward as they most always do, he kept coming directly for the ship, only much faster than he was coming before he was darted at. When he got within thirty feet of the ship he saw or heard something and tried to go under the ship but he was so near and was coming so fast he did not have room to get clear of her.
>
> He struck the ship forward of the mizzen rigging and about five or six feet under water. It shook the ship considerably when he struck her, then he tried to come up and raised the stern up some two or three feet so when she came down her counters made a big splash. The whale came up on the other side of the ship and laid there and rolled, did not seem to know what to do.

As the ship sank, the crew of 19, as well as Captain and Mrs. Jenkins (and Mrs. Jenkins's parrot), took to the whaleboats and headed for the island of Barbados, some 1,060 miles away. They were then taken to the Brazilian port of Pernambuco, then to Philadelphia. Captain Jenkins wrote up his adventures in a 1902 volume called *Bark Kathleen Sunk by a Whale*, which he introduced thus:

BARK KATHLEEN
Rammed and Sunk by an Infuriated Bull Whale

———

The most thrilling episode ever known in the history
of the American Whale Fisheries has just occurred.
It is full of mystery and thrill and terror of the deep sea. It is
even more wonderful than any of the stories told by
Mr. Frank T. Bullen, author of the famous
"Cruise of the Cachalot."*

Even today, with our concentrated studies on whale biology and behavior, we are at a loss to assign a motive for a whale's attack. It's too easy to assume that the whales that attacked 19th-century whaleships were retaliating for the murder of thousands of their brethren, and that Moby Dick sank the *Pequod* because he maintained a long-standing animosity toward whalemen, especially those with an ivory leg. (All those harpoons stuck in his side couldn't have put him in a particularly benign mood either, especially when the men were still throw-

———

*Frank Bullen (1857–1915), whom we will encounter many more times in this book, was a British author who went to sea as a cabin boy aboard his uncle's ship *Arabella* and later signed aboard a whaler. Although the ship was the *Splendid* out of New Bedford, for his 1898 book, he rechristened it *Cachalot* and changed the names of the captain and the crew members. Often read as a factual chronicle of a whaling voyage, *The Cruise of the "Cachalot"* is a heady mixture of fact and fiction, usually making a hero of its protagonist, a certain Frank Bullen. The word *cachalot* is an etymological mystery. The *Oxford English Dictionary* declares: "[Fr., fr Sp. Port *cachalote*, f. *cachola* big head = sperm whale." (The word may be pronounced *cash-a-lot* or *cash-a-lo*.) In *Giant Fishes, Whales and Dolphins*, British Museum scientists J. R. Norman and Francis Fraser (1938) said that the word comes from the Gascon *cachau* for "large tooth." Cousteau and Diolé (1972) wrote that the word "was already in use in the mid-eighteenth century. It probably is derived from the Portuguese *cachalotte* ('big head') or from the Spanish *cachalote*." We don't know how he pronounced it, but Bullen called his fictional whaleship *Cachalot,* and a U.S. submarine of that name was moored at Pearl Harbor when the Japanese attacked on December 7, 1941, but suffered no damage.

ing harpoons at him.) We simply don't have an idea of how the sperm whale thinks—or, for that matter, whether this animal, with its gigantic brain, thinks at all. On February 8, 2010, fishermen in Uwajima Bay, in the Bungo Channel between the islands of Shikoku and Kyushu in the southern part of Japan, spotted a 40-foot-long sperm whale lolling in the shallows. Three men in a small powerboat began clanking sticks against the sides of the boat, hoping to drive the whale back out to sea, when the whale turned and rammed the little boat with its head, capsizing the boat and dumping the men into the water. A CNN video shows that the whale's thrashing made it impossible for rescuers to get to the men, and by the time they arrived, one of the men had drowned. The whale then turned and swam quietly out to sea.*

Captain William Scoresby of Whitby was one of the first whalemen to set down his experiences and observations in the northern bowhead fishery; his *Account of the Arctic Regions* is the benchmark for everything that followed. The 20th-century whaling historian Sidney Harmer called it "one of the most remarkable books in the English language. With Beale and Bennett, Scoresby's magisterial work contributed to the public knowledge of the sperm whale, but more significantly for our literary discussion, all of the aforementioned works were known to Herman Melville and used in the creation of what Howard Vincent calls "the cetological center" of *Moby-Dick*.

Because Melville so obviously believed that the sperm whale was God's most divine creature, he unabashedly devoted chapter and verse to its glorification. There are entire chapters about the whale's head, the case, the skin, the spout, the tail, and the animal's social life. When he compares the sperm whale with any other whale—the bowhead, for example—the sperm whale emerges triumphant from the comparison. Here, for instance, is what Melville has to say about Captain Scoresby and his whale:

> But Scoresby knew nothing and says nothing of the great sperm whale, compared with which the Greenland whale is almost unworthy mentioning. And here be it said, that the Greenland whale is an usurper upon the throne of the seas. Yet, owing to the long priority of

*It would also be disarmingly convenient to assign a revenge motive to a sperm whale in Japanese waters. After all, Japanese whalers have been killing sperm whales (which they call *makko-kujira*) for centuries, so a little cetacean retribution might be in order.

his claims, and the profound ignorance which, till some seventy years back, invested the then fabulous or utterly unknown sperm whale, and which ignorance to this present day still reigns in all but some few scientific retreats and whale-ports, this usurpation has been in every way complete. . . . This is Charing Cross; hear ye! good people all,—the Greenland whale is deposed,—the great sperm whale now reigneth!

Even though one member of the species achieved a mythological status, the remainder of the tribe was presented as living, breathing creatures. One of the surprising aspects of *Moby-Dick* is the amplitude of the natural history it includes. In chapter 32, "Cetology," many other species are discussed, but using Beale, Bennett, and other contemporaneous authorities, Melville incorporates—in an emotional, biased, hyperbolic style, not at all the way we expect natural history to be written—almost everything that was known of the sperm whale at the time. It is surprising that this novel, written in 18 months, could not only emerge as one of the greatest works of fiction ever written in America, but could summarize the biology of one of the world's least-known large animals. Melville, while categorizing various whale species according to size, introduces the sperm whale thus in "Cetology":

This whale, among the English of old vaguely known as the Trumpa whale, and the Physeter whale, and the Anvil Headed whale, is the present Cachalot of the French, and the Pottfisch of the Germans, and the Macrocephalus of the Long Words. He is, without doubt, the largest inhabitant of the globe; the most formidable of all whales to encounter; the most majestic in aspect; and lastly, by far the most valuable in commerce; he being the only creature from which that valuable substance, spermaceti, is obtained. Some centuries ago, when the sperm whale was almost wholly unknown in his own proper individuality, and when his oil was accidentally obtained from the stranded fish; in those days spermaceti, it would seem, was popularly supposed to be derived from a creature identical with the one then known in England as the Greenland or Right Whale. It was the idea also, that this same spermaceti was the quickening humor of the Greenland Whale with which the first syllable of the word literally expresses. In those times, also, spermaceti was extremely scarce.

Once it had been established that Leviathan was a real creature, the next stage in its apotheosis was accomplished through literature, then the only available medium for popular aggrandizement. After Jonah and the whale, the earliest, and certainly the most successful, example of the animal made myth can be found in *Moby-Dick,* the story of the whale that triumphs over the puny efforts of man to kill it.

The white whale is *monstrum horrendum,* the quintessence of evil, the literary paradigm of malevolence. In the chapter in which he introduces the white whale, Melville says that he did "in the end incorporate with themselves all manner of morbid hints, and half-formed foetal suggestions of supernatural agencies, which eventually invested Moby Dick with new terrors unborrowed from anything that visibly appears." And although whiteness is usually perceived as representing goodness, Melville sees the absence of color as "the intensifying agent in things the most appalling to mankind":

> This elusive quality it is, which causes the thought of whiteness, when divorced from more kindly associations, and coupled with an object terrible in itself, to heighten the terror to the furthest bounds. Witness the white bear of the poles, and the white shark of the tropics; what but their smooth, flaky whiteness makes them the transcendent horrors they are? That ghastly whiteness it is which imparts such an abhorrent mildness, even more loathsome than terrific, to the dumb gloating of their aspect.

Moby Dick is not actually white; he is described as having "a peculiar snow-white wrinkled forehead, and a high, pyramidical white hump. . . . The rest of his body was so streaked, and spotted, and marbled with that same shrouded hue, that, in the end, he had gained the distinctive appellation of the White Whale." In chapter 42 ("The Whiteness of the Whale"), Melville lists those things that, while white, are inherently evil, including such diverse creatures as the polar bear, the white shark, the albatross, the albino man, and travelers in Lapland "who refuse to wear colored and coloring glasses upon their eyes." Melville writes, "And of all these things, the Albino whale was the symbol. Wonder ye at the fiery hunt?"*

*Except perhaps in the case of Moby Dick, white is not necessarily the color of evil. Think of "pure as the driven snow" or the white garments of hospital workers. In early Hollywood westerns, the bad guys always wore black hats, the good guys, white.

The supernatural quality of the white whale can be found throughout Melville's stirring narrative. "Some whalemen," he wrote, "should still go further in their superstitions; declaring Moby Dick not only ubiquitous but immortal (for immortality is but ubiquity in time); that though groves of spears should be planted in his flanks, he would still swim away unharmed; or if indeed he should ever be made to spout thick blood, such a sight would be a ghastly deception; for again in unensanguined bellows hundreds of leagues away, his unsullied jet would once more be seen."

Moby Dick is not only a myth and a monster; he is also the materialization of everything depraved and villainous. Even the demented Captain Ahab, his would-be conqueror, cannot compete with the power of the whale and the white whale's black heart. Howard Vincent identifies the quintessentially evil nature of the whale by equating it with the greatest evil ever known: "Through Melville, Moby Dick has been absolved of mortality. Readers of *Moby-Dick* know that he swims the world unconquered, that he is ubiquitous in time and place. Yesterday he sank the *Pequod*, within the past few years he has breached five times, from a New Mexico desert, over Hiroshima and Nagasaki, and most recently, at Bikini Atoll."

Vincent points out that it is primarily to Ahab that the whale is the personification of evil: "Ahab's reaction [to the whale] is not the normal sort of response; it is that of one scarred and maimed by life. It is the monomania of the paranoid. To Mankind in general, the White Whale in *Moby-Dick* must symbolize something else," and that is, according to Vincent, "life itself with its Good and its Evil; it is the final Mystery which no man may know and no man should pursue unrelentingly."

Although square-rigged sperm whaling ended with the end of the 19th century, the sperm whales were not spared. Mechanized whalers aboard diesel-powered catcher boats, armed with exploding grenade

---

White is the color of wedding dresses and angel's wings; a white flag is an international symbol of either surrender or truce, a sign of peaceful intent. In various cultures, real (and rare) white elephants, white buffalos, and white tigers are heralded as omens of good fortune. Every year since 1991, a pure white humpback whale named Migaloo ("white fella") has appeared among its normally colored mates, participating in the migration along Australia's east coast from the Antarctic to the Great Barrier Reef. The appearance of Migaloo is interpreted by many as a natural symbol of protest against commercial whaling.

harpoons, exponentially escalated the hunt for *Physeter macrocephalus*. First in the Antarctic, then off the coasts of South America, South Africa, Australia, and New Zealand, the whalers pursued the spermaceti whales for their oil-rich blubber. (The oil of their heads was no longer needed for the manufacture of high-quality smokeless candles.) The high point of sperm whaling history was not, as many people might assume, during Melville's lifetime (1819–1891) but rather in the 1960s. Flotillas of Soviet and Japanese catcher boats attached to gigantic factory ships killed the whales of the North Pacific—a population unknown to Yankee whalers—in numbers that could only have been accomplished by state-supported technological whaling juggernauts. In 1959, for the first time, the annual kill of sperm whales throughout the world's oceans, including the Antarctic, rose to over 20,000, and by 1964, it was over 28,000.

In chapter 32, "Cetology," Melville discourses on the known varieties of whales and dolphins. He categorizes them by size—not comparing them to other animals, but rather to the various sizes of books. Thus the largest whales are the Folio Whales (Sperm, Right, Fin-Back, Humpbacked, Razor Back, Sulphur Bottom). Next come the Octavo Whales (Grampus, Black-fish, Narwhale, Killer, Thrasher). At the end of the library are the Duodecimo Whales, the various dolphins and porpoises (Huzza Porpoise, Algernine Porpoise, Mealy-mouthed Porpoise). Many of these animals are familiar to contemporary cetologists—the Razor Back is another name for the finback; the Sulphur Bottom is the blue whale; the Grampus is actually a species of beakless dolphin (*Grampus griseus*) known as Risso's dolphin; the Black-fish is the pilot whale; the Huzza Porpoise is probably the common dolphin, known for its acrobatic leaps; the Algernine Porpoise—which Melville says is "very savage. . . . provoke him and he will buckle to a shark"—is an unknown commodity; and the Thrasher, "famous for his tail, which he uses for a ferule in thrashing his foes," is probably not a cetacean at all but the thresher shark.* Melville's cetology was limited by the available litera-

*Melville willingly admitted the limitations of his cetological expertise when he wrote, "There are a rabble of uncertain, fugitive whales, which, as an American whaleman, I know by reputation and not personally. I shall enumerate them by their forecastle appellations; for possibly such a list may be valuable to future investigators, who may complete what I have here but begun. If any of the following whales, shall hereafter be caught and marked, then he can readily be incorporated into the System, according to his Folio Octavo, or Duodecimo magnitude:—the Bottle-Nose Whales;

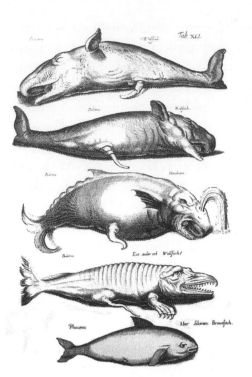

With few real whales available for inspection, 17th-century naturalists often resorted to sea monsters. Two sperm whales appear at the top, and a harbor porpoise is at the bottom. (Richard Ellis Collection)

ture and his own observations, and even though the sperm whale is the heart and soul of the novel, Melville's knowledge of this leviathan fell far short. Nothing he read—not Beale, not Bennett, not Scoresby—could have possibly prepared him for the astonishing reality of the sperm whale's biology.

It's not surprising that Melville knew little about the subjects of his Leviathanic masterwork; nobody else knew very much either. For example, here's what Thomas Beale, one of Melville's acknowledged mentors, wrote in 1835:

About the habits of the Sperm Whale we have hitherto possessed but a scanty knowledge, and that little even not yet published in a connected form. It is a matter of great astonishment, that the consideration of the habits of so interesting, and in a commercial point of view of so important an animal, should have apparently excited so little curiosity among the numerous, and many of them competent observers, that of late years must have possessed the most abundant, and the most convenient opportunities of witnessing their habitudes.

The whalers, who had more than ample opportunity to examine the sperm whale, were almost totally preoccupied with figuring out how to find and kill it. They had hardly any interest in its biology or habits—

the Junk Whale; the Pudding-Headed Whale; the Cape Whale; the Leading Whale; the Cannon Whale; the Scragg Whale; the Coppered Whale; the Elephant Whale; the Iceberg Whale; the Quog Whale; the Blue Whale &c. From Icelandic, Dutch, and old English authorities, there might be quoted other lists of uncertain whales, blessed with all manner of uncouth names. But I omit them as altogether obsolete; and can hardly help suspecting them for mere sounds, full of Leviathanism, but signifying nothing."

which, if they had bothered to learn, would have made their hunt considerably more efficient. By 1839, when Robert Jardine had published his multivolume *Naturalist's Library*, knowledge of the sperm whale had not increased, but the number of species had: "Desmaret but a few years ago admitted three subgenera and seven species; and Lacépède has three genera and eight species, including the Cachalots, physalus, and physeters. . . . Till the time of Cuvier, there was the greatest confusion regarding the alleged species, so, till a much more recent date, there was an almost unaccountable paucity of information regarding its real habits and history."

By the time Melville began his search for information on sperm whales, the story of the sinking of the *Essex* in 1820 had become embedded in the mythology of the cachalot, but the whale that stove in Captain Pollard's ship was as malicious as Moby Dick, and certainly his paradigm. As Owen Chase observed:

I saw him about one hundred rods directly ahead of us, coming down apparently with twice his ordinary speed, and to me at the moment, it appeared with tenfold fury and vengeance in his aspect. The surf flew in all directions around him, and his course towards us was marked by a white foam a rod in width, which he made with the continual violent thrashing of his tail; his head was half out of the water, and in that way, he came upon and struck the ship. He struck her to windward . . . and completely stove in her bows.

With the help of Beale and Bennett, Melville filled in many of the gaps in sperm whale biology, but the preponderance of published information emphasized the hunt for the cachalot and the inherent dangers thereof. In F. D. Bennett's 1840 *Narrative of a Whaling Voyage around the Globe*, we find this discussion of particularly aggressive bull sperm whales:

A few Cachalots have been noted individually as animals dangerous to attack. One was thus distinguished on the cruising ground off the coast of New Zealand, and was long known to whalers by the name of "New Zealand Tom." He is said to have been of great size; conspicuously distinguished by a white hump; and famous for the havoc he made amongst the boats and gear of ships attempting his destruction. A second example, of similar celebrity, was known to whalers in the Straits of Timor. He had so often succeeded in repelling the attacks of his foes as to be considered invincible, but was at length dispatched by a whaler, who, forewarned of his combative temper, adopted the expedient of floating a cask on the sea, to withdraw his attention from the boats; but notwithstanding this ruse, the animal was not destroyed without much hard fighting, nor until the bow of one of the boats had been nipped off by his jaws.*

The story of Ahab and the white whale was not a critical or commercial success during Melville's lifetime, and although he continued to write—*Pierre, or The Ambiguities* in 1852; *Israel Potter* in 1855; and *The Piazza Tales* in 1856—he could not support his family, and took a job as a New York City customs inspector. (*Billy Budd,* written shortly before his death, was not published until 1924.) He died in 1891, virtually unremembered. Upon his death, the *New York Times* wrote, "There has died and been buried in this city, during the current week, at an advanced age, a man so little known, even by name, to the generation now in the vigor of life that only one newspaper contained an ordinary obituary for him, and that of only three or four lines."

Melville's reputation languished. But in 1893, two years after his death, the tide began to turn. An anonymous reviewer (in *The Critic*) called the book a "remarkable romance," noting that "the author's extraordinary vocabulary, its wonderful coinages and vivid turnings and twistings of worn-out words, are comparable only to Chapman's translations of Homer. The language fairly shrieks under the intensity of his treatment, and the reader is under an excitement which is hardly controllable. The only wonder is that Melville is so little known and so poorly appreciated." By 1913, he had been promoted to the ranks of "minor fiction writers," and he was hailed as "a great figure in shadow; but the shadow is not of oblivion."

Moby-Dick was on the ascendant; rising from the depths, as it were. In 1917, Carl Van Doren wrote:

> Ahab, not Melville, is to blame if the story seems an allegory, which Melville clearly declared it was not; but it contains, nevertheless, the semblance of a conflict between the ancient and scatheless forces

*According to Howard Vincent, Melville changed "Timor Jack" to "Timor Tom" and also changed "New Zealand Tom" to "New Zealand Jack." Most significantly, he changed "Mocha Dick" to "Moby Dick," although "this change has never been satisfactorily explained."

of nature and the ineluctable enmity of man. This is the theme, but description can hardly report the extraordinary mixture in *Moby-Dick* of vivid adventure, minute detail, cloudy symbolism, thrilling pictures of the sea in every mood, sly mirth and cosmic ironies, real and incredible characters, wit, speculation, humour, colour. The style is mannered but often felicitous; though the book is long, the end, after every faculty of suspense has been aroused, is swift and final. Too irregular, too bizarre, perhaps, ever to win the widest suffrage, the immense originality of *Moby-Dick* must warrant the claim of its admirers that it belongs with the greatest sea romances in the whole literature of the world.

In 1919, the centennial of Melville's birth was celebrated by Raymond Weaver, who wrote, "If he does not eventually rank as a writer of over-shadowing accomplishment, it will be owing not to any lack of genius, but to the perversity of his rare and lofty gifts." Writing in 1923, the British critic Leonard Woolf accused Melville of "execrable English" and an overabundance of semicolons, which he used "without regard to meaning or convention," but Woolf concluded that he "must leave Melville see-sawing between his semi-colons on the one side and great-ness on the other." The final apotheosis came from the pen of Lewis Mumford, who published an appreciation of Melville in 1929, ensuring his permanent position in American literature. He wrote:

Melville's instrumentation is unsurpassed in the writing of the last century; one must go to a Beethoven or a Wagner for an exhibition of similar powers: one will not find it among the works of literature. Here are Webster's wild violin, Marlowe's cymbals, Browne's sono-rous bass viol, Swift's brass, Smollett's castanets, Shelley's flute, brought together in a single orchestra, complementing each other in a grand symphony. Melville achieved a similar synthesis in thought; and that work has proved all the more credible because he achieved it in language, too. Small wonder that those who were used to the elegant pianoforte solos or barrel-organ instrumentation were deaf-ened and surprised and repulsed.

In an essay included in his 1923 *Studies in Classical American Litera-ture*, D. H. Lawrence wrote that *Moby-Dick* was "a great book." (Among the other American masterworks that he discussed were Hawthorne's

The Scarlet Letter, Dana's *Two Years before the Mast,* Fenimore Cooper's Leatherstocking novels, and Melville's *Typee* and *Omoo.*) Lawrence wrote:

> A hunt. The last great hunt. For what? For Moby Dick, the huge white sperm whale, who is old, hoary, monstrous, and swims alone; who is unspeakably terrible in his wrath, having so often been attacked; and snow-white. Of course he is the symbol. Of what? I doubt if even Melville knew exactly. That's the best of it.

After describing the sinking of the *Pequod,* Lawrence continued:

> So ends one of the strangest and most wonderful books in the world, closing up its mystery and its tortured symbolism. It is an epic of the sea such as no man has ever equalled; it is a book of esoteric symbolism and profound significance, and of considerable tiresomeness. But it is a great book, a very great book, the greatest book of the sea ever written. It moves awe in the soul.

Subsequent critics have further extolled Melville and his masterpiece; there have been comparisons to Homer (Ahab as Odysseus); Shakespeare (Ahab as Lear); Mark Twain (*Moby-Dick* as the only rival for *Huckleberry Finn* as the greatest work in American literature); and Marlowe, Goethe, and virtually every other writer in the history of heroic fiction. In their annotated version of *Moby-Dick,* Harrison Hayford and Hershel Parker list 21 scholarly books written about the novel between 1953 and 1969. Perhaps the most useful of these is Vincent's *The Trying-out of "Moby-Dick,"* in which Melville's debt to earlier authors is discussed in depth, and virtually every cetological, historical, and literary element is meticulously examined. There is a Melville Society, whose members publish a journal in which (among notes on scholarship and history) the universities that have generated PhD dissertations on Melville are ranked. Between 1924 and 1980, doctoral candidates at American universities wrote a total of 246 such theses, of which Yale graduate students submitted 56.

Ernest Hemingway, born in 1899, was taken by his mother to Nantucket when he was 11, and among other diversions, he visited the whaling exhibit at the local museum. According to Susan Beegel, writing in 1985, "Perhaps Hemingway remembered lessons learned in Nantucket when, in October, 1934, a pod of sperm whales surfaced near his fishing boat off Havana." Certainly he could not resist this opportunity to try to harpoon a whale himself, and in *Esquire* magazine in 1936 he described this adventure, which was entitled "There She Breaches! Or, Moby Dick off the Morro":

There, a little way ahead, was the whale. He was very impressive. He would swim a little way under water then his broad back would come out and he would go along with the slanted top of his back out, seemingly unconcerned, but when we speeded up the engine to come up on him close enough to fire the harpoon into him he would submerge. We tried coming up on him from the back, but he would go down each time before we were in range. Then we tried coming up on him from an angle, but down he would go again to be out of sight, only to reappear ahead of us, varying his course very little. Time after time we came within thirty feet of him only to have him go down. The speeding up of the motors seemed to frighten him and put him down and only by speeding up the motors could we come up on him. He was about forty feet long and as we came up close to him we could see the indentations along the side of his blunt head running back toward the body, as though someone had made them by rubbing a finger in warm wax. Again and again we were so close to him you could have hit him with a beer bottle, but I knew for the harpoon to hold we should be almost touching him with the boat when fired.

"Shoot! For God's sake shoot!" Bob screamed. . . .

"Shoot!" yelled Carlos. . . .

"It's no use to shoot unless it's close enough," I yelled back. "The gun can't carry the weight of that hawser."

The next time we came close and they began yelling to shoot, I said "All right. I'll show you what I mean," and fired when we were not quite thirty feet from the whale and he had lowered his head to sound. The black powder roared, the wire shot out, came taut with the weight of the hawser, and the dart was short. The whale went down and this time he came up a long way ahead and it was hard to see him in the sun.

Melville's novel has also inspired many nonliterary works, such as composer and performer Laurie Anderson's 1999 modern opera, *Songs and Stories from "Moby-Dick,"* in which she compared two great American sagas: *Moby-Dick* and *Star Trek*. Also on the opera stage, there was the 2004 *Moby-Dick* composed by Princeton University professor Peter Westergaard, and another opera by composer Jake Heggie and librettist Gene Scheer, which premiered at the Dallas Opera House in 2010. On their 1969 album, the heavy metal rock band Led Zeppelin included a track called "Moby Dick," which consists mostly of a virtuoso drum solo by John Bonham that has been called the best drum solo in the history of percussion. There is a German "nautic doom" band called Ahab, whose albums feature images of nautical disasters, such as the wreck of the *Medusa* and whales sinking ships. Their playlists include titles like "Redemption Lost," "Old Thunder," and "The Hunt." Then there is Moby, the popular musician whose real name is Richard Melville Hall. He claims Herman Melville as a distant relative, but his music, as far as I can tell, has little to do with whales or whaling.

So too with the heavy metal rock band Mastodon, which, according to Craig Bernardini, of Hostos Community College, found their name in Melville's "salt-sea mastodon,"* and whose first album was entitled *Leviathan*. The tracks on this album are "Blood and Thunder," "I Am Ahab," "Seabeast," "Island," "Iron Tusk," "Megalodon" (a gigantic extinct shark), "Naked Burn," "Aqua Dementia," "Hearts Alive," and "Joseph Merrick" (the so-called Elephant Man). Although most of these titles evince a connection with *Moby-Dick*, the music does not, and as Bernardini has written (in a 2009 article in the Melville Society's journal *Leviathan*), "In fan reviews of *Leviathan* [the album] in web 'zines and on sites like Amazon.com, *Moby-Dick*, understood as the canonical great work, is used either to legitimize the album or to enable it to transcend the heavy metal genre. . . . *Leviathan* thus garners prestige by evoking a canon of prestigious music, 'music for the thinking individual,' echoing the discourse that legitimizes the literary canon."

*In his chapter on the origins of the Nantucket whaling community, Melville wrote: "What wonder then, that these Nantucketers, born on a beach, should take to the sea for a livelihood! They first caught crabs and quahogs in the sand; grown bolder, they waded out with nets for mackerel; more experienced, they pushed off in boats and captured cod; and at last, launching a navy of great ships on the sea, explored this watery world; put an incessant belt of circumnavigations round it; peeped in at Bhering's Straits; and in all seasons and all oceans declared everlasting war with the mightiest animated mass that has survived the flood; most mountainous and monstrous! That Himmalehan, salt-sea Mastodon, clothed with such portentousness of unconscious power, that his very panics are more to be dreaded than his most fearless and malicious assaults!"

American composer Bernard Herrmann (1911–1975), best known for his film scores, especially those of Orson Welles, Alfred Hitchcock, and Martin Scorsese, began *Moby Dick,* a dramatic cantata for two tenors, two basses, male chorus, and orchestra, in February 1937 and completed it in August 1938. Herrmann based his work on Melville's novel, but he claimed a more intimate connection with the subject: his father had served on two whaling ships and had often regaled his sons with stories of his adventures. Premiered by the New York Philharmonic and greeted positively by the audience and the press, *Moby Dick* seemed to portend a career as a classical composer for Herrmann, but the success of his music for Welles's famous broadcast of *War of the Worlds* and the subsequent acclaim of the score for *Citizen Kane* moved Herrmann and his music to Hollywood. (And of course, Orson Welles appeared in Huston's *Moby-Dick* as Father Mapple.)

There is no question that Alison Baird's 1999 *White as the Waves* comes directly from Melville—its subtitle is "A Novel of Moby Dick"—but it is unusual not only in that the story is told from the whales' point of view, but also because the whales are able to talk. The protagonist is called Whitewave; in fact, except in the subtitle, Moby Dick is never mentioned, presumably because that name was used by men, and this is the whale's story. Here is a conversation between Whitewave and a dolphin named Feliki (for some reason, dolphins talk like Indians in an old western):

> "What are you doing here?" the white whale demanded. "I took you back to the other dolphins."
> "I no like," she said. "Were too many males. One want mate with me, and I no want. Then he swim off, come back with more males. I know they going to surround me, try and take me by force. I once know female dolphin die when some males chase her and she go onto sandbar. So I swim away fast, look for you."

As in the original, the white whale wins in the end. After sinking the *Pequod,* he attacks the whaleboat with Ahab aboard:

> He could see the silhouette of the one-legged man's boat pulling about on the heaving hillocked surface above. Rocketing upward, Whitewave breached—directly beneath the boat. His blunt brow struck the boat's bottom with such force that it flew into the air, soar-

ing in a long parabola before dropping again. Men and oars were scattered as it fell.

Whitewave too was airborne for an instant. He fell, foaming, directly on top of the wreckage. With his tail, he lashed out once more, smashing even the minutest floating fragment. He was determined nothing should remain for his foes to use in a second attack.

"I have now done this," muses Whitewave. *"A ship—a whole ship! Sunk, destroyed, defeated. No other whale will die from its attacks. I did not fail! I won!"*

In *Moby-Dick*, a tale told by a man, the sole survivor of the whale's attack on the *Pequod* is the narrator Ishmael, buoyed up by Queequeg's floating coffin. In *White as the Waves*, a tale told by cetaceans, Ishmael again survives by hanging on to "a piece of floating wood," but Feliki the dolphin protects him from the sharks until another ship arrives to pick him up.

I seem to have missed it in my several readings of *Moby-Dick*, but you will probably be surprised as I was to learn that Ahab had a wife. In the chapter "The Ship," Captain Peleg tells Ishmael: "And once for all, let me tell and assure thee, young man, it's better to sail with a moody good captain than laughing bad one. . . . Besides, my boy, he has a wife—not three voyages wedded—a sweet, resigned girl. Think of that; by this sweet girl, that old man had a child; hold ye then there can be any utter, hopeless harm in Ahab? No, no, my lad; stricken, blasted if he be, Ahab has his humanities!"

A century and a half later, in celebration of these humanities, Sena Jeter Naslund, a professor at the University of Louisville, wrote *Ahab's Wife*, a novel about the woman Ahab left behind. Una (named for the heroine of Edmund Spenser's *Faerie Queene*) flees to New England from Kentucky to escape her father's oppressive Puritanism. Fetching up in New Bedford at 16, she disguises herself as a cabin boy and signs aboard the whaleship *Sussex*.* Somewhere in the Pacific, she sights the

---

*Down to the Sea in Ships was Clara Bow's first movie, made in 1922 when she was 17, where she too disguised herself as a cabin boy aboard a whaleship. Clara's stovepipe hat is knocked off, putting an end to her disguise, but she continues the voyage aboard the ship. As a historical document, *Down to the Sea in Ships* is vital to the study of whaling because the whaling scenes, which were shot in the West Indies, contain the only footage ever made of actual square-rigged Yankee sperm whaling, from the lookouts spotting the whales to the harpooning, lancing, "Nantucket Sleigh Ride," towing, and cutting in.

great *black* whale that will sink her ship: "No island, but an enormous black sperm whale, with a head steep as a cliff, erupting not volcanic smoke, but a huge spume of water vapor." The boats are lowered: "The black whale seemed unaware of us. He lay on the water like a slope of coal. I imagined his tiny eye, the wrinkles around the socket. *King!* The word came strangely to mind, startling as the retort of a rifle. King of all lunged creatures, this whale; king thus, of all mankind? And this dark idea spelled its way across my mind: *We shall never take him.*"

They don't, and the whale rams and sinks the ship. Una and other crew members escape in a whaleboat, heading—they hope—for Tahiti. After a horrendous episode of enforced cannibalism (reminiscent of the fate of the survivors of the *Essex*), they are rescued, and in a remarkable turn of events, Una then ships aboard the whaler *Alba Albatross*, which encounters the *Pequod* at sea. Una intends to marry Kit, one of her fellow survivors of the sinking of the *Sussex*, but when Captain Swain refuses to marry them, Una implores the captain of the *Pequod* to perform the ceremony aboard his ship, which he does. (The opening sentence of *Ahab's Wife* is "Captain Ahab was neither my first husband nor my last.") Kit Sparrow and his new wife, Una, then board the *Pequod* and return to Nantucket. Kit goes mad and runs away, leaving Una behind. Because he married Kit and Una aboard the *Pequod*, they board Ahab's ship at the dock, where he announces, "What I did join together, I now put asunder. . . . and here I wed Una and take her for my bride." Shortly thereafter, Mrs. Ahab, now pregnant, watches her new husband sail off in the *Pequod*, which is more or less where Melville took up the story. He relegated Ahab's wife and child to the single mention above, but Naslund continues Una's story beyond the fatal confrontation of her husband and the white whale. *Ahab's Wife* is a brilliant piece of storytelling, incorporating and reworking so many elements of Melville's masterpiece that it might rightly be considered a sequel, or at least a companion volume.*

*Because Naslund was writing in the late 20th century, she had access to considerably more information than Melville did, but she didn't always interpret it correctly. In *Moby-Dick*, we learn that ambergris is "an essence found in the inglorious bowels of a sick whale. . . . By some, it is supposed to be the cause, and by others the effect, of dyspepsia in the whale." That was pretty much all Melville wrote, but Naslund quotes a letter to Una, in which Ahab tells her that "the ambergris forms very much as the nacreous pearl forms in an oyster. Ye know that the chief food of the sperm whale consists of giant squid, which the whale fetches from its lair in the greatest depths. . . .

Ultimately, Melville was—and still is—recognized as the most powerful, enigmatic, and, yes, misunderstood of all 19th-century American writers, selected by fate and a predisposition for greatness to write a novel in which the most powerful, enigmatic, and, yes, misunderstood of all living animals is celebrated. Who better than Herman Melville himself to explain his—and by extension Ahab's—obsession with the great white whale?

Not even at the present day has the original prestige of the Sperm Whale, as fearfully distinguished from all other species of the leviathan, died out in the minds of the whalemen as a body. There are those this day among them, who, though intelligent and courageous in offering battle to the Greenland or Right Whale, would perhaps—either from professional inexperience, or incompetency, or timidity, decline a contest with the Sperm Whale; at any rate, there are plenty of whalemen, especially among those whaling nations not sailing under the American flag, who have never hostilely encountered the Sperm Whale, but whose sole knowledge of the leviathan is restricted to the ignoble monster primitively pursued in the North; seated on their hatches, these men will hearken with a childish fireside interest and awe, to the wild, strange tales of Southern whaling. Nor is the pre-eminent tremendousness of the great Sperm Whale anywhere more feelingly comprehended, than on those prows which stem him.

The story of Moby Dick's rise to celebrity does not end with the resurgence of interest in the novel or the various screen interpretations. As alternative devices for publicity evolved, they were used (sometimes inadvertently) in the canonization of the great white whale. Where there is an area of *Moby-Dick* or Melville scholarship to be investigated, there will surely be someone to investigate it. In the case of *Moby-Dick*

But sometimes, the horny beak of the squid lodges all indigestible within the whale. There's impediment! The whale cannot disgorge the final sword of his foe, it festers in its new scabbard, and despite the internal bandaging of the irritant with a material that is rather golden and cheeselike in texture, the whale dies." Melville's limited description is more accurate. The chief food of the sperm whale is not giant squid; ambergris does not form like a pearl; it is neither "golden" nor "cheeselike"; and the whale does not die from the "impediment."

in popular culture, it is M. Thomas Inge. In *A Companion to Melville Studies* (1986), edited by John Bryant, Inge contributed a chapter entitled "Melville in Popular Culture" in which he discusses film, comics, radio, television, recordings, children's literature, and adult fiction—that is, "popular novels that employ [Melville] and his work." In his introduction, Inge wrote, "It has inspired so many other works of literature, from science fiction and pulp novels to mainstream works for fiction, poetry and drama, that the story of Ahab's pursuit of the white whale seems permanently imbedded in the national consciousness. . . . So widely known is the basic plot structure of *Moby-Dick* and so frequently has it been adapted to the media that there are people who think they have read the novel without having gone near the actual text."

In chapter 55, "Of the Monstrous Pictures of Whales," Melville bemoans the absence of decent pictures of whales and says, "It is time to set the world right in this matter, by proving such pictures of the whale all wrong." He then proceeds to critique those "monstrous" images where the whale is depicted wrongly, from the Dutch engravings of bloated, stranded whales; to poorly realized renditions in contemporaneous works of "science"; to illustrations of the skeleton ("it is one of the more curious things about this Leviathan that his skeleton gives very little idea of his general shape"); and finally, to what is probably the most unusual portrayal of any animal anywhere: the image of the whale reproduced on a piece of the whale itself. I speak of scrimshaw, where whalemen incised the image of the whale on the very teeth of the whale they had killed.

Between bouts of frantic whaling and obligatory trying-out activities, there was only so much eating, drinking, fighting, holystoning of decks, repairing of sails and rigging, and yarn spinning to occupy the men on their seemingly endless voyages. To pass the time, some of them created what Clifford Ashley called "the only important indigenous folk art, except for that of the Indians, we have ever had in America; the art of scrimshaw." Although there are few contemporaneous accounts of scrimshanders at work—probably because the craft was too commonplace to mention—we assume that the teeth were carved during periods of sailing or while waiting in port for provisions or repairs. First the teeth were smoothed and polished. Next they were engraved with sailmaker's needles or what Melville referred to as "dentistical implements." Finally, ink or lampblack was rubbed into the etched surface.

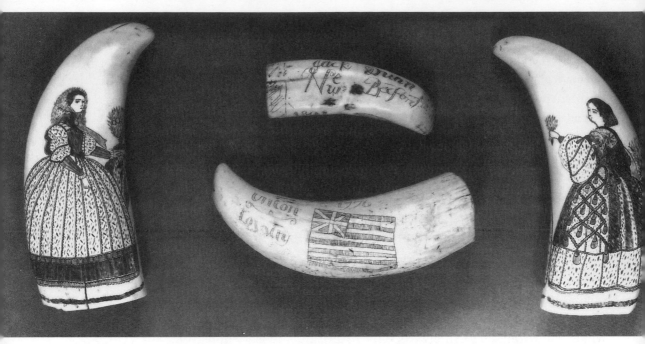

Scrimshaw was the art of decorating sperm whale teeth with images that often reminded the whalemen of home. (New Bedford Whaling Museum)

The predominant subjects carved onto the teeth were whales, ships, and whaling scenes, along with reminders of sweethearts, family, and home. The Yankee scrimshanders were best known for their decorated whale's teeth, but they also fashioned belaying pins, corset stays, canes, knife handles, dominoes, pie crimpers, and all sorts of tools and boxes. Whales have bones like any other mammals, but with the exception of the teeth and the lower jaw (known as the pan bone), whale bones are too porous for carving. Other cultures have also recognized the decorative possibilities of whale teeth. Certain Polynesian natives made necklaces of dolphin teeth, and the premissionary Hawaiians crafted the beautiful *le niho palaoa*, a gracefully curved sperm whale tooth that was worn by royalty on a necklace of braided human hair.

Melville's admonition—that "you had best not be too fastidious in your curiosity touching this Leviathan"—did not stop Leonard Baskin, Jackson Pollock, Frank Stella, Barry Moser, and many other artists, even though only a few (one of whom was me) tried to depict the white whale himself, on canvas, paper, or metal. Rockwell Kent (1882–1971) illustrated a 1930 Random House edition of *Moby-Dick* with a series of black-and-white drawings, which, in my opinion, brilliantly captured the essence of Melville's work; he managed to picture the various whale species remarkably accurately, even though hardly anything was known about the appearance of some of these leviathans when Kent was working. Also in 1930, the Lakeside Press published an oversized three-volume aluminum-slipcased edition of 1,000 copies, with the same typography and illustrations as the Random House version, but because everything is much larger, the Lakeside volumes come closer to the monumentality of Melville's original work.*

An oversized version of *Moby-Dick* was issued in 1979 by Arion Press in an edition of 265. It was described as "the most majestic presentation of America's most monumental novel." Subsequently, this 15 × 10–inch leather-bound volume was reissued by the University of California Press, at a reduced size, but still, according to the introduction, "a page with handsome dimensions, a page that, although smaller than the original, is still generous and suitable to the scope of the text." This book was profusely illustrated by Barry Moser (1940–), a well-known printmaker and student of Leonard Baskin, whose wood engravings and pen-and-ink drawings enhance Melville's narrative, but whose whale portraits are somewhat impressionistic. Of Moser's drawings, Elizabeth Schultz, a professor of English at the University of Kansas, says, "Moser has drawn his sperm whale in the manner of such well-known contemporary marine artists as Richard Ellis, who succeeded in rendering 'the true form of the whale' by combining scientific accuracy with artistic sensibility."

That there has been a succession of special, limited, and often expensive editions of *Moby-Dick* is a testimony to the importance and durability of the novel in American literature. Characteristically for its period, however, the original *Moby-Dick* contained no illustrations, relying instead on the author's descriptive powers and the imagination of the reader to portray the novel's events, personnel, or equipment. Of the whales themselves, Melville wrote, "I shall ere long paint to you as well as one can without canvas, something like the true form of the whale as he appears to the eye of the whaleman when in his own absolute body the whale is moored alongside the whale-ship so that

*Melville's novel, originally titled *The Whale*, was published in three volumes in September 1851 by Richard Bentley in England. Two months later, with its title changed to *Moby-Dick; or, The Whale*, it was brought out by Harper & Brothers in New York. Nowadays, a clean copy of the Random House edition sells for about $1,500; the Lakeside Press edition for $10,000 to $15,000, depending on condition; and you can buy a copy of the rare first American edition (many of the existing copies were destroyed in a warehouse fire in 1856), in the original blue cloth binding, for between $60,000 and $85,000—if you can find one for sale.

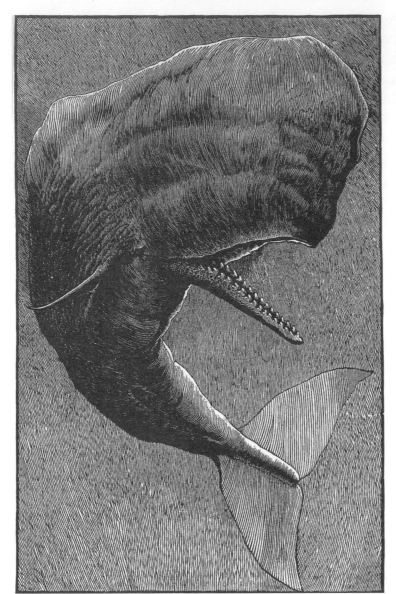

Barry Moser's excellent portrait of a sperm whale, from his 1978 illustrated edition of *Moby-Dick* published by Arion Press. (Courtesy of Barry Moser)

he can be fairly stepped upon there." Melville's vivid descriptions and the dramatic nature of the events and the dramatis personae cry out to be illustrated, and it was not long before artists began enhancing Melville's story. According to Elizabeth Schultz, "Since 1996, seventy diverse editions of *Moby-Dick* have been printed in English, the most recent in 1994; included in this figure are eight comic-book versions with variant reprint editions, four editions with photographs from movie versions of the novel, and five editions with maps, diagrams, historical engravings, and photographs. Illustrated editions of the novel have appeared in thirty-one languages, with as many as 132 artists having translated it into pictures."

In 1995, the University Press of Kansas published Schultz's *Unpainted to the Last: Moby-Dick and Twentieth-century American Art,* a comprehensive work (for which my sketch for the New Bedford mural was used as the cover illustration) that incorporates the work of many well-known artists—in every medium from watercolor to metal sculpture, and in every genre from realism to abstract expressionism—who have tried to capture the majesty and tragedy of Melville's great work in visual terms. The title of the book comes from Melville's observation that

> any way you may look at it, you must needs conclude that the great Leviathan is that one creature in the world which must remain unpainted to the last. True, one portrait may hit the mark much nearer than another, but none can hit it with any very considerable degree of exactness. So there is no earthly way of finding out precisely what the whale really looks like. And the only mode in which you can derive even a tolerable idea of his living contour, is by going a whaling yourself; but by so doing, you run no small risk of being eternally stove and sunk by him.

An oversized, profusely illustrated book of 382 pages, *Unpainted to the Last* validates the monumental importance of *Moby-Dick* in American arts and letters. It seems that almost everybody wanted to have a go at Melville's magnum opus, perhaps because they wanted to make contact with the enduring greatness of the novel. I know I did; when asked to paint a mural for the New Bedford Whaling Museum, I jumped at the opportunity to include the single most famous animal in American literature.

Beyond art and literature—or at least alongside them—is the cinema, another art form in which Melville's novel has been reinterpreted, rearranged, recast, and even rewritten. *Moby-Dick* first appeared in movie houses in 1926 as a silent film called *The Sea Beast,* with John Barrymore in the lead role. The only thing it had to do with Melville's novel was the inclusion of Ahab, a peg-legged whaling captain. He now has a last name ("Ceely"), a half-brother, Derek, and a girlfriend, Esther Harper, played by Barrymore's wife to be, Dolores Costello. The moguls at Warner Bros. obviously believed that Melville's story was too tame (and underpopulated), so they provided an assortment of ancillary

characters, a ridiculous papier-mâché whale, and a plot that empha-
sizes Ahab's love life (and Barrymore's famous profile) far more than
his pursuit of the white whale. In this version, brother Derek pushes
Ahab overboard into the jaws of the whale, neatly justifying Barrymore's hatred of his brother and the offending whale. Later, during a
typhoon, Ahab returns the earlier favor by pushing Derek into the sea,
and because Barrymore, then known as the world's greatest actor, was
so popular that audiences obviously would not stand for his being killed
off, the poor whale has to die instead.* Playing at the Warner Theater in
New York, the movie was an enormous success and took in $20,000 a
week as thousands of people a day were turned away.

Barrymore made another film version of *Moby-Dick*, and although
the story was no closer to the original than *The Sea Beast*, at least Melville's title was used. By 1930 they had obviously not been informed of
the resuscitation of Melville's reputation, so they followed the screenplay of *The Sea Beast*, basing it again on the traditional Hollywood
chestnut, the love triangle. Once again, Ahab Ceely woos a starlet,
but Dolores Costello, now married to Barrymore and pregnant with
his child, was replaced by Joan Bennett, whose role as Faith Mapple
replaces that of Costello's Esther Harper. Brother Derek appears again
in the 1930 version, as do other characters who have less than nothing
to do with Melville's novel, such as Whale Oil Rosie. Once again, Barrymore flashes his profile, the whale loses, and peg-legged Ahab returns
to New Bedford to his true love. In her biography of the Barrymores,
Margot Peters wrote,

> Compared to his appearance in *The Sea Beast* only four years before,
> Jack is shockingly older, something which his opening acrobatics up

*In 1934, Whitman Publishing Company of Racine, Wisconsin, issued one of the
strangest versions of *Moby-Dick* I have ever seen. Its full title is *The Story of Moby
Dick, the Great White Whale*. This Big Little Book (about 5 inches square) acknowledges no author, but the title page tells us that it was "*Adapted from the Novel by* Herman Melville" and "*Illustrated with Scenes from* 'The Sea Beast,' A Warner Brothers
Picture *Featuring* John Barrymore." As with the 1926 movie whose stills are reproduced on every other page, the book's narrative bears only the faintest resemblance
to the original novel, but it follows Ahab's adventures in whaling, love, and madness,
with brother Dereck [*sic*] betraying him, and his fiancée, Esther Harper, awaiting
his return. In a startling reversal of Melville's story, as Ahab enters a tavern after his
voyage, he proudly announces, "I have killed Moby Dick."

The first movie version of *Moby-Dick* was shot in 1926 but called *The Sea Beast*. In this silent movie, John Barrymore ("The Great Profile") starred as Captain Ahab, and Dolores Costello played Esther Harper, Ahab's love interest. (Richard Ellis Collection)

in the crow's nest only emphasize. This Ahab is more a caricature of John Barrymore as a drunk and a make-out artist; Jack seems drunk from beginning to end. . . . Early sound was notoriously poor, but that does not explain why Ahab talks like an old cowhand: "Beg pardon, ma'am. Aint'cha friends anuf ta intraduce me?"

Almost lost in the shuffle is the eponymous whale, which appears in the distance as a humpback and in the leg-chomping and whale-stabbing scenes as a large, shapeless lump. (Despite several references to "the white whale," this whale is always black.) Probably having more to do with Barrymore's popularity than Melville's, this version, too, was an enormous success, and a reviewer for *Theatre Magazine* (displaying a certain unfamiliarity with literature) wrote, "Altogether *Moby-Dick* is a highly creditable, moving record of Melville's stunning tale."

Remember the way book-derived movies used to open? A hand turns the pages of a book, so we will know that we are seeing a work of literature transposed to the silver screen. But what are we to make of the omission of the most famous opening sentence since "In the beginning . . ."? In place of "Call me Ishmael," we get this: "There never was, nor ever will be, a braver life than the life of a whaler." Good-bye Melville, hello Hollywood. Not only are Daggoo and Tashtego absent, but Ish-

The 1930 talking version, now named *Moby Dick*, again starred Barrymore as Ahab, but because Dolores Costello was pregnant with Barrymore's child, she was replaced by Joan Bennett. To eliminate any confusion, Esther Harper was renamed Faith Mapple. (Courtesy of Stephanie Guest)

mael also seems to have missed the boat. They probably had to leave them out because the book was so long.*

The film opens with a lone figure hanging from the masthead. It is our hero, looking for all the world like Errol Flynn, as he swashes, buckles, and hurls harpoons onto the deck below him for the sheer fun of it. When he finally swings down, he comes ashore and meets the demure Faith, who immediately falls in love with him. She had been keeping company with Ahab's stuffy brother, Derek, but upon meeting Ahab, she rejects wimpy Derek's proposal and pledges to marry the dashing whaleman when he returns in three years.

The *Mary Anne* sets sail, and we meet the only other character retained from the original novel, Queequeg. But he is not the cannibal that Melville described; nor is he a harpooner, for that occupation is reserved for Ahab. He is a black man, played by actor Noble Johnson, with a haircut like Mr. T, who is given to worshipping heathen idols and playing a conga drum. Cut to the ship at sea. High in the cross-trees, Ahab cries "Thar' she blows!" and shinnies down a line to take his place in the whaleboat. We see a whale spouting in the distance, but instead of a sperm whale, it has become a humpback. (Obviously, someone had filmed humpbacks at sea and the footage was inserted here under the assumption that nobody would know the difference.) In the scenes that follow, Ahab harpoons the whale, and after his boat is overturned, he is attacked by the whale, which looks remarkably like a giant potato. Hauled back on deck, Ahab's mangled leg is operated on by the blacksmith, a burly fellow who needs only a spreading chestnut tree to round out his character. Then the unfortunate Ahab lies in his bunk (*for two years?*) until the ship is ready to return to New Bedford and his lady love.

*Curiously, the 1922 silent movie, *Down to the Sea in Ships*, which has nothing whatever to do with Melville's novel (except that it is about whaling), opens with this stirring—but slightly modified—quote from *Moby-Dick:* "We count the whale immortal in his species. . . . He swam the seas before the continents broke water. . . . In Noah's flood he despised the Ark; and if the world is again to be flooded, then the eternal whale will still survive, and rearing upon the topmost crest of the equatorial flood, spout his frothed defiance to the skies."

As the crew gets ready for their shore leave, the carpenter brings Ahab a new leg that he seems to have made without Ahab's knowledge—a sort of homecoming present. Of course the loyal Faith is waiting on the dock. She watches everyone else leave the ship. Wondering what has become of her betrothed, she comes on board to look for him. He emerges from the fo'c'sle and tells her that she cannot possibly love him because he is not all there. When he clumps onto the deck, she takes one look at him and runs off in horror. The despondent Ahab then goes to a dockside tavern to drown his sorrows, where his crafty brother (still in love with Faith) meets him and tells him that Faith cannot love him because he is deformed. Of course nasty Derek has lied to his brother; the faithful Faith really loves him and rejects Derek again, leading to his drinking, which gets him into the dockside tavern, which gets him into a whole lot of trouble. Ahab goes back aboard the ship (which has now mysteriously become the *Boston Lass*), and sails around the world looking for the white whale.

The ship ends up in Singapore, and in another tavern, Ahab buys his own ship so that he can pursue the whale as a proper captain. We see the bill of sale for the *Shanghai Lily,* signed by Ahab Ceely. Having failed to find the white whale, Ahab returns to New Bedford, where his entire crew promptly deserts. He orders the first mate to shanghai another crew from the taverns and brothels, and sets sail again. Ahab, now deranged and bearded, asks Queequeg, "What does the wooden god say now, you jungle worshipper?" As he genuflects to his drum and his idol, Queequeg rolls his eyes like Stepin Fetchit and answers, "Him say we find Moby Dick soon."

But not before the obligatory typhoon. Amid buckets of water thrown onto the decks, Ahab hysterically steers the ship in the direction of what appears to be a waterspout but is identified as a whale. They do not lower the boats in this force-10 hurricane, but instead, the crew mutinies. (The gang of cutthroats that has been shanghaied is far and away the most disreputable crew ever to set sail, and one of them bears a distinct resemblance to Quasimodo. However, when the replacement crew watches Ahab being attacked at the film's denouement, they are the very same people that saw him the first time. I guess the more people change, the more they stay the same.) The mate tries to wake up one of the sleeping crewmen, who turns out to be none other than brother Derek, shanghaied in New Bedford, and evidently ignorant of the captain's identity for however long it has taken them to get from New Bedford to the coast of China. Upon finding out that it is his brother who is steering the ship to its doom, Derek runs on deck and confronts him. A soggy battle royal ensues, and as Derek throttles his brother, Ahab inserts his peg leg into a convenient hole—a plot device actually borrowed from the novel. Thus leveraged, he hurls the unfortunate Derek to the deck. Derek throws a knife into Ahab's back, which the stalwart captain casually removes as he takes the wheel again. Trusty Queequeg picks up brother Derek and flings him down, breaking his back. After shouting to the storm clouds that "Lucifer is my skipper now!" Ahab steers the ship into calmer waters, where, just as he is about to give up the hunt, the lookout spies Moby Dick, who appears to be the same humpback that appeared earlier in the film.

They lower the boats, and Mr. Potato Whale attacks again. Clambering aboard the whale, Ahab stabs it repeatedly, drenching himself in the whale's blood. With Ahab maniacally stabbing the whale, my copy of the film ran out. (I had taped it in the middle of the night, and had mistimed the ending.) I could have assumed that Warner Bros. adhered to Melville's ending, where the ship is sunk by the enraged whale and lost with all hands save Ishmael, but given the wild deviations from the original—and the absence of Ishmael from the screenplay—I could not safely make such an assumption. If they omitted Ishmael and added Faith and Derek, why would they bother with Melville's downbeat ending?

I was in serious trouble. I could fake it and make up an ending, but someone who had seen the entire film was bound to say that I had gotten it wrong. I contacted Warner Bros. to ask them for a copy of the film, but they couldn't help me. How could I find out how it ended? I could try to find someone who remembered the film, but 58 years is a long time. "Aha," says I, "I can ask Joan Bennett herself." Her daughter is a friend of mine, and I called her to ask if she would ask her mother if she remembered how the film ends. But when I called the daughter, she said, "My mother doesn't remember anything." Uh oh. Somehow, I had to find the ending. I took an ad out in the *Village Voice*, in which I said, "If anyone recorded 'Moby Dick' last Tuesday night, please contact me." If you know the *Village Voice*, you'll realize why this was not such a good idea. A couple of people contacted me, but they were responding more to the name of the film than to its cinematic history. One guy called me from Dannemora (do they allow you to record films in prison?) and said that he had a copy of the film, and he would send it to me if I subscribed to some porn magazines for him. Another person offered to meet me in a bar on Christopher Street. Finally somebody called and said he had recorded the film, and he would give it to me only if I brought Joan Bennett's daughter along. I did and he did.

It seems that Ahab succeeds in killing the whale, rejoins the ship, and returns to New Bedford to be reunited with Faith. Once again he swings down from the rigging (how he managed to climb up there with a peg leg is not explained), and he hobbles off to find Faith. She is demurely reading in her garden, and as she puts her book down, the two lovers embrace. "Why Ahab Ceely," she says, "you're crying." Never at a loss for the bon mot, our hero turns his famous profile to the camera, and the film ends with the immortal words "So are you." Music rises; fade to black. In true Hollywood tradition, everybody lives happily ever after—including me, who almost lost this story because of a failure of 20th-century technology.

It was not until 1956 that Melville's novel was properly brought to the screen. Produced again by Warner Bros., this *Moby-Dick* was directed by John Huston, with a screenplay written by the director and Ray Bradbury, the acclaimed science fiction writer. (Bradbury has written a "novel" called *Green Shadows, White Whale,* in which he circuitously discusses his volatile relationship with Huston, but he sheds little light on the making of the film.) The first unit worked in Ireland, and the actual whaling scenes were shot in Madeira and the Azores, while the models of the white whale—there were three in all, one of which was lost at sea—appeared in an 80,000-gallon tank at Elstree Studios outside London.

The cast consisted of Gregory Peck as Ahab, Richard Basehart as Ishmael, Leo Genn as Starbuck, Dublin drama critic Seamus Kelley as Flask, and Harry Andrews as Stubb. Friederich Ledebur, an Austrian sportsman, was made up as the cannibal harpooner Queequeg; Edric Connor, a Trinidadian calypso singer, was Daggoo; and Orson Welles played Father Mapple. There are no women in this version except those of the Irish village of Youghal (got up to look like New Bedford by removing all power lines and modern signage and refurbishing the quay) who silently watch the *Pequod* set sail in the film's opening. The whaleship was a rebuilt schooner that Alan Villiers, who supervised the seafaring aspects of the film, called "small, strained, and decrepit." The

In the 1959 version of *Moby-Dick*, Captain Ahab (Gregory Peck) prepares to hurl his harpoon at the white whale. (New Bedford Whaling Museum)

ship was decked over with a false deck, under which the actual crew sailed the ship while the actors played their roles, often in genuinely foul weather. In his biography of Huston, Alex Madsen tells the story of the filming of the typhoon scenes during an actual storm, and when Gregory Peck informed Huston that such a thing was impossible, he is quoted as saying, "It's a mistake to tell John that something is impossible." Of this film, Inge wrote:

> Huston's *Moby Dick* is clearly the most devoted effort to translate into film the letter and spirit of Melville's novel, which in its stylistic and philosophic depths ultimately defies any sort of adaptation. Bradbury obviously stood in awe of the original and fashioned a script that often uses Melville's own language, or a semblance of it anyway, and remains faithful to the plot. The earlier producers of the Barrymore versions found it impossible to bring the film to its original tragic conclusion, but Huston hesitated not a moment.

Given the total lack of love interest and the pessimistic mood of the film, it is a wonder it did well at the box office. Perhaps the exciting chase scenes carried the film. Certainly the unusual use of color to establish mood and the expert photography of Oswald Morris and Freddie Francis make viewing the film an engaging experience.

If Herman Melville was the 19th century's champion of marine life (remember, in *Moby-Dick*, the whale wins), Jacques Cousteau was his 20th-century counterpart. By 1970, Cousteau had become enormously popular as a spokesman for marine life, and on the ABC TV series *The Undersea World of Jacques Cousteau*, viewers could see for themselves the wonders of sharks, fishes, whales, and other colorful denizens of the deep. Cousteau was the inventor of the aqualung and one of the first underwater filmmakers, but he was not a marine biologist. His films were underwater adventure stories that often played fast and loose with scientific accuracy. Quoted in Brad Matsen's 2009 biography of Cousteau, Genie Clark, who *is* a marine biologist, said, "Cousteau's films are misleading in a way because they portray him as a scientist. I can't think of any particular scientific contributions he's made because he just doesn't have the time." To this, Cousteau replied, "To show the truth about nature and give people the wish to know more, I do not stand as a scientist giving dry explanations." *Undersea World* was based on a three-year (1967–1970) round-the-world cruise that Cousteau *et fils* took aboard the "research vessel" *Calypso*, filming as they went. At various locations, they encountered fin whales, gray whales, humpbacks, killer whales, and sperm whales, and they filmed these "mighty monarchs" for their television series.*

They filmed humpbacks off Bermuda, gray whales in the lagoons of Baja California, and fin whales and sperm whales in the Indian Ocean. Except for what Melville and his contemporaries had written, little was known about sperm whales at that time, and nothing at all was known about filming whales in the wild. Whenever a whale was spotted, the crew quickly launched a Zodiac and gave chase. They developed a tech-

---

The Whale: Mighty Monarch of the Sea was the title of the 1972 book in which Cousteau and Philippe Diolé recounted their adventures with the whales, and incidentally reproduced the first underwater photographs of these gigantic cetaceans. The book also includes an appendix that purports to present a summary of the natural history of whales and dolphins, and another that describes the history of whaling.

nique they called *virazeou,* which consisted of zooming around the whale "with the outboard motor running wide open. . . . The whale finally is enclosed in a circle of noise and bubbles from the Zodiac's wake. His initial reaction seemed to be one of annoyance, but he is quickly confused by the noise and the wake. Little by little he becomes quiet, as though in a stupor." They would then jump into the water to film what they hoped was a stupefied whale. Sometimes the whales swam off, so the intrepid crew had to figure out a way to round them up again. They came up with another brilliant invention that they called the *kytoon,* a combination of "kite" and "balloon." When they caught up with a whale, they harpooned it with a crossbow bolt that had a weather balloon attached, which was towed behind the whale like a kite. When that didn't work, they threw an old-fashioned iron harpoon, but this time they attached an unsinkable red buoy to the line.

On one occasion in the Indian Ocean, they spotted a number of sperm whales and rushed from one group to another in the *virazeou* maneuver, but failed to slow them down. A diver named Falco circled one young whale ("weighing only about three tons") so incessantly that the panicked whale attempted to bite the Zodiac every time the rubber boat came close enough. "Twice Falco fired his special marking harpoon at the angry whale, and twice the spear bounced off the whale's skin." The whale crashed into the Zodiac, knocking one of the crew into the water and unseating the outboard motor before swimming off. The crew continued the chase and successfully harpooned the whale with a kytoon:

> By sunset the young whale has rejoined his family. The kytoon has been attached, and we are looking forward to an interesting evening. But suddenly, the buoy stops moving, and the kytoon floats quietly over the water. Bernard and Falco rush to the buoy and pull in the line. It has not broken, and, at the end of the line, they find the harpoon attached. "These are intelligent animals," Berbet declares. "Unless someone proves otherwise, I will always believe that the adult whales pulled the harpoon out of the young whale's stomach."

We learned what some whales looked like underwater, but otherwise, we didn't learn much else from Cousteau's films, except perhaps how to stupefy a whale by buzzing around it in a motorboat, or that a whale could pull a harpoon out of an injured comrade's stomach. Nevertheless, a great many people got their first look at undersea scenery and wildlife via this TV series, and it is likely that a number of real marine biologists were inspired to pursue their careers after watching the exploits of Cousteau & Co. The book *Mighty Monarchs* provides some rare and spectacular underwater shots of sperm whales (and fin whales, gray whales, and humpbacks), and although the explanatory text was often archaic, exaggerated, or just plain wrong, it was better than nothing—which is more than you can say about the early film versions of *Moby-Dick*. Not trusting Melville's depictions of the characters and story, Hollywood put its own spin on it. First, they changed the title, then they dropped Ishmael, gave Ahab a brother and a last name, and, in the ultimate anachronistic indignity, gave him a girlfriend.

Was Starbucks coffee named for the first mate on the ill-fated *Pequod?* Chapters 26 and 27 of *Moby-Dick* (both titled "Knights and Squires") are devoted to introducing the *Pequod*'s first, second, and third mates, Starbuck, Stubb, and Flask, and two of the three harpooners, Tashtego and Daggoo. (We have already met Queequeg.) A line-by-line reading of the chapter on Starbuck reveals nothing that would support the contention that he especially liked coffee. At one point, as the *Pequod* meets the *Jungfrau*, Starbuck sees the German captain holding up something that may be a coffeepot, but that's about it for Starbuck and coffee. Yet the myth persists that the Seattle-based coffee chain was named for Melville's Starbuck because he was a coffee drinker, and that the name evoked the romance of the high seas and the seafaring tradition of the early coffee traders. According to *Pour Your Heart into It: How Starbucks Built a Company One Cup at a Time,* by Howard Schultz, who became Starbucks' sole owner when he bought out the original founders in 1987, *Moby-Dick* was indeed a book beloved of one of the founders, who proposed naming the company Pequod, after the ship. This idea was mercifully rejected by his partners (who would want to drink a cup of *Pee*-quod?), and they cast about for a name with some flavor local to Seattle. They came upon the name Starbo, from an old mining camp on Mount Rainier, and liked it. Then the *Moby-Dick* fan drew a phonetic connection between Starbo and the first mate named Starbuck. Thus the company was named. Myths such as this one die hard, and many people still believe that the name of the coffee shops originated in the pages of Melville's epic.

Outside the boundaries of art and literature, and even beyond cinema, the sperm whale's story is awe-inspiring: *Physeter macrocephalus* might be the most extreme mammal on earth—perhaps the most extreme mammal in history. In virtually every respect, the sperm whale exceeds all other whales, and because whales, by their very definition, are excessive, the sperm whale represents the pinnacle of colossal cetacean evolution. It is the largest of the toothed whales, although blue and fin whales are longer, and right whales heavier. The sperm whale has the largest brain of any animal that has ever lived. It is the deepest diving and the noisiest, although we're still unable to decipher its vocalizations. It is capable of immobilizing its prey, including sometimes giant squid, with laserlike, focused sound beams that emanate from complex components in its nose—not surprisingly, the biggest nose in history. How could it happen that this great beast still swims, dives, and breaches in today's profoundly unfathomable waters?

III

Whence the Sperm Whale?

Because of their perfected adaptation to a completely aquatic life, with all its atten-
dant conditions of respiration, circulation, dentition, locomotion, etc., the cetaceans
are on the whole, the most peculiar and aberrant of mammals. Their place in the
sequence of cohorts and orders is open to question and is indeed quite impossible to
determine in any purely objective way.—George Gaylord Simpson, 1945

EVOLUTION AS WE UNDERSTAND it is the story of various species—
make that *all* species—becoming modified over time to adapt to a
changing environment. Those that fail to adapt become extinct. Probably the most anomalous adaptation in the long and convoluted history of mammalian evolution is the return of some groups to the sea
from which they came. We know that the earliest tetrapods emerged
from the water and developed traits that enabled them to function on
land. Some went further and evolved to leave terra firma and take to
the air. But it is more difficult to imagine the evolutionary sequence
that encouraged land mammals to forsake the developments that had
brought them out of the water and, essentially throwing their history
into reverse, to go back into the water.

Of the marine mammals, some, like seals, sea lions, and walruses,
might be seen as only partially returned to the sea: they mate and give
birth on land or ice but feed in the water. Their legs have become modified into flippers, but they can walk on land. Only the cetaceans (whales
and dolphins) and the sirenians (manatees and dugongs) have adopted
a completely marine existence. Like all mammals, they are air breath-
ers, but out of the water, they are immobile and helpless, and whales
and dolphins stranded on the beach usually die.

Beginning with some carnivorous (or herbivorous) mammals that reentered the water, the cetaceans evolved into an enormously varied group that includes, with some of the larger sharks, the top predators of the sea. There are only 77 or 78 kinds of whales and dolphins, and over time, they have filled some niches already occupied by sharks and colonized others that were previously vacant. Some cetaceans live over deep water, and some can dive miles below the surface, but all must spend much of their lives near the surface because their aquatic development has persistently retained one critical link to their terrestrial forebears: they all must breathe air. Despite their occasional bathypelagic descents, there are no bioluminescent cetaceans—indeed, there are no bioluminescent mammals of any kind—but there are cetaceans that have been categorized by some as the most highly developed animals on earth. Evolution does not move toward more complexity or perfect the designs for living, because it is obvious that many of the species long extinct were as complex and "advanced" as any alive today.

But when we consider the diversity of the cetaceans that now range in size from five to 100 feet, and includes little fish eaters; coastal or riverine creatures that are exquisitely attuned to sound; deep divers that can hunt squid in the inky, icy depths; fast-moving ocean rangers that live in huge schools; and 100-ton giants that have achieved a body mass attainable only by creatures rendered neutrally buoyant by the displacement of the water they live in, we get an inkling of the incredible subtlety and richness of the evolutionary process. From a terrestrial hoofed hyena, a miscellanea of sleek water dwellers has evolved, breathing from the top of their heads, communicating with one another in ways that we have not been able to decode or understand, living in a watery world so alien to us we that we can barely enter it without benefit of artificial aids, such as face masks, snorkels, scuba gear, or submersibles.

The three major groups of cetaceans—the baleen whales (Mysticetes), the toothed whales (Odontocetes), and the extinct whales (Archaeocetes)—have more homologies than differences. They all have (or had) dense ear bones, space around the bones for fat deposits, air sacs to isolate the ear from the skull, a long palate, and nostrils located on top of the snout. They all have a long body and a short neck (often with the cervical vertebrae fused), reduced hind limbs (usually completely absent), paddle-shaped front limbs, and a long tail designed to move up and down. All the living whale species are completely aquatic and perform all their activities in the water, including feeding and giving birth to live young. In recent years, more ancestral whales have been identified, and although some of them appeared to have been at least partially aquatic, some were terrestrial, and this has led to a certain degree of confusion and controversy about how whales developed.

Even if we assume that sperm whales did not arrive several million years ago in a whale-sized spaceship that crashed in the middle of the Pacific Ocean, releasing a leviathanic cargo bent on oceanic domination, the question of where *Physeter* came from (and when) is still largely unresolved. The earliest vertebrates lived in the sea and ultimately evolved into land animals. The earliest known whale ancestors were quadrupeds with fur, four legs, and nostrils at the end of the nose, so in order to become fully aquatic, they would have to become hairless; lose at least two legs (the hind ones, which would be replaced by a horizontal tail fin, while the forelegs turned into flippers); have their nostrils migrate to the top of their head (the better to breathe at the surface without having to slow down with every breath to poke one's snout out of the water); and, at least in some whales, lose the teeth altogether and replace them with a unique feeding system, the baleen plate apparatus, which would enable them to take advantage of a heretofore unexploited food source.* The baleen whales lost their teeth, but *Physeter's* ancestors did not. Still, the pathway from furry terrestrial quadruped with teeth in the upper and lower jaws to hairless, big-headed cetacean with a narrow lower jaw armed with peglike ivory teeth is not easy to discern in the paleontological understory of whale evolution.

*In a 2007 study, David Lindberg and Nick Pyenson of the University of California, Berkeley, hypothesized that toothed whales developed sonar to chase squid swimming at night at the surface. The toothed whales, say Lindberg and Pyenson, first occupied the ocean and responded to the diel cycle of cephalopods—they come close to the surface at night and descend to the depths by day—and then developed the wherewithal to follow them and echolocate the prey in the depths. The "evidence" for this theory can be found in the development of a scooped-out forehead (to house the sound-focusing material) and of bone structures that might have been used to generate sounds. The animal with the most scooped-out skull of all is the sperm whale, and although Lindberg and Pyenson do not discuss the development of the sound-generating elements in the whale's spermaceti organ (which would not fossilize), *Physeter* is the living proof that certain odontocetes echolocate their cephalopod prey in the depths.

Remington Kellogg (1892–1969) entered the University of California, Berkeley, in the fall of 1916 in pursuit of a PhD. He began by studying zoology, but the following year his academic career was interrupted by World War I. He became a sergeant in the 20th Engineer Battalion, where, because he was an incipient naturalist, he was assigned the job of rat control in the trenches of France. Kellogg returned to Berkeley in 1919, where Professor John C. Merriam of the paleontology department encouraged him to study the rich record of fossil marine mammals being unearthed on the Pacific coast. He described fossil pinnipeds and cetaceans from the Sharktooth Hill in Kern County (California), a fossil sea cow and humpback whale from Santa Barbara County, and marine mammal fossils from other parts of North and South America. His 1928 PhD dissertation, "The History of Whales—Their Adaptation to Life in the Water," was considered so important that it was published as a two-part article in the *Quarterly Review of Biology*. In 1928 Kellogg left the Biological Survey and transferred to the division of mammals at the U.S. National Museum (now the National Museum of Natural History) at the Smithsonian Institution. His PhD thesis had established him as an authority on cetaceans, and with concern growing over the need to protect whales from overexploitation, he was invited to speak at a 1930 conference on whaling held by the League of Nations. Further conferences followed, and Kellogg was appointed as a U.S. delegate to the International Conference on Whaling, held in London in 1937, which resulted in the International Agreement for the Regulation of Whaling. Kellogg was head of the U.S. delegation in two further conferences, in 1944 and 1945, and was chairman of the 1946 conference, after which he became the U.S. commissioner to the International Whaling Commission from 1949 to 1967. He served as vice chairman of the commission from 1949 to 1951 and chairman from 1952 to 1954. Remington Kellogg died before Gingerich, Thewissen, and other cetologists published their revolutionary discoveries of the ancestral whale fossils from India and Pakistan, but his monograph is still considered a benchmark summary of whale adaptations. In the introduction, Kellogg defined his subject matter:

Cetaceans are air-breathing, warm-blooded mammals, generally having pointed heads, torpedo-shaped bodies, fin-like forelimbs, and horizontal caudal flukes. All external structures which might offer resistance to the water have been either eliminated or sunk below the surface level. This group embraces the extinct zeuglodonts (Archaeoceti), the whalebone whales (Mysticeti), and the toothed whales (Odontoceti), including porpoises, dolphins, beaked whales, and sperm whales.

There are no whale fossils older than the middle Eocene, 53 million years ago, and although such a negative statement cannot be used to prove that whales did not exist before then, it is currently believed that whales arose around that time, descended from terrestrial mammals known as mesonychids, also considered to have been the ancestors of modern ungulates, such as horses and pigs. Analysis of postcranial fossil material (Zhou et al. 1992; O'Leary and Rose 1995) has clearly shown that they were hoofed, omnivorous, running animals, designed for endurance rather than speed. In 1968, Leigh Van Valen, a research associate at the American Museum of Natural History in New York, wrote,

> Only two known families need to be considered seriously as possibly ancestral to the archaeocetes and therefore to the ancestral whales. These are the Mesonychidae and Hyaenadontidae (or just possibly some hyaenadontid-like palaeoryctid). No group that differentiated in the Eocene or later need to be considered, since the earliest known archaeocete, *Protocetus atavus,* is from the early middle Eocene and is so specialized in the archaeocete direction that it is markedly dissimilar to any Eocene or earlier terrestrial mammal. It is also improbable that any strongly herbivorous taxon was ancestral to the highly predaceous archaeocetes. . . . Diverse and apparently equally valid objections exist for the various groups of Paleocene insectivores, one common to all being their small size. All marine mammals are large or rather large mammals.

The mesonychids lasted for about 20 million years, from the middle Paleocene to the Oligocene. They are among the most enigmatic of fossil mammals, not only for their unexpected appearance in the genealogy of whales, but also for the confusion about what they actually were. We have unimpeachable evidence of their existence, but the fossils send conflicting messages, and interpretation differs widely. When Edward Drinker Cope published a description of an early mesonychid called *Pachyaena* in 1884, he said that its hind legs were higher than its forelegs, that it was a good swimmer, and that its teeth appeared to be designed for catching fish—all of which were probably incorrect. Cope also said that when it walked, its soles rested flat on the ground—the type of gait known as plantigrade and seen today in bears—but when Maureen O'Leary reexamined the *Pachyaena* fossils, she concluded that they had hooves and therefore belonged to the mammalian group known as ungulates. (A hoof is a modified claw that consists of a dorsal, scalelike plate, known as the *unguis* [Latin *unguis*, "fingernail"], and a softer ventral plate, the *subunguis*. In a hoof, like that of a horse, the unguis is usually curved so that it encloses the subunguis laterally, but the ventral portion of the digit—the part the horse walks on—is not covered or enclosed.)

Now classified as ungulates because they had hoofs, the mesonychids did not behave at all like today's herbivorous ungulates, which include the even-toed artiodactyls (pigs, giraffes, sheep, goats, deer, antelopes, hippos, cattle) and the odd-toed perissodactyls (tapirs, rhinos, horses). They were carnivorous ungulates that resembled big-headed dogs or skinny bears. Probably the best documented is *Mesonyx*, meaning "middle claw," which is also the source of the name *mesonychid*, its adjectival form. It was about the size of a wolf, with a disproportionately large skull equipped with powerful jaws and teeth. Where canids have claws, the mesonychids had a hoof on each digit. They ranged in size from the fox-sized *Haplodectes*, to the bear-sized *Pachyaena*, all the way up to *Andrewsarchus*, a 13-foot-long mesonychid with huge crushing and tearing teeth—the largest carnivorous mammal that ever lived on land.

Andrewsarchus mongoliensis, "the giant mesonychid of Mongolia," was described by Henry Fairfield Osborn in 1924 and was first thought to be a huge omnivorous pig. When the skull reached the American Museum in New York, vertebrate specialist W. D. Matthew immediately recognized it as a gigantic version of *Mesonyx*. In his awed description, Osborn wrote, "This is the largest terrestrial carnivore which has thus far been discovered in any part of the world.* The cranium far

*When Osborn (1924) described *Andrewsarchus mongoliensis* as "the largest terrestrial carnivore which has thus far been discovered," either he or the proofreaders left out a critical word. It was (and still is) the largest *mammalian* terrestrial carnivore, but the dinosaur *Tyrannosaurus rex*, which had been described by Osborn himself in 1905, was 40 feet long compared to *Andrewsarchus*'s 13. Where the mesonychid's skull was 35 inches long, that of *Tyrannosaurus* was 52 inches long, and the

The whale ancestor *Andrewsarchus mongoliensis,* "the giant mesonychid of Mongolia," was first thought to be a huge omnivorous pig. Further study revealed that it was a terrestrial ancestor of cetaceans. (Richard Ellis)

surpasses in size that of the Alaskan brown bear (*Ursus gyas*), which, when full grown, weighs 1,500 lbs; in length and breadth of skull, *A. mongoliensis* is double *Ursus gyas* and treble the American wolf (*Canis occidentalis*)."

Immediately noticeable in the reconstructions of the mesonychids is the size of the head. *Mesonychus* was about the size of a wolf, but its head was as big as a bear's. (Its gigantic skull was probably the reason that Osborn thought it was a much larger animal than it actually was.) The mesonychids were probably hyena-like in their behavior, scavenging everything they found, from dead mammals to dead fish. Somehow, these terrestrial scavengers that looked like big-headed hyenas metamorphosed into short-legged semiaquatic animals that spent half their time in the water and half out. The carnivorous, terrestrial mesonychids are usually placed in the ancestry of whales, but they do not reside there comfortably. As Robert Carroll wrote in the 1988 textbook *Vertebrate Paleontology and Evolution,*

> The appropriate taxonomic rank of the mesonychids is difficult to judge. They were less diverse and long lived than most of the placental orders, but they are anatomically very different from the remainder of the early ungulates. . . . The problem is complicated by the fact that early mesonychids were almost certainly close to the ancestry of

---

skull of *Carchodontosaurus,* another gigantic carnivorous dinosaur, was even longer. The skull of the sperm whale, however, the largest carnivore that has ever lived, can be 20 *feet* long.

whales. Despite the extreme difference in habitus, it is logical from the standpoint of phylogenetic classification to include the mesonychids among the Cetacea.

In his 1998 summary of the paleobiology of cetaceans, Philip Gingerich observed that there are conflicting theories about the rise of whales, "cast[ing] doubt on our ability to reconstruct past evolutionary history from living animals." (The alternative, of course, is to reconstruct evolutionary history from fossils, but they are considerably less numerous, less complete, and often less revealing than living animals.) Although the details are far from clear, said Gingerich, "most authors now accept as a working hypothesis Van Valen's idea that Mesonychia gave rise to Archeoceti." But because of the similarity of mesonychid teeth to those of primitive whales, it is treated as fact (or as much "fact" as is available to paleontological theory) that the mesonychids are ancestral to whales and dolphins. Of the early cetaceans, Rice (1998) wrote, "some of them are so similar to mesonychids that it is difficult to decide whether to call them mesonychids or cetaceans." Gingerich (1998) compares the form of mesonychids to dogs: "Mesonychians are usually interpreted as solitary carrion feeders and scavengers that spent many of their waking hours trotting in search of dead animals and were best able to chew flesh after it was partly decomposed. This is plausibly the kind of animal from which the archaeocetes evolved."

Like the process it describes, the story of the evolution of whales occurred in stages. Before 1983, when Gingerich published the description of *Pakicetus,* the first ancestral whale with hind legs, all accounts of cetacean evolution were devoted to legless whales, like the well-known *Basilosaurus.* (Actually, *Basilosaurus* did have hind legs, albeit small ones, but the evidence was not found until 1990.) The first of what came to be known as archaeocetes, *Basilosaurus,* which reached a length of 70 feet and had a small head with a mouthful of sharp teeth, might also have had horizontal tail flukes and a dorsal fin. Until *Pakicetus,* however, the fossil record seemed devoid of any creature that could have become a whale; all we knew was that there were extinct whales like *Basilosaurus,* and then there were fossil whales that were directly referable to living cetaceans. Because it was obvious that whales were descended from terrestrial mammals, and because it was equally clear that they could not have developed spontaneously, the absence of any

When it was discovered in 1832, *Basilosaurus* was named "king of the lizards" because it was thought to be a giant reptile. It was later shown to be an early whale, but the rules of zoological nomenclature mandated that the original name be applied. (Richard Ellis)

transitional forms made cetacean evolution the most enigmatic and troublesome of studies.

Whales and dolphins are descended from land mammals, and they have the anatomy to prove it. They are warm-blooded, have lungs to breathe air, suckle their young, and have rudimentary facial hair. In addition, the flippers of cetaceans contain the characteristic five-fin-gered *manus,* and many large whales have vestigial hind limb bones buried deep within their bodies. Paleocetologists differed on what sort of creatures might have led to the development of whales. Not only was there no evidence for the transition of whales from land to water, but the only evidence we had seemed to be an evolutionary dead end. The study of cetacean evolution was at an impasse. Vertebrate paleontolo-gists knew there had to be something connecting the ancestral whales to their living descendants, but they were stuck with what Darwin called the "imperfection of the geological record."

Basilosaurus, discovered in Louisiana in 1832 by Richard Harlan, was first thought to be a reptile—hence its name, which means "king of the reptiles," from the Greek *basilikos,* for "royal." But nine years later, Richard Owen realized that it was a mammal and renamed it *Zeu-glodon,* a reference to its yoked teeth, from the Greek *zeugos* for "pair." However, the rules of scientific nomenclature dictate that the first name has priority, no matter how misguided. This early whale reached a length of 70 feet, but it was considerably less bulky than the levia-thans of today. Reconstructed from numerous fossils found in Louisi-ana, Mississippi, Alabama, Australia, and Egypt, *Basilosaurus* was a stretched-out animal, with a proportionately tiny head, greatly elon-gated vertebrae, powerful foreflippers, and hip bones that supported tiny, fully formed hind legs. It had a mouthful of differentiated teeth, cone-shaped in the front of the jaw for catching its prey, and multi-cusped molars with double roots ("zeuglodont") in the rear of the jaw.

The first *Basilosaurus* specimens, found and inappropriately named by Harlan, were believed to have no hind legs. A second species (*Basilosaurus isis*), found by Gingerich, Smith, and Simons (1990) in Egypt, was equipped with complete, tiny hind legs, but they were so small that they had to have been useless. Puzzled by these tiny, inadequate limbs, Gingerich and his colleagues wrote, "Hind limbs of *Basilosaurus* appear to have been too small relative to body size to have assisted in swimming, and they could not possibly have supported the body on land." In their discussion of the hind limbs, Gingerich et al. suggested that "the hind limbs of *Basilosaurus* are most plausibly interpreted as accessories facilitating reproduction. Abduction of the femur and plantar flexion of the foot, with the knee locked in extension, probably enabled hind limbs to be used as guides during copulation, which may otherwise have been difficult in a serpentine aquatic mammal." Annalisa Berta (1994) wrote that they "could just as reasonably be interpreted as vestigial structures without a function."

Hind legs appear to be an artifact of the evolution of four-legged animals to an aquatic existence because they appear not only in early cetaceans, but also in the marine reptiles such as ichthyosaurs, plesiosaurs, and mosasaurs, which are also believed to have descended from terrestrial forms. There are a few recorded instances where living whales were found with what have been described as vestigial hind legs, including a humpback (Andrews 1921), a sperm whale (Ogawa and Kamiya 1957), and a striped dolphin (Ohsumi 1965). These atavisms are usually cited as evidence that whales are descended from terrestrial mammals—which they surely are—but the tiny bones embedded in the muscle of the pelvic region seem a much stronger argument than protruded hind limbs. After all, mammals are occasionally born with anomalous mutations—think of two-headed calves—and nobody argues that cows are descended from animals that had two heads.

If we speed up the tape of history, we will see fishlike creatures with stubby legs emerge from the water; their legs will grow longer and sturdier as the tetrapods disperse to colonize the terrestrial habitats, and then, in a surprising move, they stand at the water's edge and look hungrily at the fishes swimming in the shallows—and return to an aquatic existence, their legs either transformed into flippers or eliminated altogether. The cetaceans, therefore, after evolving for several hundred million years, are shaped not unlike their piscine forebears, but they have persistently retained a trait that is a serious handicap for life in an aquatic environment: they have to breathe air.

In a 1998 study, Gingerich wrote, "We now know, thanks to the fossil record, that the modern orders Mysticeti and Odontoceti have a fossil record extending back to the Oligocene, and that they are thought to have diverged from each other sometime in the late Eocene or early Oligocene, some 50 million years ago. Whales that are known from the earth's Eocene rivers and oceans all belong to a third suborder, Archaeoceti, which is a group with much more specialized morphology."

The oldest known whale fossil identified to date is *Himalayacetus*, described by Bajpai and Gingerich in 1998. They found only a partial jawbone with teeth, but the animal is said to be an archaeocete because its molars closely resemble those of the mesonychids, and there was a small canal in the mandible that suggests underwater hearing. *Himalayacetus*, which is 53.5 million years old, was found in the Simla Hills of northern India, which area was underwater (as the Tethys Sea) when India was an island off the Asian continent. *Himalayacetus*, which was probably seal-like in size and habits, is believed to represent the beginning of the transition from terrestrial mesonychid to aquatic archaeocete. It is, however, more than a little difficult to postulate its complete life history and affinities from a single jawbone.

When *Pakicetus* was first discovered in the Himalayan foothills in 1979, it was hailed by paleontologists—and almost everyone else except the creationists—as the missing link between terrestrial mammals and whales. Because all that was found was a part of the skull, the appearance of the rest of the animal was not obvious, and although Gingerich and his colleagues named it and described it in 1983 as an ancestral whale, there was no way of telling whether it had hind legs, flippers, or flukes. Its teeth were not unlike those of the giant mesonychid *Andrews-*

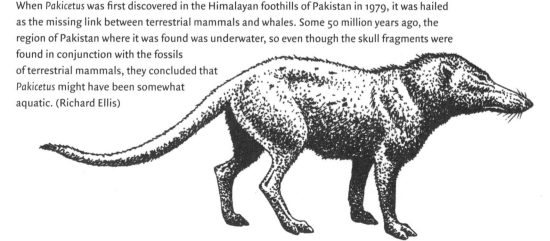

When *Pakicetus* was first discovered in the Himalayan foothills of Pakistan in 1979, it was hailed as the missing link between terrestrial mammals and whales. Some 50 million years ago, the region of Pakistan where it was found was underwater, so even though the skull fragments were found in conjunction with the fossils of terrestrial mammals, they concluded that *Pakicetus* might have been somewhat aquatic. (Richard Ellis)

archus, and it was probably a completely terrestrial animal. The find also included ear bones, and because they were similar to those of a known archaeocete, it was concluded that it was a protowhale because it could probably hear well underwater. Some 50 million years ago, the region of Pakistan where it was found was underwater, so even though the skull fragments were found in conjunction with the fossils of terrestrial mammals, they concluded that it might have been somewhat aquatic.

When Jerold Lowenstein (1983) read of this find, he wrote an article in *Oceans* magazine entitled "Very Like a Whale" (taken from a quote in *Hamlet,* where Polonius is describing cetacean-shaped clouds), in which he suggests that *Pakicetus* is not only *not* an ancestral whale, but it was probably a creature whose lineage died out long before it could give rise to the cetaceans. His was a lonely voice, however, because there is almost complete unanimity on *Pakicetus*'s place in the lineage of cetaceans, and since its discovery, more fossil whales have been found in the same location.*

Found later but believed to have lived earlier is the archaeocete *Nalacetus,* named by Thewissen and Hussain (1998) for the Urdu word *nala,* which means "seasonally dry riverbed," the location of the fossil find in the Kuldana Formation of Pakistan. As shown by measurements of the teeth (all that was found of this species was a mandible with fragmentary teeth), *Nalacetus* was intermediate in size between the other pakicetids, *Ichthyolestes* and *Pakicetus,* and like them, it dates from the early Eocene at approximately 49.0 to 49.5 million years ago. It is therefore considered among the oldest of all archaeocetes. In O'Leary and Uhen's (1999) review of the origin of cetaceans, they say, "The fragmentary fossil *Nalacetus* was found to be the most primitive cetacean, based on improved understanding of character transformation in the dentition of cetaceans."

Next in the proposed evolutionary sequence of the development of whales is *Indocetus ramani,* found by Gingerich and colleagues after

---

*Lowenstein's response was not that far off, especially since the later species had not yet been found. In an article in *Discover* magazine in January 1995 ("Back to the Sea," by Carl Zimmer), Gingerich is quoted as saying, "We were making it up before; now we don't have to." Lowenstein was probably correct anyway, because the path from the mesonychids through *Pakicetus* and to the whales of today is more likely a succession of false starts rather than a continuum.

the *Pakicetus* fossils but placed earlier in the sequence. Only the leg bones were found, but they were clearly from an early whalelike animal, and their large size encouraged Gingerich et al. (1993) to erect the new genus *Indocetus* and place it between *Pakicetus* and *Basilosaurus*. Of the bones they wrote:

> The pelvis has a large and deep acetabulum, the proximal femur is robust, the tibia is long. . . . All these features, taken together, indicate that *Indocetus* was probably able to support its weight on land, and it was almost certainly amphibious, as early Eocene *Pakicetus* is interpreted to have been. . . . We speculate that *Indocetus*, like *Pakicetus*, entered the sea to feed on fish, but returned to land to rest and to birth and raise its young.

Upon closer inspection, *Indocetus* was renamed *Remingtonocetus harudiensis* (Gingerich et al. 1995), a species for which the skull was already known. *Remingtonocetus* (named for Remington Kellogg) was originally described by Kumar and Sahni in 1986, from fossils found in western Kutch (now known as *Kachchh*) on the India-Pakistan border. The remingtonocetids are 43 to 49 million years old and are characterized by a narrow and elongated skull with correspondingly long and narrow mandibles filled with sharp teeth. As Kumar and Sahni wrote,

> The overall dentition of *Remingtonocetus* is well adapted for a dominantly carnivorous food habit. Its premolars probably served both purposes, i.e., grasping the prey as well as its initial cutting. The extremely elongated premolars could have enabled the animal to grasp small as well as large prey. A long-snouted animal like *Remingtonocetus* would face much difficulty in retaining large prey in its mouth. This problem has been overcome by sharp and elongated anterior teeth, which will kill the prey rapidly and not allow the prey to slip out.

Fordyce and Barnes (1990) described the "bizarre" remingtonocetids as "a short-lived archaeocete clade characterized by a long narrow skull and jaws with cheek teeth placed relatively far forward of the eyes." Because they are both freshwater inhabitants with long, narrow, tooth-filled jaws, it seems not unreasonable to compare *Remingtonocetus* to *Platanista*, today's Ganges River dolphin. It is believed that the remingtonocetids died out without leaving any descendants, and the living dolphins, which live in the turbid, murky waters of the Indus and Ganges rivers, have greatly reduced vision, and specialized bony processes in the skull that enhance their echolocating abilities.

Again in Pakistan in 1992, Hans Thewissen, then a student of Gingerich's, found a skull, ribs, and leg bones of a fossil that appeared to him (and his colleagues, Madar and Hussain) to be an animal that might fall between the mesonychids and *Pakicetus*. It was fully described in an 86-page paper published in 1996. The skull of "the walking whale that swims" was huge and the jaws powerful; it had strong forelegs, flipperlike hind legs, and a long, thick tail. Thewissen, Hussain, and Arif (1994) wrote that it "was the size of a male of the sea lion *Otaria byronia* (approximately 300 kg)," but its long tail accounted for much of its 12-foot length. The forelimbs were similar to those of a sea lion, and in the 1994 reconstruction, it is shown with its forelimbs arranged like those of a sea lion, where "the semipronated elbow left the hands sprawling when the shoulder was abducted and the wrist extended," meaning that its hands were splayed to the sides or even slightly tailward, as in sea lions.

In accordance with its name, *Ambulocetus natans* could walk on land and swim in the water. A careful examination of its leg bones and vertebrae suggested to the authors that it was not a fast swimmer, but that it probably swam by moving its webbed hind feet up and down through the water like a sea otter. It didn't look much like a whale; the reconstruction drawing in Thewissen et al. (1996) shows it looking like a giant, web-footed shrew, and other illustrators have portrayed it

Known only from fossil evidence, *Ambulocetus natans*, the "walking whale that swims," had powerful forelegs, webbed feet, and a long, thick tail that accounted for much of its 12-foot length. (Richard Ellis)

as a sort of furry crocodile—and it probably fed crocodile-fashion by ambushing its prey in the shallows. Like a crocodile, its jaws and teeth were powerful and its head disproportionately large. In the drawing of the skeleton, it looks as if the head of a crocodile has been affixed to the body of an otter.

As to the importance of *Ambulocetus* fossils in the history of mammalian evolution, Thewissen et al. (1996) wrote,

> The acquisition of aquatic adaptations by cetaceans resulted in a number of major morphological changes. This kind of revolutionary change is common in the long history of life, as new morphologies were typically acquired early in the diversification of major clades. However, few of the transitional forms that arose are preserved in the fossil record. The origin of whales has become an exception to this rule; it is better documented than most other morphological transitions. As such it forms one of the best examples of evolutionary change across a strongly constraining environmental threshold.

The discovery of a whale that walked prompted many scientists to reexamine the very definition of the term *whale*. The living odontocetes and mysticetes are clearly whales because they are all vertebrate mammals that are fully adapted to a marine existence. Modern cetaceans do not walk because they have no legs and they never leave the water. They all have flippers in place of forelegs, no visible hind limbs, and they propel themselves through the water by flexing their horizontal flukes up and down. They breathe air through single or paired blowholes on top of the head, and they give birth to live young that they nurse underwater. The only other completely aquatic mammals are the sirenians (manatees and dugongs), which fulfill all of these criteria except they breathe through nostrils at the front of the head, and they are descended from a completely separate line of mammals. In a 1994 article in *Science* that accompanied Thewissen, Hussain, and Arif's (1994) discussion of the first walking whale, Annalisa Berta wrote:

> Whales can be defined in several different ways, emphasizing either possession of certain characters (character-based definition) or with respect to certain ancestry (stem-node- and apomorphy-based definitions). Character-based definitions are problematic. For example, how can whales be defined as lacking hindlimbs, since some whales

(for example, several archaeocetes) possess them? Another problem arises considering that discoveries of ostensible whales occur quite frequently with new combinations of characters making it difficult to decide whether they are whales following a strictly character-based definition. A more reasonable solution is to use a phylogenetic definition, that is, one based on a common ancestry. In the previous example, because archaeocetes are more closely related to modern whales than they are to mesonychids, *Ambulocetus* is a whale by virtue of its inclusion in that lineage. Evidence to support the inclusion of *Ambulocetus* in the whale lineage comes from derived characters it shares with modern whales. . . . Although its relationship with other whales is uncertain, *Ambulocetus natans* is a whale.

Eight million years after the pakicetids, along came *Protocetus,* much more whalelike, with paddlelike forelimbs, functional hind limbs, and tail vertebrae that suggest the musculature necessary for supporting flukes. This creature had a long snout and teeth arranged in a zigzag pattern in the front of the jaws. Whereas the nostrils of *Pakicetus* and *Rodhocetus* were at the tip of the snout, those of *Protocetus* had begun to move rearward. (Eventually, this modification would enable whales and dolphins to inhale and exhale without slowing down, but it is difficult to imagine the advantage of a slight rearward movement of the breathing passages, unless the evolution of this character was somehow designed to become advantageous in the distant future.) Found in the Egyptian desert, *Protocetus* had dentition that suggested that it chewed its food before swallowing, dense tympanic bullae (ear bones), and "windows" in the lower jaw like the ones known to be used for transmission of sound in modern dolphins.

One of the more difficult adaptations to explain in cetaceans is the horizontal tail flukes. Because the flukes have no bones, they do not lend themselves to fossilization, so there can be no record of half-formed flukes—indeed, half-formed flukes would probably have served no useful purpose. They are not modifications of the hind legs, because many whales have vestigial hind leg bones, and some of the proto-cetaceans may have had reduced hind legs *and* tail flukes. In a 1998 analysis of the origin of cetacean flukes, F. E. Fish wrote, "Although we understand the evolution of terrestrial locomotion because of available skeletal remains and footprints, no such record exists for swimming by cetaceans as no fossilized imprints of the flukes have been unearthed and the sea leaves no tracks." The flukes consist of collagenous tissue attached to the numerous short caudal vertebrae and intervertebral disks. In simple terms, the flukes represent a unique, flattened appendage attached to the tail bones, which is used for propulsion. As with so many of the modifications that enabled cetaceans to return to the water, it is difficult to understand how natural selection could have produced a horizontally flattened instrument that is the whale's sole means of propulsion.*

What was the purpose of even a slightly flattened tail in a land animal? In their description of *Ambulocetus*, Thewissen and colleagues (1996) wrote, "Like modern cetaceans—it swam by moving its spine up and down, but like seals, the main propulsive surface was provided by the feet. As such, *Ambulocetus* represents a critical intermediate stage between land mammals and cetaceans." For Stephen J. Gould, this somehow translates into the development of flukes. In a 1994 essay, he concludes that the discovery of *Ambulocetus*'s swimming technique of "forward motion supplied primarily by extension of the back and subsequent flexing of the hind limbs [was] directly responsible for the development of flukes." Thewissen et al. actually wrote, "*Ambulocetus* shows that spinal undulation evolved before the tail fluke. . . . Cetaceans have gone through a stage that combined hind limb paddling and spinal undulation, resembling the aquatic locomotion of fast-swimming otters," and from that sentence, Gould somehow decides that "the horizontal tail fluke . . . evolved because whales carried their terrestrial system of spinal motion to the water." Gould makes this point by misunderstanding or misreading the propulsive actions of otters when he writes that they move in water by powerful vertical bending of the spinal column in the rear part of the body: "This vertical bending propels the body forward both by itself (and by driving the tail up and down) and by sweeping the hind limbs back and forth in paddling as the body undulates." This would be most convenient for Gould's theory if it were true, but otters move by paddling with their legs, not their tails, although there is an attendant

*There is, of course, a semiaquatic mammal with a greatly flattened tail that doesn't use its tail for propulsion. The beaver (*Castor canadensis*) uses its webbed hind feet for propulsion, and its paddlelike tail may be used as a rudder in the water, or as a prop when the animal sits upright on land to cut down a tree. Beavers also slap their tails to express alarm or fear. Like that of the beaver, the tail of the platypus (*Ornithorhynchus anatinus*) is also flattened, although furred where that of the beaver is naked, but the platypus also swims by paddling with its webbed feet.

flexion of the spine that causes the tail to undulate. Besides, no matter how the spine undulates, flukes are not modified legs; they are a modified tail, an adaptation that evolved separately in two kinds of marine mammals, the sirenians and the cetaceans. And however efficient this modification has turned out to be for fast-swimming cetaceans, a flattened, horizontal tail has provided the manatees and dugongs with little power and even less undulating flexibility.

After the important discovery of *Ambulocetus natans* in 1994, Thewissen and Fish (1997) tried to figure out how it moved on land or in water, and what this meant to the future development of whales. On the basis of a careful analysis of its structural morphology, they wrote that it was probably not a quadrupedal paddler, a caudal oscillator, or a pelvic undulator—in other words, it did not paddle with four legs, it did not move its tail up and down, and it did not flex its pelvis. The length of the tail implied that "a fluke at the distal part of the tail would have a poor lever arm and would not be an efficient hydrofoil." The closest analogue seems to be the Brazilian river otter (*Pteronura*), which has a somewhat flattened tip on its heavily muscled tail. But after noting that "lutrines [otters] are the best extant functional models for early cetacean locomotion, and that the locomotor morphology of *Ambulocetus* may have been most similar to that of *Lutra* or (less likely) *Pteronura*," Thewissen and Fish were careful to point out that "this does not imply that *Ambulocetus* was ancestral to all later cetaceans," or that "all early cetaceans swam similarly." *Ambulocetus,* a four-legged amphibian mammal, walked on land and probably swam like an otter, but it vouchsafed us little additional information on the swimming motions of the early or later whales.

The archaeocete genus *Dorudon,* named in 1845 by Robert Gibbes of South Carolina from jawbones, cranial fragments, and caudal vertebrae of a specimen found "in a bed of Green sand near the Santee Canal," was smaller than *Basilosaurus* but similar in shape and habits. Their hydrodynamic streamlining, shortened neck, reduced hind limbs, and powerful tail—but lacking the strangely elongated vertebrae of the basilosaurs—made them what Gingerich (1998) called "good candidates for the ancestry of modern whales." Another, similar species known as *Dorudon atrox* (previously known as *Prozeuglodon atrox*) flourished in the shallow seas that covered what is now northeast Africa some 40 million years ago. One of the many skeletons collected from the marine deposits at Wadi Hitan in the deserts of western

Egypt was particularly well articulated, and it was prepared for exhibit at the Exhibit Museum of Natural History of the University of Michigan at Ann Arbor. Under the supervision of Mark Uhen, who studied under Gingerich and wrote his doctoral dissertation on *Dorudon atrox*, a mold of the fossil was made, and the 19-foot-long skeleton was put on exhibit in the museum.

Like *Basilosaurus*, *Dorudon* had a streamlined shape, paddles for forelimbs, and tiny hind limbs, but it was only 20 feet long to *Basilosaurus*'s 70, and fossils of both species have been found together. Another early archaeocete was *Zygorhiza*, serpentlike in form, and found on the Atlantic coast of North America. When Barnes and Mitchell published reconstructions of *Basilosaurus* and *Zygorhiza* in 1978, they gave both species tail flukes that are not notched, and therefore look not unlike the flukes of a beaked whale. They wrote:

> In most modern mammals with tails modified for aquatic propulsion, the tails are expanded in the transverse plane. There is clearcut evidence for horizontal caudal flukes on the tail based on analogy with modern cetaceans. This analogy depends on the abrupt change in length versus diameter and the change from flat-faced to round-faced vertebral centra within the tail. . . . Our restoration shows that while the osteological evidence only shows the length of the base of the caudal flukes, if a wide caudal fluke of about the same proportions as that on a large balaenopterid whale is restored in a basilosaurine, it makes a reasonable appearing propulsive organ, even though the base of the fluke is much shorter.

In other words, they didn't know whether the basilosaurines had flukes, but because modern whales propel themselves this way and this was certainly a whale, they put flukes on our reconstruction. The beaked whale analogy is even stronger in these reconstructions when we look at the body proportions of these basilosaurines, particularly *Zygorhiza*. The authors have added a dorsal fin, probably for the same reason they added flukes: it is useful for swimming, and most modern cetaceans have them. The animal thus looks quite like a beaked whale, but with more teeth. With this body plan, *Zygorhiza* looks as if it might have been the ancestor of the beaked whales, but there is not enough material on primitive ziphiids to make the necessary comparisons.

Although paleontologists like to believe that successful fieldwork results primarily from research, geological analysis, and experience, they willingly admit that luck also plays an important part. Fossil whales had been found during surface surveys in Pakistan in 2000 and 2004 and were brought back in plaster jackets to Gingerich's lab at Ann Arbor. When the plaster was removed from one of the specimens, it was seen to be an adult female of a previously unknown species, with the skull and partial skeleton of a single near-term fetus—a startling and valuable find. In describing their discovery, Gingerich and colleagues (2009) wrote, "The fetal skeleton is positioned for head-first delivery, which typifies land mammals, but not extant whales, evidence that birth took place on land. The fetal skeleton has permanent molars well mineralized, which indicates precocial development at birth." They christened the whale *Maiacetus innuus, Maia* for "mother," and *inuus,* a Roman fertility god. The other specimen was an adult male of the same species, 8.5 feet long (the same size as the female), but with proportionally larger teeth: "Relatively large canines," they wrote, "are commonly found in male mammals that use their teeth in threat displays and fighting."

Maiacetus differed in size and proportions from basilosaurids such as *Durodon atrox* but was close in size and proportions to *Rodhocetus.* It had cheek teeth with shearing crests, like its mesonychid ancestors; flexible elbow joints; and well-formed hind limbs with feet and toes. Calling their finds "extraordinary fossils," Gingerich et al. wrote that *Maiacetus* "has a piscivorous dentition, and, like many other early protocetids, is interpreted as an amphibious, semi-aquatic, foot-powered swimmer that came ashore to rest, mate and give birth. While the hind limbs were capable of bearing the weight of the body on land, the proportions of the limbs and the long phalanges of both hands and feet would have limited terrestrial locomotion and prevented *Maiacetus* from traveling any substantial distance from water." It is truly amazing how much you can learn from fossils.

Living beaked whales are almost all pelagic species, and it is not unreasonable to suggest that their ancestors were too. Deep ocean sediments, although plentiful, do not yield many fossils, largely because they are so inaccessible. It is probably the exception rather than the rule that the body of a marine mammal is preserved at all, because those that die in the open sea are likely to be fragmented by scavengers such as sharks, and the most propitious circumstances for fossilization occur when an animal strands or dies close to shore and is washed up on the beach.

There are paleocetologists who believe that modern odontocetes and mysticetes had different origins. They cite the symmetrical skull, vestigial femur, paired blowholes, and lack of proper teeth as characteristics that only the mysticetes possess. Others, such as Lawrence Barnes, Leigh Van Valen, and David Gaskin (quoted in Barnes's 1984 article), believe that both groups of living whales had a common archaeocete ancestor, and according to Barnes (1984), "When we look at fossil cetaceans of the Oligocene, about 25 to 30 million years ago, we find odontocetes and mysticetes that are remarkably similar to each other and to archaeocetes. They represent exceptionally good intermediate stages." An example cited by Barnes is *Aetiocetus* from the Oligocene of Oregon, which had many features of baleen whales, such as a loosely articulated lower jaw, but it also had a full complement of mammalian teeth. Many cetologists disagree, however, and believe that *Aetiocetus* was the ancestor of the mysticete whales. In her 1994 "What Is a Whale?," Annalisa Berta wrote that "Archaeocetes are a 'scrapbasket' group of extinct Eocene whales that together with the two living whale lineages, the toothed whales (Odontoceti) and the baleen whales (Mysticeti), comprise the mammalian order Cetacea."

In 1883, William Flower published a discussion he entitled "On Whales, Past and Present, and Their Probable Origin," in which he suggested that cetaceans were the direct descendants of hoofed mammals known as ungulates. When Flower's thesis was published, it was generally derided because anyone could see that whales and dolphins bore no resemblance to cows, horses, and pigs. However, in recent years, the idea has not only been revived but is now accepted by at least a portion of whale evolutionists. Those who follow the traditional route of classifying animals on the basis of similarities in bones and flesh are inclined toward the mesonychians as the ancestors of whales because there appears to be a traceable pattern leading from these terrestrial carnivores through the protocetids, the basilosaurids, right up to the modern whales and dolphins.

Fossil or extant, a whale is a whale and cannot be taken for anything else. All whales are built along a stretched-out, horizontal body plan. The neck vertebrae are shortened and at least partly fused into a single bony mass. The vertebrae behind the neck are numerous and similar to one another; the bony processes that connect the vertebrae are greatly reduced, allowing the back to be flexible and to produce powerful thrusts from the tail flukes. The flippers that allow the whale to steer are composed of flattened and shortened arm bones; flat, disklike wrist bones; and multiple elongated fingers. The elbow joint is virtually immobile, making the flipper rigid. In the shoulder girdle, the shoulder blade is flattened, and there is no clavicle. A few species of whales still possess a vestigial pelvis, and some have greatly reduced and nonfunctional hind limbs. The rib cage is mobile—in some species, the ribs are entirely separated from the vertebral column—which allows the chest to expand greatly when the whale is breathing in and allows the thorax to compress at depth when the whale is diving deeply.

The skull also has a set of features unique among mammals. The jaws extend forward, giving whales their characteristically long head, and the two frontmost bones of the upper jaw (the maxillary and premaxillary) are telescoped rearward, sometimes entirely covering the top of the skull. The rearward migration of these bones is the process by which the nasal openings have moved to the top of the skull, creating blowholes and shifting the brain and the auditory apparatus to the back of the skull. The odontocetes have a single blowhole, while the mysticetes have a pair.

In the odontocetes, there is a pronounced asymmetry in the telescoped bones and the blowhole that provides a natural means of classification. Although teeth often occur in fetal mysticetes, only odontocetes exhibit teeth as adults. These teeth are always simple cones or pegs; they are not differentiated by region or function, as teeth are in other mammals. Whales have no tear glands, no skin glands, and no sense of smell. Their hearing is acute, but the ear has only a minute external opening. Hearing occurs via vibrations transmitted to a heavy, shell-like bone formed by fusion of skull bones (the periotic and auditory bullae).

The evolution of whales is a complex subject, and many questions remain unanswered. Nevertheless, the basic chronological sequence is known—as much as anything can be known from fragmentary evidence that is far from conclusive. Philip Gingerich, since his discovery of *Pakicetus* one of today's best-known paleocetologists, summed it up succinctly in a popular article written in 1997:

> What history tells us is that the teeth of whales changed first, before their ears or limbs. The transition to whales started when the mesonychians went into the water to feed, possibly first on dead or dying fish, and later on healthy fish. The ears changed next, as hearing

replaced vision as the dominant sensory mode, enabling archaeocetes to communicate and find their prey. Archaeocetes, however, never developed the sonar or echolocation found in modern odontocetes. The next and final modification involved the reduction of the sacrum and pelvis for tail-powered swimming, which enabled archaeocetes to disperse widely in the world's oceans.

However they developed, and wherever they came from, it is obvious that the mysticete and odontocete branches of the cetacean family tree diverged millions of years ago, while the archaeocete branch was destined for extinction. The mysticetes, now represented by the baleen whales—blue, fin, sei, Bryde's, minke, gray, humpback, right, and bowhead—are the product of one of the most bizarre evolutionary paths imaginable: they lost all their teeth, which were eventually replaced by a unique system of hanging, fringed plates that act as a sieve to trap the small marine organisms on which they subsist. All cetaceans eat living prey, but the method used by the baleen whales is closer to grazing than it is to hunting. Along with the development of this unique feeding apparatus, some of the mysticete species evolved to gigantic sizes. As far as we know, the blue whale is the largest animal ever to have lived on earth.

Some explanation seems required when discussing the vast size that has been achieved by some of today's whales. *Basilosaurus* was a long, skinny creature that may have reached a length of 70 feet, but today's blue and fin whales can exceed that, and although the blue whale is long, it is anything but skinny; an adult female was weighed at 150 tons. Many marine creatures were larger in earlier times. There were crocodiles and turtles much larger than their living decendants, and although they left no heirs, some of the ichthyosaurs and mosasaurs were also giants. Freed from the restraints imposed by gravity (a 150-ton whale is neutrally buoyant, and in effect weighs nothing in the water), it is possible for aquatic animals to reach a huge size, but why the large crocodilians and monster turtles died out while the whales were getting larger and larger is one of the historical mysteries of life in the sea.

As to when the divergence between baleen whales and toothed whales occurred—never mind *why*—the fossil evidence is inconclusive and often contradictory. In *Marine Mammals: Evolutionary Biology* (1999), Berta and Sumich wrote, "Estimates of the divergence time for the mysticete-odontocete split based on gene rates are clearly in conflict with the known fossil record. According to the fossil record, mysticetes and odontocetes diverged from a common archaeocete ancestor about 35 million years ago . . . [but] on the basis of rates of evolution in mitochondrial ribosomal genes of ungulate mammals, Milinkovitch et al. (1993) postulated a 10–13 million years ago split between sperm and baleen whales. This figure has been revised to 25 million years ago based on a slower rate of evolution for whale mitochondrial DNA and the recognition that sperm whales and baleen whale fossils are older than 15 million years."

Baleen probably developed from teeth, but as Berta and Deméré (2009) wrote, it "appears to have been a stepwise transition from an ancestor with teeth only to an intermediate state with functional teeth and baleen to the derived condition with baleen only." Baleen is composed of keratin, the same substance as human hair and fingernails, and rarely fossilizes, so the shape of the skull and the palate is used to demonstrate the presence of baleen plates. Fordyce and Barnes (1994) wrote that "primitive toothed mysticetes are more widespread than formerly understood. Their teeth may have been used, in the mysticete fashion, for bulk feeding rather than for selecting individual prey."

The oldest described mysticete is the toothed *Llanocetus denticrenatus,* whose description is based on a mandibular fragment found on Seymour Island in the Antarctic (Mitchell 1989). Another early mysticete is *Aetiocetus,* from the Upper Oligocene (23 million years ago) of Oregon. Although this species had teeth, the structure of the skull is otherwise typical of primitive mysticetes, and as Douglas Emlong wrote in 1966, "It is very unlikely that this animal would possess so many mysticete features if it were not directly in mysticete lineage." The Aetiocetidae appear to be the rootstock of all subsequent mysticete families, including the cetotheres (now extinct), the Balaenidae (right whales and bowheads), and the Balaenopteridae, which includes all the rorquals: the blue, fin, sei, Bryde's, and minke whales. The small skull of the Eschrichtiidae (gray whales) is a primitive feature that may suggest an earlier origin. Eventually, the early mysticetes lost their teeth altogether and became the great filter feeders we now know as baleen whales, but the branch that led to the physetererids was trimmed and trimmed again, until only one species remained: *Physeter macrocephalus,* the great and powerful sperm whale.

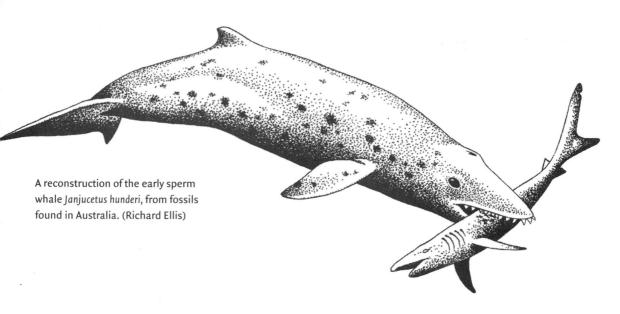

A reconstruction of the early sperm whale *Janjucetus hunderi*, from fossils found in Australia. (Richard Ellis)

Examining a fossil described as "A Bizarre New Toothed Mysticete from Australia," which included an almost complete skull, mandibles with teeth, two ribs, and three cervical vertebrae, Erich Fitzgerald (2006) of Monash University, Melbourne, wrote, "Unlike all other mysticetes, this new whale was small, had enormous eyes, and lacked derived adaptations for bulk filter-feeding. . . . It refutes the notions that all stem mysticetes were filter-feeders, and that the origins and initial radiation of mysticetes were linked to the evolution of filter feeding." Fitzgerald named the species *Janjucetus hunderi,* after Jan Juc township in central coastal Victoria, and for Staumn Hunder, the surfer who discovered the specimen. It would have been about 12 feet in length, much smaller than any other baleen whale, but despite the teeth, the structure of the skull unequivocally identifies it as a mysticete. There were no structures that suggest echolocation, but the large eyes and powerful teeth and jaws indicate that *Janjucetus* hunted large prey by sight—rather like some of the extinct marine reptiles, and even like today's leopard seal.

A 25-million-year-old mysticete with teeth calls into question earlier estimates of the time of the split between toothed and baleen whales, usually held to have been about 34 million years ago. It now appears that during the Late Oligocene there were many different types of toothed whales, some of which, like *Janjucetus,* can be placed in the ancestry of the baleen whales, whereas others, whose skull shape was radically different, gave rise to many forms of toothed whales, including dolphins, beaked whales, and the quintessential toothed whale, *Physeter macrocephalus.*

Remington Kellogg's 1922 description of "Two Fossil Physeteroid Whales from California" points up the difficulties and complexities of analyzing whale phylogeny. Before getting to the new species (of old whales), he devotes 18 pages of a 34-page paper to a discussion of the state of whale paleontology, reviewing "at least 25 genera [that] have been proposed for the remains of fossil cetaceans which at one time or another have been referred to as physeteroids," including some questionable specimens that were eventually classified as baleen whales, dolphins, sea lions, or even manatees. *Idiophyseter merriami,* the first of Kellogg's new species, is represented only by a skull, small enough to be either a fetus or perhaps a pygmy sperm whale ancestor, but definitely a physeteroid because of the scooped-out cranial basin, which indicates the presence of a spermaceti organ.

But *Ontocetus oxymycterus* is something else.* The fossil, found in a sea cliff north of the Santa Barbara lighthouse, consisted of the end of the rostrum, the extremities of both mandibles, the roots of 10 or 11 teeth, and several other teeth found in the adjoining matrix. Kellogg wrote: "Comparative measurements indicate that a complete skull of this species will measure between 12 and 15 feet in length. If the estimate is correct, then the skull of this species is twice as long as that of *Idiophorus patagonicus,* from a lower Miocene tuff formation on the coast of Chubut Territory, Patagonia, and probably represents the largest Miocene physeteroid thus far described." The large teeth were found in the left and right mandibles (the lower jaw), but there were three teeth on each side of the extremity of the rostrum—the tip of the upper jaw—suggesting that the upper teeth of some of the early sperm whales were on the way out by the Lower Miocene, some 25 million years ago.

In whales, as in no other living mammals, the trend has been toward gigantism. Just as the blue and fin whales are the giants of the mysticetes, the sperm whale is the largest of the odontocetes, and it is considered the largest predatory animal in history. (Of course, blue and

---

Oxymycterus is also the name of a genus of South American burrowing mice known locally as *hocicudos.* But *Ontocetus* simply means "living whale"—from the Greek *ontos,* meaning "living thing" or "being" (*ontology* is the study of things that exist or may exist)—and Kellogg (1922) tells us that the name was first used by Joseph Leidy, a Philadelphia paleontologist, in 1868.

fin whales, which are longer than sperm whales, are predators too, in that they consume live prey such as krill, but their feeding methods are more like grazing, so sperm whales are the biggest large-prey predators that ever lived.) Sperm whales are almost as enigmatic in the fossil record as they are today. A late Miocene physeterid was *Ferecetotherium,* found in the Caucasus and originally described as an archaic mysticete (Mchedlidze 1984), but the published drawings clearly show a creature that resembles a sperm whale, with peglike, homodont teeth and a large pan bone in the lower jaw. *Ferecetotherium* is probably from the Late Oligocene, between 23 and 30 million years ago. Fossils have also been found in early Miocene deposits (22 million years ago), and by that time, the fossils show characteristic sperm whale features, such as the deep cranial basin; large, tusklike teeth; and an enlarged periotic (ear bone). It is likely that the early physeterids were deep divers like their descendants, and they were probably squid eaters (teuthophages) as well. Primitive sperm whales had upper and lower jaw teeth, but members of the sole surviving genus (*Physeter*) have visible teeth in the lower jaw and none in the upper jaw, suggesting some adaptive advantage to losing half of your teeth. Of today's toothed whales, most of which feed on squid, only the sperm whale uses one set of teeth, but it actually has two; those in the upper jaw remain unerupted in their sockets throughout the whale's life. Remington Kellogg described *Diaphrocetus poucheti* in 1928; it is one of the oldest physeterids. Like *Physeter,* its skull had a distinctive basin that presumably held a large, fatty melon and spermaceti organ.

The recently discovered early cetacean fossils—Thewissen (1998) refers to their number as a "cascade"—may give us some idea of the changes that were required for a terrestrial mammal to become an aquatic one (migration of nostrils, loss of legs, etc.), but we still have no idea why a group of animals would leave one environment for another. Thewissen (1998), summarizing cetacean evolution, wrote,

> A question that cannot be answered satisfactorily is why cetaceans took to the water. It has been suggested that they took to the water to take advantage of a plethora of resources that had gone untapped since the extinction of the Mesozoic marine reptiles. This explanation is too simplistic. The earliest cetaceans lived in or near freshwater, and it is unlikely that they profited from extinctions in the nearshore marine realm. Lack of competition for food is also an unlikely

The pygmy sperm whale differs from its larger cousins by the presence of a dorsal fin and the strange, gill-like pattern forward of the flippers. (Richard Ellis)

reason for the subsequent shift to the seas. The earliest marine cetaceans did not live like modern cetaceans, but resembled crocodiles ecologically. Crocodiles were not greatly affected by the K-T extinctions, and they are abundant in the same deposits that yield the earliest cetaceans. If the early cetaceans were generalized feeders, they must have suffered considerable competition and predation from crocodiles.

Physeter macrocephalus is unlike any other animal on earth, past or present. There are two mini–sperm whales in existence today, the dwarf (*Kogia simus*) and the pygmy sperm (*Kogia breviceps*), but neither gets longer than 12 feet. They do not have a disproportionately enlarged nose, but both have a miniature version of the great sperm whale's spermaceti organ and smallish teeth in the underslung lower jaw. Both sport a dolphinlike dorsal fin and a pattern on the head that strangely resembles gills.* The dwarf and the pygmy are classified in the genus *Kogia,* making them different both taxonomically and physically from *Physeter,* the great sperm whale. They are usually classified as "relatives" of the sperm whale, but this term is a convenience for taxonomists and does not really mean they would acknowledge one another at a family gathering.

Sperm whales, shaped somewhat differently than the ones we know today, have been around at least since the Late Oligocene, about 25 million years ago. (That's the date of the earliest known sperm whale

*When a mother and a calf dwarf sperm whale were trapped in a tuna net in the eastern tropical Pacific, the mother released a cloud of feces whenever one of the dolphins (also trapped in the net) approached her or the calf, and appeared to hide herself in the opaque cloud. At one point the mother rammed the purse seiner with her head and then released another reddish cloud. "This suggests behavior similar to that of other physeterids," wrote Scott and Cordaro (1978), "that release of feces occurs in response to threatening situations, and that one possible use of this behavior may be for concealment."

relative, *Ferecetotherium kelloggi,* but there are surely older species and fossils that we haven't found yet—and may never find.) *Ferecetotherium,* the precursor of the known Physeteridae (sperm whales), was relatively small, about 15 feet long, the size of a large dolphin. It had a small head and pincerlike narrow jaws, but it was definitely a sperm whale because it had that diagnostic scooped-out, asymmetrical skull. Over time, the Physeteridae grew larger, and *Zygophyseter varolai,** described from an almost complete skeleton unearthed in a quarry in southern Italy by Giovanni Bianucci and Walter Landini (2006), was 22 feet long. Large teeth in both jaws were reminiscent of the extant killer whale (*Orcinus orca*), an active, large-prey predator, so the authors suggested the English common name "killer sperm whale." Bianucci and Landini included a hypothetical reconstruction of *Zygophyseter,* which depicts a long-beaked cetacean with a bulging forehead, not unlike a bottlenose whale (*Hyperoodon*), but where the bottlenose whales—which are completely different from bottlenose dolphins—have only a small, ineffectual pair of teeth at the tip of the lower jaw, *Zygophyseter* had both jaws filled with an impressive array of large, pointed teeth.

At 22 feet, the killer sperm whale was closing in on the size of today's sperm whale. (From the skeleton, it is not possible to ascertain the sex of the specimen.) Bottlenose whales, where only the male has the characteristic bulging forehead, can reach a length of 30 feet. They are among the largest of the beaked whales and are now thought to be a distant branch of sperm whale developmental evolution. But even today, beaked whales, known as ziphiids or mesoplodonts, are the least known of all living cetaceans.

From a partial mandible found in Apulia, southern Italy, Bianucci, Landini, and Varola (2004) described a specimen of a North Atlantic sperm whale that they referred to as *Orycterocetus quadratidens,* originally described in 1867 by Joseph Leidy, the 19th-century Philadelphia paleontologist, who wrote,

*The generic name *Zygophyseter* was derived from the Latin adjective *zygomaticus,* referring to the zygomatic process, a protrusion from the rest of the skull, rather like the bumper of a car. "The extreme elongation of the zygomatic process of the squamosal," wrote Bianucci and Landini (2006), "and the circular supracranial basin (probably for housing the spermaceti organ), delineated by the peculiar anterior projection of the supraorbital process of the right maxilla [half of the upper jaw], are the most distinctive features of this bizarre sperm whale." The animal was named for Angelo Varola, who discovered the specimen.

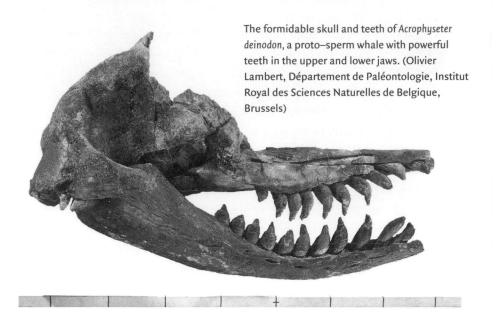

The formidable skull and teeth of *Acrophyseter deinodon*, a proto–sperm whale with powerful teeth in the upper and lower jaws. (Olivier Lambert, Département de Paléontologie, Institut Royal des Sciences Naturelles de Belgique, Brussels)

Quite lately, I received from Prof. Holmes fragments of both sides of a lower jaw, two teeth and a portion of a rib of a cetacean from the Miocene formation of Virginia. . . . The more perfect of the teeth appears to have been about five inches in length, and is curved conical. The fang is quadrate and hollowed, and the surface of the tooth, nearly to the end of the crown, which appears not to have been covered with enamel, is annularly and longitudinally corrugated. The greatest circumference of this tooth is three inches in length [sic] and nearly straight. For the animal I propose the name of *Orycterocetus quadratidens.**

In 2008, Olivier Lambert, Giovanni Bianucci, and Christian de Muizon described *Acrophyseter deinodon,* new "stem-sperm whale from the Latest Miocene of Peru." The rostrum was sharply pointed, so they named the species *Acrophyseter* from the Greek *akros,* "acute," and *deinodon* from the Greek *deinos,* "terrible," a reference to its sharp and powerful teeth. These teeth, swollen at the root and narrowing to a sharp point, suggest some degree of shearing ability, indicating "predation on large prey." In contrast to most interpretations of recent sperm

*Leidy's name for the Virginia specimen seems particularly uninspired, albeit literal. In Greek, *oryctos* means "dug out" (*oryctology* is an antique term for the study of fossils, or paleontology), so *Orycterocetus quadratidens* is simply "fossil whale with quadrate teeth."

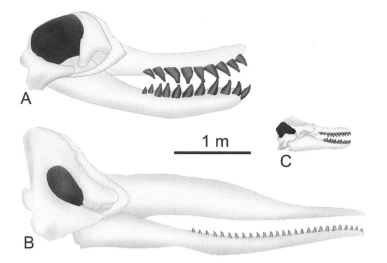

Comparison of the outline of the skull and mandibles of *Leviathan* (A) with a modern sperm whale *Physeter* (B) and the killer whale *Orcinus* (C). (Richard Ellis)

A

1 m

C

B

whales feeding by debilitating their prey with sound and then gobbling it up, Lambert, Bianucci, and de Muizon have followed the 2006 suggestion of A. J. Werth, who believed that the loss of teeth in the later odontocetes (such as the modern sperm whale) indicated that they relied primarily on suction to capture and ingest prey. Shearing, said the authors, "would contrast strongly with the suction feeding demonstrated in recent sperm whales, related to the loss of the upper teeth." Suction is indeed the primary feeding method used by walruses as they plow along the bottom sucking up clams, but clams are not particularly elusive. Werth (2000) videotaped captive long-finned pilot whales (*Globicephala melas*) sucking in frozen, thawed herring in a tank at the New England Aquarium, but he pointed out that "suction feeding has not been documented in adult *G. melas* feeding on live, elusive prey in its natural environment."

Then in July 2010, Lambert, Bianucci, and de Muizon, along with several other paleontologists, published a description of the most formidable marine predator ever, comparable only to the contemporaneous giant shark, *Carcharodon megalodon*. Discovered in the same Miocene Peruvian formation as *Acrophyseter*, the skull, mandible, and teeth of *Leviathan melvillei** proclaimed it as "one of the largest

*When naming a new species, it is customary to include an etymological explanation of how and why the name was chosen. Lambert et al. (2010) wrote: "From Hebrew 'Lvyatan' ('Leviathan' in Latin); name applied to large marine monsters in popular and mythological stories. Species is dedicated to novelist Herman Melville (1819–1891)." For me, the naming of *Leviathan melvillei* was a godsend.

raptorial predators [with] the biggest tetrapod bite ever found." The skull of this whale was 10 feet long, and its overall length has been estimated at between 45 and 57 feet, approximately the size of today's *Physeter*. In the above discussion of *Acrophyseter deinodon*, Lambert et al. discussed the suction feeding of today's sperm whale, but the massive, sharp teeth—up to 14 inches long!—in the upper and lower jaws of *Leviathan melvillei* proved conclusively that it was a large-prey predator that probably fed on medium-sized baleen whales. "This sperm whale," they wrote, "could firmly hold large prey with its inter-locking teeth, inflict deep wounds, and tear large pieces from the body of the victim." The scooped-out skull, described by the authors as "a vast supercranial basin," shows that this Miocene superwhale also had a spermaceti organ, which, if the analogues to *Physeter* are accurate, suggests that *Leviathan melvillei* could add echolocation and perhaps even directed sound blasts to its predatory arsenal. But despite the similarities between *Leviathan* and *Physeter,* the fossil appears to be a distant relative of today's sperm whale, not an ancestor.

The largest animals ever to have lived on land were the sauropod dinosaurs, such as *Brachiosaurus,* but even at an estimated weight of 70 tons, they were dwarfed by the baleen whales of today. A 100-foot-long blue whale can weigh 150 tons, but even if some of the dinosaurs reached that length, they would not weigh nearly as much because of the heavy body and legs, accompanied by a long skinny neck and an even longer, even skinnier tail. Animals that live in water can grow to extreme sizes because their great weight, which would otherwise have to be borne by legs, is buoyed up by water. In the world's ancient oceans (some of which have themselves disappeared), there were huge jawless fishes, such as the 30-foot-long *Dunkleosteus,* and the Early Jurassic plankton-eating fish known as *Leedsichthys,* which might have been 100 feet long. Throughout the history of marine life—which substan-tially predates terrestrial life—there has been a succession of large and often formidable predators. There once were huge, carnivorous marine reptiles, such as the 50-foot-long mosasaur *Tylosaurus,* or the pliosaur *Kronosaurus,* approximately the same length. According to big-game fishermen, there are indeed giant fishes today, but the biggest bluefin tunas weigh about three-quarters of a ton, and the heaviest black mar-lins weigh in at about 2,000 pounds. (The largest living fish is actually a shark: the whale shark [*Rhincodon typus*] is the current titleholder at 50 feet and 15 tons.) The great shark *Carcharodon megalodon,* with its human-hand-sized, serrated teeth, was long considered the definitive marine predator—that is, until they found the fossil remains of *Leviathan melvillei.*

Even though we don't quite understand how it works, extinction is an integral part of the evolutionary process. (Of course we understand extinction caused by the hand of man, as with the dodo or the passenger pigeon.) Species develop, live and prosper for a while, and then die out, sometimes adapting to changes in the environment, but more often not. Most of the animal species that have ever lived on earth are gone, whether predator or prey. The great sharks such as *C. megalodon* are gone (despite silly novels that postulate their continued existence), and *Leviathan melvillei* has not spouted for about 13 million years. Sometimes the extinction of a predator can be attributed to the disappearance of its prey, but the baleen whales (or ones very much like them) are still with us. What happened to *Leviathan?* We don't know. Once there were giant raptorial predators, taking chunks out of baleen whales, and then they were gone. Something might have replaced them as top predators, but the giant sharks died out too. The water they inhabited may have cooled, warmed, or become saltier to an extent that *Leviathan melvillei* could not tolerate it. Their prey animals might have died out or moved elsewhere. A random DNA mutation might have rendered them nonfunctional or susceptibíe to a previously unknown disease. Or they might simply have reached the end of their line, for reasons beyond comprehension. Whatever the reason, *Acrophyseter, Orycterocetus, Leviathan,* and all the other early sperm whales are gone, and what remains of the physeterid line are the true sperm whale, the pygmy and dwarf sperm whales, and the closely related beaked whales.

There are some 20 species of beaked whales, generally characterized by a spindle-shaped body, protruding upper and lower jaws (the "beak"), no central notch in the tail fin, and teeth only in the lower jaw of the males. Very little is known about the habits of ziphiids (mesoplodonts), and some species are represented only from animals that have stranded. In those cases where the stomach contents have been examined, it was seen that beaked whales feed mostly on squid. The smaller species may be around 15 feet long, but the largest, Baird's beaked whale (*Berardius bairdi*), can reach a length of 42 feet. They may be the least understood group of large animals in the world, and in recent years, two unsuspected new species have been described, the

Peruvian beaked whale (*Mesoplodon peruvianus*) in 1991 and Bahamonde's beaked whale (*Mesoplodon bahamondi*) in 1995.

For many years, the least known of the little-known beaked whales was *Indopacetus*, whose entire catalog consisted of two skulls, one found on a beach in Queensland in 1882, and the other in Somalia, on the east coast of Africa, in 1955. In his 1971 *Field Guide of Whales and Dolphins*, W. F. J. Mörzer Bruyns wrote (of *Indopacetus*) of "the possibility that the author observed these animals in the Gulf of Aden and the Sokotra area, being very large beaked whales and certainly not *Ziphius* [another genus of beaked whales]." Mörzer Bruyns also wrote that "Mr. K. C. Balcomb of Pacific Beach, Washington USA took a photo of a school of 25 beaked whales on the equator at 165° West [in the vicinity of the Gilbert Islands], which were almost certainly this species." In Balcomb's and other photographs, the animals looked more like bottlenose whales (*Hyperoodon*) than beaked whales, but the two species of bottlenose whales (*H. ampullatus* and *H. planifrons*) are from high northern and southern latitudes, respectively, and had no business in tropical waters.

In 1999, Pitman et al. published an article in *Marine Mammal Science* with the intriguing title "Sightings and Possible Identity of a Bottlenose Whale in the Tropical Indopacific: *Indopacetus pacificus?*" The article incorporated a collection of photographs (one of which was Balcomb's 1966 picture), and when the photographs and eyewitness descriptions were compared, it was clear that the whale was a bottlenose whale, and it was neither the northern nor southern version. Pitman et al. note that the whale's body color "has been variously described as tan, light brown, acorn brown, gray-brown or just gray." When an adult female beaked whale stranded in late 1999 in the Maldives in the northern Indian Ocean, it was identified as *Indopacetus*, and comparison with the museum specimens showed once and for all that *Indopacetus* was actually the tropical bottlenose whale. In a 2009 article, in the *Encyclopedia of Marine Mammals*, Pitman wrote,

> For a hundred years, cetologists had nothing to work with but two skulls on the shelf. We now have specimen material from six individuals (including five skulls and one complete skeleton), records of more than two dozen sightings, numerous photographs of large animals in the field, recordings of their vocalizations, and (welcome to the twenty-first century) eight minutes of digital video footage.

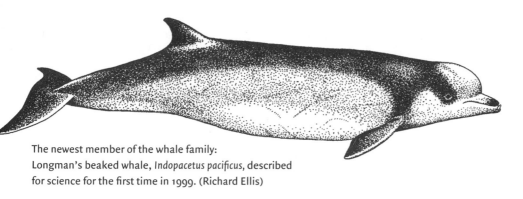

The newest member of the whale family:
Longman's beaked whale, *Indopacetus pacificus*, described
for science for the first time in 1999. (Richard Ellis)

Then, in July 2003, a complete description of *Indopacetus pacificus* appeared in *Marine Mammal Science*. It was written by nine authors who came from New Zealand, Australia, the Maldives, South Africa, Kenya, and California (Robert Pitman was the California contributor). They examined the two original skulls from Queensland and Somalia and added four new specimens to the list: a skull that was found in the National Museum of Kenya; several ribs and vertebrae that had been found on a Natal beach in 1976; a skull, mandible, teeth, ribs, and ear bones of a specimen mistakenly identified as *H. planifrons* in the Port Elizabeth (South Africa) Museum; and the adult female that was collected in the Maldives in January 2000. DNA sequencing showed that all these specimens belonged to the same species: *Indopacetus pacificus,* now known as Longman's beaked whale. The range of this species is now known to incorporate "the western reaches of the tropical Pacific Ocean . . . and the western, northern, and southern latitudes of the tropical Indian Ocean" (Dalebout et al. 2003).*

"Ziphiids in general," wrote James Mead in 2009, "have reduced their teeth to the point that teeth in the upper jaw are vestigial or absent and teeth in the lower jaw are reduced to one of two pairs that usually erupt only in adult males." (Female beaked whales have no teeth at all.) For all intents and purposes, then—except perhaps for fighting among males—the vestigial teeth of beaked whales are useless. But because their deep-water hunting dives are effective for collecting enough squid and fish to sustain their high mammalian metabolism, it must be concluded that teeth are unnecessary for capturing the beaked whales' prey.

*As if to point up the esoteric nature of the beaked whale family, yet another new species was added in 2002. On the basis of the DNA analysis of five animals stranded on the coast of California between 1975 and 1997, *Mesoplodon perrini,* named for dolphin specialist William Perrin, was introduced as a new species (Dalebout et al. 2002).

The ziphiids probably don't incapacitate their prey with sound; only the bottlenose whale has a bulging forehead, but in the males, it contains dense bony crests that are used, according to Gowans and Redell (1999), for head-butting contests between males. A large bulging "forehead" in the sperm whale (it is actually the end of the whale's nose) is evidently used for sound generation, and although no one would refer to an adult sperm whale as "toothless," the peglike teeth are probably not used to capture prey, as most of the various squid species examined from sperm whale stomachs are unmarked. "Toothless capture" would therefore represent a convergent adaptation for the ziphiids and the physeterids, which diverged on the cetacean family tree millions of years ago.

In 1994, Kiyoharu Hirota and Lawrence Barnes published a description of *Scaldicetus shigensis,* a primitive sperm whale with teeth in its upper and lower jaws. Found in the central Japanese prefecture of Nagano at Shiga-mura (hence its generic name), the skeleton was nearly complete and was a rare record of an early physeterid in Japan. The genus *Scaldicetus* was originally proposed in 1867 by B. A. L. de Bus for a specimen found in Antwerp, Belgium, which was named *S. grandis.* Other specimens have subsequently been assigned to the genus, but a comparative study of the specimens has not been undertaken. That the Japanese specimen was a primitive sperm whale was clearly demonstrated by its sloping, asymmetrical skull with the characteristic long, slender rostrum, and the "very asymmetrical external nares, comparable to the living sperm whale." There were 12 equal-sized, conical teeth in each of the mandibles, fewer by half than the number in living sperm whales, as well as functional upper teeth, suggesting "a different manner of feeding than is employed by living sperm whales"—that is, a diet of fish instead of squid. The authors concluded, "*Scaldicetus shigensis* represents an interesting and important stage of sperm whale evolution, and adds to the known diversity of the family in Miocene time. Its primitive morphology is in some ways transitional between primitive Oligocene odontocetes and more derived sperm whales."

Conventional evolutionary theory would explain the existence of today's sperm whale as the culmination of a series of adaptations to changing circumstances over time, where an ancestral species with teeth in both jaws lost the uppers (or at least the visible uppers), and a small-nosed species, not unlike today's dwarf and pygmy sperm whales, gave rise to the gigantic *Physeter macrocephalus,* a whale with

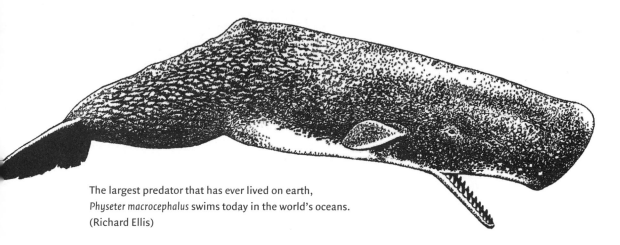

The largest predator that has ever lived on earth,
Physeter macrocephalus swims today in the world's oceans.
(Richard Ellis)

the largest nose in history, able to dive to depths heretofore unavailable to any cetacean, there to feed on the giant squid that had presumably gone unpredated in their deep-ocean sanctuary until the arrival of the greatly modified whale.

The acknowledged forebears of our *Physeter—Orycterocetus, Zygophyseter, Ferecetotherium, Scaldicetus, Aulophyseter*—probably fit comfortably into our idea of what an early carnivorous toothed whale might look like, but today's sperm whale represents such an enormous leap of paleontological faith that we are hard-pressed to imagine how it might have evolved. Where (or why) did its upper teeth go? Was the disappearance of the upper teeth somehow related to a developing taste for ammonia-filled giant squid?* Did it develop its great size to enable it to battle these giants? Why that monstrous nose? And how about the spermaceti organ, equipped with sacs, valves, lips, and fluids found nowhere else in nature? Did that complex organ develop to make sounds? Or are

*Evidently there were giant squid long before there were whales to eat them. In the mid-Cretaceous period, some 85 million years ago, the Western Interior Seaway covered most of midwestern North America from the Gulf of Mexico to the Arctic Circle, incorporating all of Saskatchewan, North and South Dakota, Kansas, Nebraska, and Oklahoma, as well as most of Texas. The waters of the seaway were not particularly deep, rarely exceeding 1,000 feet, but they still provided a habitat for large squid. The fossil remains of a large squid known as *Niobrarateuthis* was first described by Richard Green in 1977 from a specimen found in Kansas. It was named for the Niobrara Chalk Formation and *teuthis,* which is Greek for "squid." The Western Interior Seaway, which closed up some 70 million years ago, was home to various giant marine reptiles, such as mosasaurs, which fed on the squid of the Cretaceous seaway, but there were no whales then.

the sounds a secondary function of an organ that might be related to buoyancy control? The sperm whale is an evolutionary conundrum. Somehow, all these disparate, exaggerated parts were combined into what is unquestionably one of the weirdest of all mammals—perhaps the weirdest animal on earth. The sperm whale is a testimony to and a validation of the theory of adaptive evolution: every aspect of its development—even those that we don't understand—enables it to pursue its unique lifestyle with consummate competence. However mysterious and misunderstood the process, the great sperm whale has come down through time to its present form: a living, breathing, spouting, diving testimony to the miraculous process we know as evolution.

If there are many variations within a particular genus, something that happens to one species does not portend the extinction of the genus. But if there is only one species in the genus—for example, we belong to the only living species of the genus *Homo*—a catastrophe such as epidemic contagious disease or massive environmental change can wipe out the entire genus. Thus Hirota and Barnes (1994) wrote, "Sperm whale taxonomic and phyletic diversity was much greater in the Miocene than it is at the present time. With only one living species today, the family Physeteridae appears to be at an all time taxonomic low. This may have critical implications for the evolutionary survival of the family." As if sperm whales didn't have enough problems!

IV

The (Un) Natural History
of the Sperm Whale

IT IS NOW BELIEVED that there is only one worldwide species of *Physeter,* but numerous other variations have been identified over the years, all given new, and at the time seemingly appropriate, names. In an 1835 publication called *The Natural History of the Order Cetacea,* H. W. Dewhurst listed the following species: *P. macrocephalus, P. cetadon* [*sic*], *P. trumpo, P. cyhndricus,* and *P. microps.* All of these are identifiable from the descriptions and illustrations as sperm whales except for *Physeter bidens sowerbyi,* which is the beaked whale now known as *Mesoplodon bidens.* The Dutch call the sperm whale *potvisch;* the Norwegians, *spermhval;* and the French, *cachalot.* In Japanese the sperm whale is known as *makko-kujira,* which means "like the color of *ko,* a powdered perfume which is burned at Buddhist funeral ceremonies" (*kujira* is Japanese for "whale").

Until recently, the sperm whale was known as *Physeter catodon,* but *P. macrocephalus* has recently been shown to be the correct name. The specific name *macrocephalus* means "big head," and its derivation is obvious. This name was originally used by Linnaeus in 1758, and from that year to 1911, the species was known as *Physeter macrocephalus.* In a 1911 review of Linnaeus's nomenclature, Thomas "decided that *P. catodon* and *P. macrocephalus* were definitely synonymous and he accepted the name *P. catodon* for the species" (Husson and Holthuis 1974). The name *P. macrocephalus* had been uninterruptedly in use from 1758 to 1911, and Husson and Holthuis, in a comprehensive study of the synonymy of the species, concluded that "there seems to be no good reason not to apply the rules [of the International Code of Zoological Nomenclature] strictly here, and to adopt the name *P. macrocephalus.*" Rice, in the 1998 *List of the Marine Mammals of the World,* uses the name *macrocephalus,* citing Husson and Hoithuis for authority. The junior synonym *P. catodon* comes from the Greek *kata* ("down") and *odous* ("tooth") and refers to the prominent teeth found in the lower jaw.

The sperm whale looks very like a whale; it is extremely large and has a blowhole, paddlelike pectoral fins, and a horizontal tail. But from its communications, migratory patterns, and internal structure, it just might be an envoy from the planet Cetacea. Hardly anything the sperm whale does is comparable to what other large whales do. It is the only species with teeth, and then only in the lower jaw; the only one that dives and hunts in the abyssal depths; the only one that is able to subdue and swallow large prey items; and the only animal on earth with a noseful of oil, not to mention a nostril shaped like the f-hole of a violin, a monkey's muzzle in its nose, and the ability to generate—and aim—the loudest sounds made by any living creature. If the sperm whales are reporting back to their home planet (maybe that's what the sounds are all about), they are probably saying that this planet is not good for the health of their species. A certain bipedal creature has been killing them for centuries, to the point that the entire population is threatened. Perhaps the time has come for *Physeter* to return home. Colonization of Earth's oceans doesn't seem to have worked out so well.

Whereas most whales—indeed, most vertebrate animals that spend most of their lives in the ocean, from fishes to penguins, from seals to eels—are more or less pointed at the front end, the better to pass through the thick, unforgiving medium in which they live, the sperm whale pushes forward with the flattened forepart of its great nose, probably the worst design imaginable for an animal trying to move through the water with the least resistance. Yet somehow the design serves the sperm whale well. It slides downward into the abyssal depths of the ocean, catches itself a bite of food, and returns to the surface apparently unaffected by the depth, duration, and immense pressure of these prodigious dives. (That it accomplishes most of its foraging in total darkness is a subject that will be addressed in time.)

There are fishes that are profusely festooned with spines, spikes, warts, lumps, and bumps that counterindi-

In 1973, the publishers of the *Encyclopedia Britannica* thought my painting of a blue whale would look better upside down. (Richard Ellis)

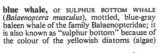

blue whale, or SULPHUR BOTTOM WHALE (*Balaenoptera musculus*), mottled, blue-gray baleen whale of the family Balaenopteridae; it is also known as "sulphur bottom" because of the colour of the yellowish diatoms (algae)

Blue whale (*Balaenoptera musculus*)
Painting by Richard Ellis

A drawing of a sperm whale by somebody who probably never saw one. This fellow has a prominent dorsal fin, convoluted flukes, and a silly grin that would belie the harpoon stuck in his back. (New Bedford Whaling Museum)

cate the concept of hydrodynamic streamlining, which would seem to be a requirement for moving through the water. But for every goose-fish, anglerfish, batfish, scorpionfish, or toadfish—there is even a fish known as a lumpfish—there is a graceful tuna, mackerel, mako shark, swordfish, or marlin that is perfectly designed to slip through the water unimpeded by any drag-inducing protrusions. The bluefin tuna, gener-ally regarded as the epitome of piscine design, has eyes that are flush with its head, fins that can be tucked into slots, and scales so small that they cannot be seen by the human eye. The pinnipeds—the seals and sea lions—have fur coats, but when wet, the fur becomes smooth and slick, facilitating an easy passage through the water. So it is with almost every cetacean: dolphins are slick-skinned torpedoes that can rocket through the water at amazing speeds (the Dall porpoise can probably swim 35 miles per hour, but the killer whale can catch it), and even the hulking, slow-moving right and bowhead whales have no impedi-ments, except occasional infestations of barnacles, to slipping through the water. In synchrony with its contrarian design scheme, however, the sperm whale, unlike any other cetacean, does not have the slick skin that might be helpful in its swimming and diving pursuits. Rather,

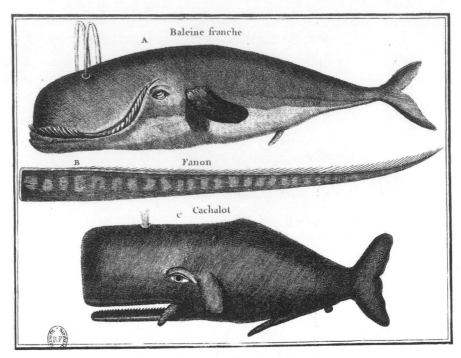

In their 1972 book *The Whale: Mighty Monarch of the Sea*, Cousteau and Diolé reproduced an old illustration of a sperm whale (*Cachalot*) and a right whale (*Baleine franche*), where the right whale was drawn upside down, with its mouth curving up over the eye, and a pair of spouts emerging from the "top" of its head (actually the underside of its lower jaw). *Fanon* is French for "baleen plate" (from the right whale). (Bibliothèque Nationale, Paris)

the hindmost half of its body is covered with prunelike wrinkles. These corrugations probably don't slow the whale down, but as with so many aspects of sperm whale physiology, their purpose remains a mystery.

The great rorquals—the blue whale, fin, sei, Bryde's, and minke—all have a dorsal fin. The humpback does too, although it is a peculiar two-step affair. In place of the dorsal fin, the gray whale has a series of "knuckles" running from the middle of the back to the base of the tail. Bowheads and right whales have no dorsal fin at all. Naturally (or unnaturally), the sperm whale conforms to none of these standards; it has its own style of fin that is not quite a fin and not exactly a knuckle. In their detailed discussion of the sperm whales of the southeast Pacific, Clarke, Aguayo, and Paliza (1968) described the "dorsal fin" thus:

> The dorsal fin of the sperm whale varies a good deal in size and its limits are ill-defined. Measurements of the length and height can be approximate only, and observations vary between different observers, so that, as will be seen in the paper on body proportions now

in preparation, comparison of these measurements in whales from different regions is really impracticable. Because the fin is always so broad transversely, the old whalemen often called it the "hump," giving the impression that it scarcely merited the name of "fin."

As identified by Kasuya and Ohsumi (1966), female sperms have a "callus" on the dorsal fin or hump of the cow, which has been referred to as a "secondary sexual characteristic." Unlike the callosities on the right whale, this growth does not appear in the unborn calf, so it must be acquired. Because "it occurs in females at every sexual condition," wrote its discoverers, "it is presumed that [the] male sexual hormone inhibits the manifestation of the callus, and some female hormone, probably estrogen, stimulates it." Tailward of the hump, there are a series of smaller "posterior dorsal humps," numbering from three to nine, culminating at the tailstock in a vertical, narrow rising that smoothly intersects with the flukes.

The sperm whale is black in color—or brownish gray, slate gray, or iron gray, depending on the circumstances. When photographer William Curtsinger had the opportunity to see sperm whales underwater in the open ocean, he described the color this way: "The black was deep and rich next to the white pigments of their jaws, blacker than anything I'd ever seen or imagined" (Fadiman 1979). Most specimens, male and female, have some white on the body, usually on the underside in the genital and anal regions and on the lower jaw. These white or unpigmented areas are irregular and not necessarily symmetrical, consisting of spots, flecks, streaks, patches, and whorls. In many specimens, the underside is lighter than the back or sides, but this trait is not consistent. Matthews (1938) illustrated a head-on view of an adult male sperm whale with a lighter area on the flattened foremost portion of the head, calling it a "head-whorl." Some observers were of the opinion that lighter color on the head was an indication of age; Scammon (1874) referred to these animals as "vicious, gray-headed old Cachalots." Best and Gambell (1968) recorded head whorls in all sizes of animals killed off South Africa, but they noted that "size seems to be an important factor in the development of the head whorl."

Adult male sperm whales are much larger than females, reaching a maximum length of 62.32 feet (19 meters), whereas the females do not get much beyond 40 feet (12.2 meters). Most contemporary writers do not extend their figures much beyond the 60-foot limit, but there were no such constraints on the earlier historians of the mighty cachalot. Clifford Ashley (1926) mentioned a 90-foot specimen "taken by the bark *Desdemona* in the late seventies"; Beale (1835) said that a full-grown male may be "about 80 feet"; and Bullen (1898) told of a whale "over seventy feet long." There is no question that the intensive hunting of sperm whales in recent years has resulted in a reduction of the average length of those whales taken (Jonsgård 1960), but it seems unlikely that the largest bulls were much more than 62 feet long. Adult females, which are much slimmer than males, rarely reach a length of 40 feet. At that length, they would weigh about 30 tons. In 1835, Beale wrote that they "are more slenderly formed which gives them that appearance of lightness and comparative weakness which the females of most species possess," and Scammon (1874) said that the female has "an effeminate appearance."

A 25-year-old bull will be about 40 feet in length, and at this time the animal will engage in what Ohsumi (1971) called the "struggle for joining a nursery school in the breeding season." This struggle probably results in some of the scars, lacerations, missing teeth, and broken jaws found in older bulls. Many of the scars, especially the circular ones, are thought to result from the suckers of giant squid, and the broken jaws are here differentiated from the deformed jaws mentioned earlier, which are presumed to be birth defects. Although battles between sperm whales have been seen only infrequently, there are a few eyewitness descriptions of these titanic contests. Zenkovich (1962) observed "fights between two big male sperm whales on two occasions," and in another situation, described by Hopkins (1922), a bull tried to join an established herd and a battle between two bulls ensued, wherein "great pieces of flesh were torn from the animals' heads" and the jaw of one of them (later captured) was badly broken and hanging by the flesh.* In

---

*Whitehead includes this battle in his 2003 book, *Sperm Whales,* but with this disclaimer: "It is not clear whether this account is fact or fiction, or (perhaps most likely) something between the two." A careful reading of the account suggests that it is much closer to fiction than to fact. Hopkins's (1922) book, *She Blows! And Sparm at That!* is a novel written for younger readers, and as such, it contains any number of episodes that the author obviously included for entertainment, not veracity. For example, there is an account of a fatal attack on a whale by a swordfish, an attack on the crew by cannibals, and an attack on the crew by Chinese pirates. It is obvious that Hopkins knew a lot about whaling, but like Melville, he also knew how to liven up a story.

TABLE 4.1. Length and Weight of Sperm Whales Caught Near Japan			
Animal No.	Body Length (feet)	Weight Metric Tons	Short Tons
1	30	6.82	7.55
2	35	11.14	12.28
3	40	17.03	18.77
4	45	24.77	27.30
5	50	34.36	37.86
6	55	46.89	51.67

After Omura (1950).

The Year of the Whale (1969), Victor Scheffer gave the following fictionalized account of a battle between two bull sperm whales:

> The young whale turns on his left side and charges, clapping his jaw violently, forcing each tooth with a smash into the firm white socket of the upper gum. The old bull turns deliberately on his back, belly up, responding in kind with a racket that carries through the sea for a league in all directions. His great jaw swings at right angles to his body, tip waving in the air. The first impact of the bodies with a total mass of a hundred tons throws a geyser of green water high in the sky. Within seconds the movements of the whales are lost in a smother of foam. Each infuriated beast is trying to engage the other's jaw, or to seize a flipper—the action is all confused.

An old whaler's formula held that bull sperm whales weighed "a ton a foot," but this estimate seems to have been based on guesswork as much as anything else, because the whales were certainly not weighed at sea, where most of them were tried out during the 19th century's deep-water whaling. Japanese scientists weighed sperm whales caught in the "adjacent waters of Japan" and produced the figures (after Omura 1950) shown in Table 4.1.

These whales were weighed in parts, and a percentage was added for body fluids lost in the cutting process. In 1970 a 43.7-foot (13.32-meter) male was weighed in its entirety using a flatcar and a weigh bridge at a Durban whaling station. It came to 69,300 pounds (31,450 kilograms), or nearly 31 long tons.* From Table 4.1, it becomes apparent that the larger whales begin to approach the ton-per-foot figure of the whalers. Scheffer (1969) mentioned a 50.5-foot (15.4-meter) male that weighed 44 short tons, without blood and intestinal fluids. The figure given by Gambell (1970) for the percentage of the total weight accountable to these fluids was 12 percent. Therefore, adding 10,560 pounds to 88,000 pounds, we get 98,560 pounds, or 49.28 tons, almost exactly a ton per foot.

The mouths of the great whales are, like the whales themselves, more than a little unusual. In most of the baleen whales—blue, fin, sei, Bryde's, minke, humpback, right, and bowhead—the flattened upper jaw sits atop the lower like a (somewhat misshapen) lid on a pot. (The gray whale's upper and lower jaws are similar in size, and uniquely, the upper protrudes beyond the lower, like the beak of a parrot.) In most mammals, the upper jaw is larger than the lower—imagine the head of a deer or a dog—the arrangement in baleen whales at first looks wrong. In the past, those who would illustrate, say, the bowhead turned the whale upside-down so the jaws would look right.

Top or bottom, nothing about the sperm whale's mouth is ordinary. The upper jaw is part of the skull, of course, but the lower jaw, narrowed to a rod, slots into a cavity, not unlike a switchblade fitting into its housing. Now things begin to get interesting. All the other so-called great whales—those that reach a length of 30 feet or more—have baleen plates suspended from the upper jaw, which are used to strain small organisms from seawater.† But the sperm whale has nothing visible in its upper jaw but empty sockets. In the sperm whale's lower jaw are two rows of great ivory pegs that, because they have no enamel, provided

*A "long" or "metric" ton is 2,240 pounds, or 20 hundredweight. A hundredweight, abbreviated "cwt," equals 50.8 kilograms.

†Correction: There is one species of Octavo whale that can be longer than 30 feet, and it most assuredly has teeth. The killer whale (*Orcinus orca*), of which Melville said "little is precisely known to the Nantucketer, and nothing at all to the professed naturalist," is, after our nominal subject, the largest predator on earth. Found throughout the oceans of the world, from the poles to the tropics in both hemispheres, males can reach a maximum length of 30 feet, females 26, which makes them by far the largest of the dolphins. They live all their lives in tightly knit family groups. Sometimes known as wolves of the sea, they feed in packs on everything from sharks, fish, and squid to seals, sea lions, dolphins, and even the great whales. They are fast, powerful, and intelligent, but usually docile in captivity. There are no records of killer whales

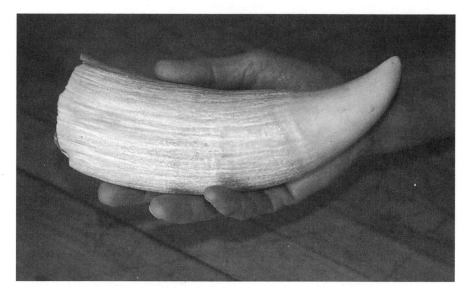

This is what a whale tooth looks like before it is polished and decorated with scrimshaw drawings. (Richard Ellis)

a ready surface for scrimshanders, and which may—or may not—serve the usual functions of teeth for the whale.

The teeth are gracefully curved items, rounded rather than pointed at the tip, and at their largest, they are as big around at the base as a man's wrist. (Although the Yankee whalers hunted the sperm whale for the oil in its head, the teeth were collected and decorated by whalemen to provide an activity during slack periods, which provided an additional source of income when the whalers returned home.) There are 20 to 30 teeth per side, protruding from the gum for a third of their length. Ashley (1926) recorded a pair of teeth that weighed 8 pounds, 7 ounces (3.8 kilograms), and were 11 inches long, but they were from a whale that was supposedly 90 feet long. When the mouth is closed, the teeth fit into sockets in the upper jaw, where there are also teeth, but these are rudimentary and usually unerupted. Because these upper teeth are not subject to wear, as are the lower teeth, they can be used in the determination of the age of the whales from the annual layers of dentine that are laid down (Nishiwaki et al. 1958). It might appear that the function of the sperm whale's teeth is obvious enough: they ought to be used in the procuring, holding, or perhaps even chewing of prey. But like so many aspects of sperm whale biology, it is not nearly that

---

attacking people in the wild, but in March 2010, at Sea World in Orlando, Florida, a male orca named Tilikum dragged trainer Dawn Brancheau into the tank and killed her.

Sperm whale skull in the Whale World Museum, Albany, Western Australia. Notice the scooped-out cranium and the toothed lower jaw. (Richard Ellis)

simple. From the examination of many specimens, it has been deduced that the lower teeth are not cut until the animal is between 28 and 31 feet long (8.54–9.45 meters), some time after weaning (Clarke 1956). This means that the whales have to feed themselves with no teeth for several years.

Occasionally, sperm whales of both sexes are found with a twisted or otherwise deformed lower jaw, but in most cases, these animals were found to be well fed and otherwise healthy (Nasu 1958). (Moby Dick was described as having a "sickle-shaped lower jaw," and this did not appear to affect his biting abilities, either the first time he encountered Ahab or the second.) It is apparent, therefore, that the teeth of the sperm whale play an insignificant role in feeding because the animals seem capable of grasping and ingesting the cephalopods that make up the major part of their diet without actually biting them. It has been noted that most of the squid from the examined stomachs of sperm whales lack tooth marks and are often whole (Okutani and Nemoto 1964).

In the chapter he called "The Battering Ram," Melville wrote this of the sperm whale's head:

You observe that in the ordinary swimming position of the Sperm Whale, the front of his head presents an almost wholly vertical plane to the water; you observe that the lower part of that front slopes considerably backwards, so as to furnish more of a retreat for the long socket which receives the boom-like lower jaw; you observe that the mouth is entirely under the head, much in the same way, indeed, as though your own mouth were entirely under your chin. Moreover you observe that the whale has no external nose; and that what nose he has—his spout hole—is on the top of his head; you observe that his eyes and ears are at the sides of his head, nearly one third of his entire length from the front. Wherefore, you must now have perceived that the front of the Sperm Whale's head is a dead, blind wall, without a single organ or tender prominence of any sort whatsoever.

All whales, from Octavo to Duodecimo, propel themselves through the water with an up-and-down motion of their tail fins, known to the whalers as flukes. Most of the 20,000-odd species of fishes move by

Although the forward portion of a sperm whale's head seems poorly designed to cut through the water, the whales have been swimming and diving successfully for millennia. (Victor Scheffer)

wagging their tails from side to side, although some members of the piscine fraternity—eels, for instance—move with a snakelike undulation of their entire bodies. From his readings of Cuvier, Beale, Bennett, and others, Melville certainly understood that whales were mammals, but perhaps with his tongue in his cheek, he wrote,

> Be it known that, waiving all argument, I take the good old fashioned ground that the whale is a fish, and call upon holy Jonah to back me. This fundamental thing settled, the next point is, in what internal respect does the whale differ from other fish. . . . lungs and warm blood; whereas all other fish are lungless and cold-blooded. Next: how shall we define the whale, by his obvious externals, so as conspicuously to label him for all time to come? To be short, then, a whale is *a spouting fish with a horizontal tail.*

And how Melville adored that tail! He devoted an entire chapter to it, which begins like this:

> Other poets have warbled the praises of the soft eye of the antelope, and the lovely plumage of the bird that never alights; less celestial, I celebrate a tail. . . . Reckoning the largest size Sperm Whale's tail to begin at that point of the trunk where it appears to taper to about the girth of a man, it comprises upon its upper surface alone, an area of at least fifty square feet. The compact round body of its root expands into two broad, firm, flat palms or flukes, gradually shoaling away to less than an inch in thickness. At the crotch or junction, these flukes slightly overlap, then sideways recede from each other like wings, leaving a wide vacancy between. In no living things are the lines of beauty more exquisitely defined than in the crescentic borders of these flukes.

So from fore to aft, the sperm whale is like a big, ungainly dolphin—it has teeth; a bluff-headed shape that would seem to make its passage through the water more than a little difficult; a flat, powerful tail that powers it through the water; and a single blowhole. (Only the baleen whales have paired blowholes on the top of the head, rather like a human nose writ large, and grafted backward onto the head of the whale, to keep water from entering its nostrils as it moves forward.) The twin towering spouts of a blue whale can be spotted from a great distance, but our subject, conforming not for an instant to the cetacean

norm, has but one nostril at the end of its nose, and upon exhaling, it sends a low, bushy plume forward at a 45-degree angle.

Most animals that live in the water—fishes, cephalopods, mollusks, echinoderms, crustaceans—breathe with gills, which extract dissolved oxygen from the water. (They do not breathe the "O" in H_2O.) Mammals, birds, amphibians, and reptiles acquire their oxygen by inhaling air via their nostrils, then transferring the oxygen to the blood. In the bloodstream, oxygen and carbon dioxide (CO_2) bind with hemoglobin, a protein in red blood cells that has an affinity for CO_2. In most mammals, the need to breathe is felt when the CO_2 level in the blood rises above a certain threshold, causing the pH to decrease and making the blood slightly more acidic. Sensors in the part of the brain known as the medulla oblongata tell the animal that it is time to breathe again. The amount of oxygen needed for everyday functions differs greatly from animal to animal. According to Slijper (1962), a horse replaces some 20 percent of the air in its lungs with every breath; a manatee, about 50 percent; and a deep-diving bottlenose whale, about 85 percent. (Humans replace about 15 percent of the air with every breath.)

Certain breath-holding diving animals, such as some seals, a few species of penguins, and many whales and dolphins, have developed an additional mechanism to cope with the CO_2 buildup that would become dangerous or even fatal during protracted or frequent deep dives. They don't have particularly large lungs, but they replace a much larger proportion of the air in the lungs with each breath. The oxygen in the lungs, which is necessary for breathing, is forced into the muscle tissue, where it binds with another blood pigment, myoglobin, until the pressure is reduced as the whale returns to the surface and the stored CO_2 is replaced with fresh oxygen. The amount of myoglobin in a diving animal's muscle tissue determines the maximum time it can spend underwater. The animal with the most myoglobin in its muscles is . . . you guessed it: the sperm whale. The high myoglobin content renders sperm whale meat inedible by humans; it is greasy and purplish black in color, and it is reputed to taste awful.

It is most unlikely that Yankee whalers ate sperm whale meat. An occasional "porpoise" was harpooned to supplement the whalemen's diet, but nobody ate whale meat. The early whalers and writers, such as Beale, Bennett, Melville, and Scammon, were acquainted with the outside of the sperm whale, and even with some of the interior elements— the skeleton, the skull, the case, the junk, the spermaceti oil—but almost a century would pass before some of the innermost workings of the great nose were revealed, and then not particularly helpfully. Dissections have uncovered the incredibly complex structure of the spermaceti organ—which occupies virtually the entirety of a male's 20-foot-long nose—but even with some of the elements laid out before us, we only have the vaguest idea of how they function, or how the whale's sounds

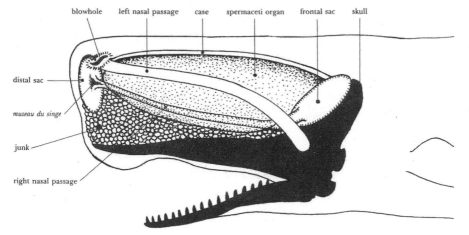

blowhole   left nasal passage   case   spermaceti organ   frontal sac   skull

distal sac

museau du singe

junk

right nasal passage

Diagrammatic cross section of the head of the sperm whale, from the left side (after Raven and Gregory 1933). (Richard Ellis)

are generated or broadcast. And we have only an inkling of what these clicks, pops, creaks, groans, and bangs are used for.

The sperm whale's unusual name is a function of its anatomy. In its nose—and it is a nose, being forward of the brain, over the mouth, with a nostril at the end—there is a huge reservoir of oil, which solidifies to a waxlike consistency when exposed to air. As long ago as Shakespeare's time (he died in 1616), spermaceti oil was a favorite remedy for bruises. In *Henry IV*, act 1, scene 3, Hotspur says that the "sovereign'st thing on earth was parmaceti for an inward bruise." This oil, rather than the oil that could have been boiled out of the blubber as with the other great whales, was the primary reason that sperm whales were hunted relentlessly during the 18th and 19th centuries by the doughty whalers of Nantucket and New Bedford. The oil, as bailed out of the "case"—as the reservoir was known—was the ne plus ultra of whaling, the most important product sought by the Yankee whalers. It was the finest oil known. It was used in the manufacture of smokeless candles and for the lubrication of delicate devices. In *Moby-Dick*, Melville suggests that it had to have been the oil used to anoint the brows of kings. (Because it does not congeal at low temperatures or liquefy at high temperatures, it is the only oil used today in satellites and space vehicles.)

After being ladled out of the case, the congealed head matter was shoveled into strong woolen or canvas bags and placed between the heavy wooden leaves of the spermaceti press. As the press beam was lowered, the pressing began. The oil drawn off "winter-strained sperm oil" was clear and considered to be the finest of all spermaceti oils. The material remaining in the bags was then reheated and molded into 40-pound chunks, called black cake. Here is the Reverend Cheever (1850) on spermaceti: "It is then a yellow viscous substance, which is afterward put into strong canvas bags, and subjected to a screw press, and next to the pressure of the hydraulic engine, whereby the oily matter is all expelled leaving the spermaceti in hard, concrete masses. This, after boiling with potash and purifying, is molded into those beautiful oilless candles which are sold under the name of spermaceti."

Physeter macrocephalus is the only great whale with a single nostril at the front of its head; all the others have paired blowholes located much further aft. *Physeter* is Greek for "blower," and *macrocephalus* means "big head." Thus its name means "big-headed blower," a fitting appellation for the creature with the largest head—and concurrently, the largest brain—of any animal that has ever lived on earth. A brain is a metabolically expensive organ, and there has to be a damned good reason for such a massive brain. But so far, our three-pound brain can't figure out what the whale does with a 20-pounder. The huge brain probably has something to do with sound production for hunting and communication. Sperm whales are noisy animals, making all sorts of clicks, bangs, pops, and grunts that are probably used for conversing with one another, and also for hunting.

Even though sperm whales were hunted from open boats for centuries, the early whalers didn't seem to notice the noises the whales made. As it turns out, the deeper the whale dives, the noisier it gets, so it is not surprising that the whalemen believed the whales to be silent—or nearly so.* In 1957, Worthington and Schevill published the first observations of underwater sounds coming from sperm whales. In the journal *Nature,* they wrote:

> While the sperm whale *(Physeter catodon)* is one of the more conspicuous cetaceans, it has not figured among the relatively few that have been demonstrated to make underwater sounds, although they have occasionally been suspected of doing so. We have now obtained reliable evidence that they, too are soniferous. . . .

*The little Arctic whale known as the beluga has always been known by the whalers as a noisemaker. While they were being hunted, or sometimes when they were just passing by, the chirps, pops, squeaks, and whistles of the "canary of the sea" could easily be heard, even by men in boats.

Three types of sound were distinguished. The first, heard before the whales were sighted, was a muffled, smashing noise, with impulses about half a second apart, increasing in intensity to the end of the series. At first it was supposed to be hammering somewhere in the ship, but it was determined that the sound was not made on board. Later, when the whales were in plain sight most of the time, this sound was less conspicuous. A second sound was a grating sort of groan, very low in pitch, which reminded some of a rusty hinge creaking. This lasted as long as five seconds at a time. By far the most common sounds were series of sharp clicks, which were loud enough to blacken the sounding recorder paper, and which usually occurred at intervals of about half a second, but occasionally as rapidly as about five clicks per second. As many as 73 successive clicks were counted. They usually came in groups of 20 or so. Different individuals chimed in from time to time, and there was no period of more than a few seconds without clicks.

We are familiar with dolphin echolocation: dolphins send out bursts of clicks, which bounce off whatever is in range, and the little whales interpret the returning echoes to determine the shape and consistency of the object in front of them. It is the audio version of vision: light rays (not necessarily generated by the sender, but they can be if emitted from, say, a flashlight) bounce off what's in front of the eyes, and return to sender to be processed by the brain. When you see a wall ahead, you don't walk smack into it; or, when you see light bouncing off a hamburger in front of you, you can pick it up and take a bite. You're a dolphin underwater, and you've just analyzed the echoes from a school of anchovies. What do you do next? If you can see them, you wade right into the school and start gobbling them up. So far, so good. But what if it's deep water and it's too dark to see the individual anchovies? You can't echolocate each one individually, so you have to find a way to slow them down. Faced with the same situation, a swordfish slashes wildly with its sword, killing or injuring enough fish so that the swordfish can gobble them up at leisure. Although a dolphin doesn't have a weapon affixed to the end of its nose, it does have a powerful sound generator. It uses bursts of sound to kill or injure the small fishes.*

---

*When I was diving with dusky dolphins in Patagonian waters, we came across a tightly balled school of anchovetas, surrounded by clicking dolphins. I saw that the

Now change the relative sizes of predator and prey: Substitute a 60-foot, 60-ton sperm whale for the 10-foot, 500-pound dolphin, and change the foot-long anchovetas to a 50-foot-long giant squid. To make things more difficult, stage the encounter in total darkness. Sperm whale sends out clicks to echolocate food items. Echoes bounce back: squid ahead! Whatever the whale does in this situation, it has to do it quickly: the whale is holding its breath, and the squid is not only a faster swimmer, it is a gill-equipped water breather, and it can stay submerged indefinitely. Bang! The whale sends out a focused laser beam of sound aimed directly at the squid, and its internal organs are ruptured by the sonic boom, as if a stick of dynamite had been dropped on it. QED Feeding time for *Physeter*.

In October 2009, when diver/photographers Tony Wu and Douglas Seifert were in the waters of the Ogasawara Islands (south of Japan, where Tsunemi Kubodera had obtained the first underwater photographs of a living giant squid in 2005), they saw—and photographed—something that nobody had ever seen before: a sperm whale feeding on a giant squid. It would have been impossible for the divers to approach the depth at which the whales normally hunt their cephalopod prey, but near the surface, they saw a large female that had obviously caught a huge squid. In their photographs, the squid, minus the long tentacles, is grasped in the mouth of the whale as she comes up to breathe near two smaller whales, one of which might be her offspring.

The nose of the sperm whale is one of the most complicated organs in nature, and one of the largest. In a full-grown whale, it may be as much as a third of the animal's total length. A 60-foot-long whale can have a 20-foot-long nose. It is shaped rather like a tanker truck: a cylinder squared off at the forward end with a single off-center nostril, the blowhole, located at the tip. (It is not smoothly rounded, however; to Hal Whitehead [2003], it "appears to consist of two barrels stacked on top of each other, rather like a double barreled shotgun.") Air taken in through the blowhole is transported to the lungs through a 15-foot-long nasal passage, the only nasal passage the whale has. The other one is a closed system, connected at one end to an air sac and at the

---

little fishes would suddenly stop swimming, and float downward as they were zapped by the dolphins, and then be gobbled up. Once the dolphins clicked on me—to figure out what I was, I think, not to eat me—and I could feel the rapid buzz go right through me, like a ticklish electric shock.

other end to a unique organ called the *museau du singe,* or "monkey's muzzle," which consists of a pair of horny "lips" inside the whale's nose that are involved in the production of the whale's varied repertoire of sounds. As T. W. Cranford (1999) wrote,

> The world's largest nose belongs to the sperm whale, yet its functional significance remains equivocal. In order to help shed light on its function, the head of a postmortem neonate sperm whale was subjected to CT scanning. Geometric comparisons between homologous cephalic structures in sperm whales and dolphins (normalized for body size) show extreme hypertrophy and size sexual dimorphism in the sperm whale's lipid spermaceti organ. Anatomic geometry, energetics, and behavior suggest that this immense nasal apparatus is a bioacoustical machine. Sexual selection *via* an acoustic display is suggested as an explanation for the size and continuous (physiologically isolated) energy investment in the construction and maintenance of the male's spermaceti organ.

All of these nasal passages, *museaus,* oil, and air sacs (there is a sac at the front in addition to the one at the back) are embedded in this oil-filled case, which occupies the upper portion of the whale's nose. Below it is another large organ known colloquially as the *junk.* It too is filled with oil, but this time, the oil is contained within a complex arrangement of cells and connective tissue. Together, the junk and the case make up the spermaceti organ. When the first sperm whalers cut open the head of a dead whale, they found gallons of a milky liquid gushing out. The old whalers didn't know what they were seeing. Even though they were aware that the nose is an unusual repository for an animal's seminal fluid, the rest of this gigantic creature was so strange—ivory teeth only in lower jaw, an asymmetrical skull, only one nostril where all other whales had two, paddlelike flippers that contained bones remarkably like those of the human hand—that sperm in the nose might not have been that strange after all. (Apparently nobody questioned the appearance of seminal fluid in the nose of the *female* sperm whale.) Thus they named this great beast the "spermaceti whale" (*spermaceti* means "seed of the whale") after its mysterious nasal seminal fluid. The name, in a slightly shortened but no more accurate version, is still in use today.

As with all mammals, the reproductive organs of a male *Physeter mac-*

rocephalus consist of testes and a penis, tucked in more or less where you would expect to find them, around where the hind legs would originate, if whales had hind legs. But the great gushing oil reservoir in the whale's head is as poorly understood today as it was when Melville called it "The Great Heidelburgh Tun" and described the moment when Tashtego tumbled in and had to be rescued by his fellow harpooner, Queequeg. In his definitive study of sperm whales, Hal Whitehead (2003) wrote,

> There have been a number of theories about the function of the spermaceti organ, ranging from the battering ram of Melville, to the buoyancy regulator of M. R. Clarke. Most can be largely dismissed. Instead, modern science has focused attention on the acoustic theory of Norris and Harvey. It now seems that the spermaceti organ is involved in the production of echolocation clicks, but the exact mechanism by which clicks are produced is not quite clear.

When Moby Dick sinks the *Pequod,* he smashes it with his head—actually, the end of his nose. The whale that smashed and sank the *Essex* in 1820 also rammed it with his head, so Messrs. Carrier, Deban, and Otterstrom (2002) decided that this sort of thing might be the reason the great nose evolved in the first place. They hypothesized "that the ability of sperm whales to destroy stout wooden ships, 3–5 times their body mass, is a product of specialization for male-male aggression. Specifically, we suggest that the greatly enlarged and derived melon of sperm whales, the spermaceti organ, has evolved as a battering ram to injure an opponent." Melville surely would have agreed: he named a whole chapter "The Battering Ram," and wrote,

> You observe that in the ordinary swimming position of the Sperm Whale, the front of his head presents an almost wholly vertical plane to the water; you observe that the lower part of that front slopes considerably backwards, so as to furnish more of a retreat for the long socket which receives the boom-like lower jaw; you observe that the mouth is entirely under the head, much in the same way, indeed, as though your own mouth were entirely under your chin. Moreover you observe that the whale has no external nose; and that what nose he has—his spout hole—is on the top of his head; you observe that his eyes and ears are at the sides of his head, nearly one third of his entire length from the front. Wherefore, you must now have perceived that

the front of the Sperm Whale's head is a dead, blind wall, without a single organ or tender prominence of any sort whatsoever.

Furthermore, you are now to consider that only in the extreme, lower, backward sloping part of the front of the head, is there the slightest vestige of bone; and not till you get near twenty feet from the forehead do you come to the full cranial development. So that this whole enormous boneless mass is as one wad. Finally, though, as will soon be revealed, its contents partly comprise the most delicate oil; yet, you are now to be apprised of the nature of the substance which so impregnably invests all that apparent effeminacy. In some previous place I have described to you how the blubber wraps the body of the whale, as the rind wraps an orange. Just so with the head; but with this difference: about the head this envelope, though not so thick, is of a boneless toughness, inestimable by any man who has not handled it. The severest pointed harpoon, the sharpest lance darted by the strongest human arm, impotently rebounds from it. It is as though the forehead of the Sperm Whale were paved with horses' hoofs. I do not think that any sensation lurks in it.

The "battering ram" of the sperm whale may be impervious to the hurled harpoon, but within that "boneless toughness" there would be found the single most extraordinary organ in the animal kingdom, a complex of tubes, air sacs, lips, and vast reservoirs of oil, the function of which is only now being investigated. If Melville had any idea what really went on inside that nose, *Moby-Dick* would have been a very different novel—wilder and deeper, perhaps, but certainly a lot noisier.

Reviewing Carrier et al. (2002), Hal Whitehead (2003) noted that the spermaceti organ of the female is also huge, and its evolution probably had little to do with male-male aggression. He wrote, "The standard acoustic function gives a perfectly reasonable, and well supported, explanation for the increased relative size of the male spermaceti organ: sexual selection on the sounds of male sperm whales, if the sounds are either used as weapons themselves or include honest signals—most likely attributes of the slow click—that determine male reproductive success." How like the sperm whale, probably the most anomalous animal on earth, to use its nose for breathing, broadcasting, and ramming into things!

When moving at the surface, the sperm whale swims with an undulating motion, "a slow, shallow and dignified porpoising movement"

(Gaskin 1964). The speed when "making a passage" has been estimated at three or four knots, and when alarmed, they can double this. The action of swimming has been described thus by Ashley (1926): "At top speed his head lifts entirely from the water until the jaw is in view and the head rises and pitches with the rapid beat of his flukes, but he does not disappear beneath the waves." Beale commented that the "narrow inferior surface [of the head] bears some resemblance to the cutwater of a ship, and which would in fact answer the same purpose in the whale." Under extreme duress—that is, when harpooned—the sperm whale has been recorded to achieve speeds of 10 to 15 miles per hour, often towing a whaleboat with a full crew—the Nantucket sleigh ride.

Sperm whales have been observed lobtailing, which is slapping their flukes on the surface, creating great noise and commotion; pitch poling, a behavior that involves raising the forward portion of the head out of the water, presumably to enable the whale to see at the surface; and settling, a peculiar action in which the whale will "sink bodily in the water with the apparent rapidity of a lump of lead . . . unaccompanied by any change in the horizontal position, or any movement of the tail or fins" (Starbuck 1878). Other unexplained behaviors include tail slapping, swimming in a circle, swimming in line, and swimming abreast (Gambell 1968). Gaskin (1964) also reported whales seen lying on their sides or submerging so that just the tip of the snout is visible as the whale hangs in a vertical position. (This position may in fact indicate that the whale is sleeping.) Sperm whales breach frequently, energetically throwing themselves almost completely out of the water and then falling back with an enormous splash. Most accounts of this spectacular behavior indicate that the whale does not emerge completely from the water, but Ashley (1926) wrote, "I have seen an 8-barrel bull sperm leap clear out of the water so that the afternoon sun was framed for an instant under his hurtling form."

Breaching may occur when the whale surfaces from a deep dive, but this is not always the case. The largest whales, the mature bulls, dive the deepest (Lockyer 1976) and spend the most time at the surface recovering from their exertions. "Among the whole order of cetaceans," wrote Scammon (1874), "there is no other which respires with the same regularity as the Cachalot." An old whaler's rule, quoted by Ashley (1926), was that "a sperm whale will spout once for every minute he has been down." They will also stay down one minute for each foot of length. Combining these figures with the whaler's ton-per-foot estimate, we can therefore assume that a 50-foot whale, weighing 50 tons, will stay down for 50 minutes, then breathe 50 times at the surface. As in the case of the length/weight ratio, the whalers were not far wrong. The whales can stay below for more than an hour. Bennett (1840) recorded a 90-minute dive; Beale (1835), a dive of 80 minutes. At the surface after a long dive, the whale pants, breathing short, rapid breaths approximately every 10 seconds. If this pattern and the number of respirations are interrupted, the whale cannot remain submerged as long on the next dive (Caldwell et al. 1966).

In preparation for a deep dive, the sperm whale gives a strong spout, rounds its back into an arch with the hump prominently displayed, and descends steeply with the flukes raised almost vertically in the air. These deep dives are presumed to be hunting expeditions, with the whales searching for food at various depths. Females and juveniles do not dive as deeply as adult males and therefore do not remain submerged as long. By acoustic tracking, Watkins (1977) observed that the initial deep dive was converted immediately below the surface to a shallower dive angle; whales that had been together fanned out underwater, but returned to the surface close together again. Sperm whales have long been known to be able to maintain contact with one another even at considerable distances, but as Beale wrote in 1835, "the mode by which this is effected remains a curious secret."

As biological acoustician Shippa Nummela (2009) wrote, "Cetaceans have succeeded superbly in aquatic hearing, and have also become crucially dependent on their hearing while adapting to the aquatic world." It is a given that an animal that depends on sound will hear well. If it sends out clicks and reads the returning echoes, it has to be able to hear them. But what do they hear with? Odontocetes have external ears, but they are no more than a pinhole located behind the eye, and "the meatus [auditory canal] and ear drum of the porpoise no longer function in the hearing process" (McCormick et al. 1979). The porpoise hears by a method known as bone conduction, which McCormick and colleagues first described in 1970. In *The Porpoise Watcher* (1974), Norris devotes an entire chapter to "The Jaw-Hearing Porpoise"* and dis-

---

*In the days of these early technical and experimental studies, the terms *porpoise* and *dolphin* were used interchangeably. Ken Norris insisted on using *porpoise* for the animal we know today as the bottlenose dolphin. Most of the experimental work was done with bottlenoses, but some other odontocetes were also used as subjects,

cusses his finding of a porpoise jawbone on a remote beach in Baja California:

> What peculiar jaws they were! Most land animals have stout jaws adjusted to the forces of chewing or tearing, or of pulling up grass or the leaves from trees. One can see strong flanges where the jaw muscles attach. Nothing like this seemed to be represented in the porpoise jaw. In fact, it was so thin towards its rear end that I could see sunlight shining through the translucent bone, which at its center was less than a millimeter thick. . . .
>
> Nothing in nature as bizarre as this is created unless the forces of survival hold out value for it. Why, I wondered, did a porpoise find it useful to have a jaw this thin and delicate as the finest porcelain, and a nerve and blood vessel canal that consumed the whole rear end of the jaw?

The hearing apparatus is structurally similar in all odontocetes,* from the smallest porpoise to the largest toothed whale, which is, of course, the sperm whale. There is an inner ear, but sound reception takes place in the lower jaw. Norris identified the actual sound receptors as pockets

---

particularly the harbor porpoise in the Soviet Union, and common dolphins and Pacific white-sided dolphins in the United States. In 1964, Schevill wrote, "Our knowledge is still depressingly limited, being based on a very few individuals of one species, *Tursiops truncatus*. We extrapolate cheerfully from this short base, with the encouragement that all other known odontocete clicks are very much like those of *Tursiops.*"

*Except perhaps the freshwater dolphins known as platanistids. The Ganges River and Indus River dolphins (*Platanista gangeticus* and *P. indi*) are constructed so differently from other odontocetes that they may have evolved separately. Living in turbid rivers, these dolphins have no functional eyes. They depend exclusively on echolocation to navigate and hunt their prey. The platanistid skull has twin crests of bone projecting forward, for which Purves and Pilleri (1973) have postulated a sound-gathering function, as contrasted with other odontocetes, where the primary hearing organ is the lower jaw. Furthermore, Pilleri (1979) wrote that although the upper nasal passage is believed to be the source of clicks in other odontocetes, in the platanistids, it has been shown to be the larynx. (As of 2008, the Chinese river dolphin or *baiji* [*Lipotes vexillifer*], a close relative of the platanistids, found only in the Yangtze River, became the first cetacean in modern times to be officially declared extinct.)

of fat in each side of the mandible, which transmit the incoming sound to the bulla (ear bone), "thence to this tiny bone strut to the inner ear and brain." When Norris.wrote that in 1974, he followed it by saying, "All of this is supposition, supported here and there with just enough fact to allow various scientists to explain what they see in different ways." Continuing experimentation over the next five years has proven Norris's supposition correct, for in 1979, he wrote, "It seems clear now that the environmental sounds enter the odontocete mandible by passing through overlying blubber through what I have termed the 'acoustic window,' penetrating the very thin 'pan bone' of the rear mandible, entering the mandibular fat body and are transmitted to the middle ear."

The posterior flanges of the lower jawbone of the bottlenose dolphin—known jointly as the pan bones—contain fat pads that are composed of triacylglycerols that are known to conduct vibrations efficiently. "Experimental evidence," wrote Nummela (2009),

> has shown that the odontocete lower jaw is very sensitive to sound, and it has been suggested that odontocetes use their lower jaw as an outer ear that collects sound energy that is then guided forward by the mandibular fat pad to the tympanic plate. The tympanic plate vibrations are moved forward to the ossicular chain and to the cochlea. It should be noted that in water, ossicles [the bones of the inner ear] of high mass can transmit high frequencies which in air can be transmitted by very light ossicles. Killer whales and mice can hear equally high frequencies, but their ossicles differ hugely in size. The jawbones of whales and dolphins differ in size too. The tooth-studded mandibles of the bottlenose dolphin are shaped like a narrow "V," and are about fifteen inches long. The lower jaw of a sperm whale, studded with much larger teeth, is a long, narrow "Y," and can be fifteen *feet* long.

Is it too much of a stretch to suggest that it looks a lot like a tuning fork?

In the hunt, after the whales were sighted and the boats lowered, it was not always a frenzied race to get the boat close enough for the harpooner to dart his iron. In many instances, a long chase was required, where the oarsmen's mission was to keep the whale (or whales) in sight until they were in harpoon range. Whaleboats carried short sails to make the pursuit less arduous, and if the oars were shipped, the chase boat made even less noise in its approach. Sometimes the oars were muffled, which consisted of wrapping them in cloth where they fit into the oarlocks, to keep quiet as they approached the whale. With their remarkably sensitive hearing, the whales could certainly hear the clanking of oarlocks, and probably even the conversations of the whalemen. Still, the hunters tried as hard as they could not to spook or "gally" the whales, which would cause them to disperse in a panic.

Off Sumatra, the *Pequod* lowered three boats, and the whales, more than a mile away, somehow became aware of the hunters:

> A general pausing commotion among the whales gave animating token that they were now at last under the influence of that strange perplexity of inert irresolution, which, when the fishermen perceive it in the whale they say he is *gallied*. The compact martial columns in which they had been hitherto rapidly and steadily swimming, were now broken up in one measureless rout.... They seemed going mad with consternation. In all directions expanding in vast, irregular circles, and aimlessly swimming hither and thither, by their short, thick spoutings, they plainly betrayed their distraction of panic.

Because the whales could not possibly have known the intentions of the lowered boats, it seems reasonable to assume that the strange clankings and shoutings that accompanied the lowering a mile distant were enough to gally the whales. The sensitive hearing of odontocete whales also serves them well in feeding.

Squid have no bones. Their only hard parts are the beak and in some species a chitinous "pen" (the cuttlebone used in birdcages to provide calcium for the bird). Echolocating an object that is mostly soft tissue might present a problem for a hunting whale or dolphin. In discussing sperm whales, acousticians Fristrup and Harbison (2002) wrote, "Squids reflect less than one-hundred-thousandth of the incident acoustic energy. These measurements exaggerate the reflectivity of squids in relation to sperm whale clicks, because they utilize much higher frequencies at which the geometry of the beaks becomes significant." "It has been hypothesized," wrote Madsen et al. (2007), "that the soft bodies of squid provide echoes too weak to be detected by toothed whale biosonars, and that only the few hard parts of the squid body might generate significant backscattter." The investigators found that the pen, beak, and lenses did not contribute significantly to the backscatter, but "the muscular mantle and fins of the squid *Loligo pealeii* (the North American longfin squid) constitute a sufficient sonar target for individual toothed whales at ranges between 25 and 335 m, depending on squid size, noise levels, click source levels, and orientation of the ensonified squid." Madsen et al. concluded that bottlenose dolphins and beaked whales could detect *Loligo* squid at ranges up to 35 meters (114 feet), but the sperm whale, "which generates the highest known sound levels in the animal kingdom . . . under noise limiting conditions should be able to detect a 25 cm [10 inch] *Loligo pealeii* at ranges between 25 and 325 m depending on squid orientation." If a sperm whale can detect a 10-inch squid at 1,000 feet, it would have no problem with echolocating *Mesonychoteuthis* or *Architeuthis,* which can be 50 feet long.

The sounds made by the sperm whale have been recognized at least since 1957, when Worthington and Schevill recorded a "muffled smashing noise," first thought to be a hammering on board ship, a "grating sort of groan," very low in pitch, reminding some of a rusty hinge creaking, and a series of clicks. A creaking noise was also heard by Hass (1959) while diving beneath a harpooned animal off the Azores: "A dying beast snapped open its lower jaw almost at right angles, and a most curious noise echoed through the water. It sounded like the creaking of a huge barn door turning on rusty hinges. It was quite a deep, harsh, vibrating tone, carried clear and powerful through the sea." The majority of sperm whale noises have been clicks, but whistles, chirps, pings, plinks, squawks, rasps, yelps, and wheezes have been reported by Perkins and associates (1966), and Norris (1974) described the sounds as "'knocks' (like someone tapping on an empty keg with a fist)."

After analyzing hundreds of hours of sperm whale sounds, Watkins and Schevill (1977) concluded that certain repeated sequences were probably an exchange between two whales, and that these "repetitive temporal pulse patterns," which they called *codas* because they came at the end of longer click sequences, were "signatures" of individual whales. Sperm whale sounds are very loud; they can be heard for miles by humans with good listening gear, and presumably at even greater distances by whales, which have been shown to have an excellent sense of hearing (Yamada 1953). Caldwell and colleagues (1966) wrote that "for exploration of the environment, this acoustic system may very well serve in the stead of primate hands and well-developed stereoscopic vision." At first it was assumed that the clicks served the same echolocation function as those emitted by the bottlenose dolphin, but it is now believed that the clicks (and many of the other sounds) are also "intelligence-bearing communicative signals" (Perkins et al. 1966). Sperm

Ken Norris lecturing aboard a whale-watching cruise in Magdalena Bay, Baja California, in 1986. The author is behind him. (Richard Ellis)

whales can "talk" to each other—or, at least, they can locate and identify animals they cannot see.*

The means by which sperm whale sounds are produced is not clearly understood, but it is obviously related to the unique structure of the head. The spermaceti organ, nasal passages, and various air sacs are all assumed to be involved with deep-diving capability, buoyancy, and sound production, but the exact relationship and interaction of these organs remain a mystery. The spermaceti organ is a huge reservoir of straw-colored oil (known as spermaceti oil or sperm oil to the whalers), which hardens into a whitish wax when exposed to air. In 1725, Paul Dudley wrote, "The *Sperma Ceti* Oil, so called, lies in a great Trunk about four or five Feet deep, and ten or twelve Feet long, near the whole Depth, Breadth, and Length of the Head, in the Place of the Brains, and seems to be the same." The organ itself is encased in a tough membrane, known as the case, which Beale described (1835) as a "beautiful glistening membrane. . . . covered by a thick layer of muscular fibres." Immediately below the case is the junk, a network of tough, spongy cells, also filled with oil. The sperm whale has two nasal passages, but only the left is connected directly to the blowhole. The right nasal passage is one-seventh the diameter of the left, and Norris (1969) believed it is involved in sound production by way of its connection with the *museau du singe* and with two air sacs, one at either end of the case.

When he dissected the head of a whale at Richmond, California, Norris (1974) wrote, "I thought I could perceive from the almost unbelievably peculiar anatomy of the sperm whale's head that sperm whales produce sounds with a pair of huge lips located inside the soft tissue of the forehead. . . and then rocket them back and forth between a pair of sacs also located inside the forehead, to produce the loud banging sounds we heard." Beale believed that the case, containing oil of a lighter specific gravity than water, "will always have a tendency to rise, at least, so far above the surface as to elevate the nostril or 'blow hole'

*Bottlenose dolphins emit both whistles and clicks. The clicks are primarily used in echolocating, while the whistles are thought to be communication devices. By these signature whistles, the dolphins identify each other and probably convey additional information as well (Caldwell and Caldwell 1966).

sufficiently for all purposes of respiration." In a 1933 analysis of the spermaceti organ, Raven and Gregory wrote,

> We infer that its main function is to act as a force pump for the bony narial passages, drawing a great quantity of air into the respiratory sacs and preventing the escape of air under pressures of great depth. It may possibly also act in part as a hydrostatic organ, since by severe contractions of part of its muscular sheath the contained oil might be squeezed toward one end or the other, while the air sacs were being inflated, thus lightening the specific gravity of that end and tending to alter the direction of motion of the animal.

Norris and Harvey (1972) suggested that "the spermaceti organ is an acoustic resonating and sound-focussing chamber used to form and process burst-pulsed clicks," and therefore the physical evidence indicates that the organ is "especially useful for long-range echolocation in the deep sea." When spermaceti oil was tested for its sound-transmitting qualities, it was found to be far more effective than water and approximately twice as good a conductor as the oil of a known echolocator, the bottlenose dolphin. Spermaceti transmitted sound at 8,857 feet (2,684 meters) per second, compared with 4,462 feet (1,352 meters) per second for the oil of the dolphin. Norris and Harvey built an apparatus that approximated the nasal sacs of the spermaceti organ and were successful in duplicating the click pulses, demonstrating at least that the sounds could be produced by such an arrangement. They further postulated that the time between clicks was a function of the distance between the frontal and distal air sacs and could therefore be used as a measure of the animal's total length.

In a comprehensive survey of the literature, Malcolm R. Clarke (1978a) listed the suggestions that have been made for the function of this extraordinary organ: sound production, movement of air between the nostrils and the lungs when the whale is at depth, control of closure of the nasal passages, absorption of nitrogen under pressure, or even a means of attack and defense. His own theory is that the organ is used for buoyancy control. Briefly, he suggested that the organ regulates the whale's density, enabling it to maintain neutral buoyancy in dives deeper than 660 feet (200 meters). The whale accomplishes this by controlling the temperature of the spermaceti oil: "If the whale needs to achieve neutral buoyancy it must be able to increase its own density. . . . The necessary increase can be achieved if the whale can lower the temperature of the spermaceti oil of its head so that it freezes." Clarke hypothesized that the whale controls the temperature of the spermaceti oil by circulating seawater through the nasal passages (especially the right one), although he admitted that "the mechanism by which the heat is lost is not irrefutably established." Because the sperm whale floats after death, it is neutrally buoyant at the surface, and some sort of buoyancy control is obviously necessary to enable the animal to dive as deeply as it does. In conclusion, Clarke wrote, "While the present work clearly shows the sperm whale *could* use the spermaceti organ for buoyancy control . . . final proof must await a measurement of temperature and density of the oil *in situ* in a diving sperm whale, a difficult but not impossible task."*

Earlier, Backus and Schevill (1966) had remarked on the regularity of the clicks ("We are dealing here with a sensory system that is generally conceded to be elaborately developed, and a motor task that the whale can perform at a rapid rate"), but they were assuming that the clicks were generated physically rather than mechanically, as a function of the distance between the two air sacs. They pointed out that the clicks might be used not only for echolocation, but for communication. "While listening at sea," they wrote, "we have heard the now familiar clicks without seeing the whale. . . . We have sometimes met cachalots within a few hours of having merely heard them. Such experiences in fine weather lead us to believe that these clicks are very loud and can be heard with good underwater listening gear up to a few miles."

Nobody has ever observed a sperm whale hunting, so much of what follows is conjecture, based primarily on theories developed by Ken Norris. In 1970, at the last American whaling station (at Richmond, California), Norris examined a sperm whale that had been brought in. In his 1974 book, *The Porpoise Watcher*, he described the head:

*Although the paper by Norris and Harvey predated the one by Clarke that is quoted above, they commented on his 1970 paper in the British journal *Nature*, which presented a preliminary version of the same theory. Norris and Harvey (1972) wrote that the spermaceti organ could not possibly function as a heat exchanger: "It seems inconceivable to us that the gossamer webwork of connective tissue and blood vessels that invades the spermaceti could be an efficient heat exchanger capable of repeated heating and cooling of such massive amounts of lipid tissue in short periods of time." They also contended that "a buoyancy organ on one end of the body seems a grotesque arrangement," because the animal would constantly be tipping up or down. The function of this organ is still unknown, although there is no shortage of hypotheses.

The head didn't look like the pictures I had seen. The bow was bluff, and to be sure, as one sees in old whalers' paintings, but from the front the upper part of the head was rounded, almost cylindrical, like the end of a boiler, and then just below it was narrow and indented, and a foot or so above the upper jaw it finally bulged again. The lower jaw was all but invisible under this peculiar snout, a long, thin rod of bone lined with teeth.

Norris dissected the head and examined the *museau du singe,* along with the spermaceti organ and the attendant air sacs and tubes. He thought he had resolved the problem of the whale's sound production, but, as he wrote, "inferences like these, made from anatomy without actual tests of true functions in life, are speculations. . . . While I thought I could see how this incredible system might work, I did not know that my speculations were right." He then did what any inquisitive cetologist would do: he built a testable model of a sperm whale. With engineer George Harvey, he developed a model of the whale's resonating chamber, complete with a spermaceti-filled plastic tube, a sound generator, and an oscilloscope to display the results. Each of the clicks was composed of a packet of diminishing sounds, exactly like the recorded sounds of the whale. "I wondered," wrote Norris,

> why the sperm whale should make such peculiar sounds in the first place. What was different about the way a sperm whale lived that would require these particular bursts of sound? The only possibility that came to mind was that sperm whales feed in the deep sea, sometimes several thousand feet down, and eat mostly very large, swift prey—very large squid, some of which may be ten or even twenty feet in length. Probably the range at which the sperm whale locates its food is very long—a mile or more.

He suggested that the sound chamber in the whale's nose not only broadcasted the sounds, but also amplified the returning echoes, enabling the whale to "hear tiny sounds that would otherwise be lost in the background noise of the deep sea."*

*Although Norris published extensively on the sperm whale's ability to send out focused sound beams to immobilize its prey, the idea was not original to him, a situation he willingly acknowledged. In 1963, Soviet cetologists Bel'kovich and Yablokov

While Norris's imaginative interpretation provided a possible explanation as to why and how the sperm whale's nose could produce "these particular bursts of sound," it did not answer the next question: even if the whale managed to locate the squid in the black depths of the ocean, how did it manage to catch them? Squid are among the fastest-moving marine creatures; one source attributes to them burst speeds of up to 55 kilometers per hour (34 miles per hour), and as Norris notes, "the sperm whale is not a particularly fast or maneuverable odontocete." Nevertheless, the majority of sperm whale stomachs examined by whalers and cetologists are filled with squid, so they have to be able to catch them somehow. (One whale had the beaks of 14,000 squid in its stomach.) There was something most peculiar about the squid in the stomachs of whales: they rarely showed evidence of having been chewed or bitten. In his 1972 monograph, Soviet cetologist A. A. Berzin wrote, "Apart from the very rare exception, even on large food items found in sperm whale stomachs, no traces of teeth are visible." The teeth of sperm whales do not erupt from the gums until long after weaning, and there are many cases of captured sperm whales with completely deformed lower jaws that nevertheless had full stomachs. The whales must obviously have developed a method of capturing the fast-moving squid that does not involve chasing them or snagging them with their teeth. Earlier theories had the whales attracting prey by dangling their white lower jaw in the water, or slurping in the squid by suction. "All these questions," wrote Norris with Bertel Møhl, a Danish zoologist, "seem to be resolved if the sperm whale is able to immobilize its prey before engulfing it." Norris and Møhl (1983) wrote, emphasizing the speculative nature of their hypothesis, "the Sperm whale may catch its swift squid prey leaving no evident tooth marks, and such prey may be alive in sperm whales' stomachs. The disparity between the speeds of sperm whale and squid and the costs of sperm whale acceleration are discussed [in the paper]. Sperm whales eat a very wide-sized range of prey, and small items will not repay the costs of the whale's locomotion. The forehead sound-beaming anatomy is postulated to allow prey debilitation."

In 2000, with several Danish colleagues, Bertel Møhl reexamined

---

published "The Whale—An Ultrasonic Projector," in which they wrote that the whale is able to focus a shock wave that can be "an effective instrument for stunning an immobilizing prey far away."

Norris and Harvey's hypothesis that the sperm whale could generate loud, aimed sound beams. They wrote:

> In sperm whales, the nose is vastly hypertrophied, accounting for about one-third of the length or weight of an adult male. Norris and Harvey ascribed a sound-generating function to this organ complex. A sound generator weighing upwards of 10 tons and with a cross section of 1 m is expected to generate high intensity, directional sounds.

And indeed it does. Off the coast of Norway, Møhl et al. suspended a five-hydrophone array to pick up the clicks of sperm whales, and because the same click was recorded at a different intensity from separate hydrophones, they interpreted their results as "evidence for high directionality."

The skull that cradles the spermaceti and other organs is as curious as the organs themselves. It swoops down from a high posterior portion to a pointed rostrum, making a hollow basin, roughly the shape of "a wheel-less chariot—indeed, the whalers used to refer to it as the 'coach' or 'sleigh'" (Norman and Fraser 1938). The nasal bones are asymmetrical—in fact, only the left one is visible at all—and the lower jaw is a shaped rod of bone, the two halves being fused for most of their length. In 1840, F. D. Bennett described it as "somewhat the shape of a plough," but that was when people were more familiar with that shape. The sperm whale's skull is the largest of any whale and, a fortiori, larger than that of any mammal. The skull of a 43.7-foot (13.3-meter) bull weighed 4,760 pounds (Gambell 1970). The brain of the sperm whale is the largest brain known. The average weight in a study done by Kojima (1951) was 17.16 pounds (7.8 kilograms), and the heaviest brain he measured, from a 49-foot bull, weighed 20.24 pounds (9.2 kilograms). The body weight of a 40-ton whale is about 5,000 times greater than its 17-pound brain, as compared with a man's 180-pound body weight and three-pound brain, a ratio of 1 to 60.

There is no question that sperm whales echolocate, even though the only evidence we have is circumstantial. The click sounds have been recorded (Worthington and Schevill 1957; Watkins and Schevill 1977; and others). Sperm whales hunt at great depths and therefore in total darkness, so a logical conclusion would be that they, like many other toothed whales, read the echoes of sounds bounced off potential prey organisms. However, what seems obvious in the case of the sperm whale is often not so. In the matter of the spermaceti organ, for example, it has been suggested that the whale is able to focus sound into a shock wave that can be "an effective instrument for stunning and immobilizing prey far away" (Bel'kovich and Yablakov 1963). If the concept of projecting sonic booms to stun the prey is not bizarre enough, consider the theory of Kozak (1974): he hypothesized the existence of a "video-acoustic system" that enables the whale to transfer sound energy into images, using the surface of the rear wall of the spermaceti organ as an "acoustic retina." What actually transpires at the great depths at which the sperm whale feeds may never be known to us, but in his comprehensive review of the sperm whale's biology, Berzin (1972) summarized the available information and produced this description of its feeding habits:

> Having dived, the animal covers several hundred meters, investigating the surrounding waters by echo sounding with ultrasound of varying frequency (for dolphins it will be from 12 to 170 kHz) and intensity. Analyzing the returning signals, it receives the most detailed information on the objects around it. If the animal dives to the bottom, it swims with wide-open mouth (underwater cables damaged by sperm whales were usually coiled around the lower jaw). When mobile squid and fish are discovered, the ultrasonic beam narrows and focuses on them, its frequency sharply increases, and the prey is stunned and then seized.

Certainly, the idea of an animal immobilizing its prey with focused sound beams is incredible, but so is the image of a huge, slow-swimming creature foraging in total darkness and capturing thousands of pounds of fast-moving squid every day.

Only the creature with the three-pound brain has consistently attacked and killed sperm whales, but Matthews (1938) reported that "some agent removes semi-ovoid pieces of blubber" (probably the 18-inch shark, *Isistius*), and stalked barnacles, *Conchoderma,* are sometimes found on the front lower teeth. These teeth are not always enclosed when the mouth is shut, and the barnacles also appear on the jaws and teeth of whales whose lower jaws are deformed (Nasu 1958; Clarke 1966b). Mocha Dick, the white whale described by Reynolds in 1839, had barnacles on his head, "clustered until it became absolutely rugged with the shells."

One of Bullen's (1898) more fanciful tales concerned a "bull cachalot and so powerful a combination of enemies that even one knowing the fighting qualities of the sperm whale would have hesitated to back him to win." The enemies were two killer whales and a 16-foot-long swordfish. The sperm whale took the torpedolike charge of the swordfish in "the impenetrable mass of the head, solid as a block of thirty tons of india-rubber"; the whale turned and caught the "momentarily motionless aggressor in the lethal sweep of those awful shears, crunching him in two halves, which writhing sections he swallowed *seriatim*." The sperm whale then smashed one of the "allied forces" with his flukes, and the other killer whale fled, "for an avalanche of living furious flesh was behind him, and coming with enormous leaps half out of the sea every time." It is wild and colorful descriptions like these that cast the balance of Bullen's work into an unfavorable light, for the exaggerations and unrealistic behavior of the combatants put the entire episode squarely in the realm of imaginative fiction.*

After being chased by men in a rowboat and then stabbed with a sharp spear, a sperm whale was likely to be more than a little perturbed, and if it failed to dislodge the harpoon by diving, it might retaliate in a frenzy by upsetting the boat to which it was painfully attached—a reportedly common occurrence during the heyday of the fishery—or even charge the ship, as in the case of the *Essex*. (Moby Dick smashes and sinks the *Pequod:* "Retribution, swift vengeance, eternal malice were in his whole aspect, and spite of all that mortal men could do, the solid white buttress of his forehead smote the ship's bow till men and timbers reeled. . . . Through the breach, they heard the waters pour, as mountain torrents through a flume.") Melville's white whale was "so incredibly ferocious as continually to be a thirst for human blood," and "when swimming before his exulting pursuers, with every apparent symptom of alarm, he had several times been known to turn round suddenly, and bearing down upon them, drive them back in consterna-

*There are records of swordfish attacks on whales—or, rather, records of swordfish swords broken off in whales—but we cannot really assign a motive to the fish. Scheffer (1969) fictionalized an incident in which a sperm whale was thus pierced, and there are reports of swordfish swords or marlin bills in sei whales, minke whales, bales of rubber, ships, and even a submarine. In 1967 the minisub *Alvin* was attacked at 1,800 feet by a 200-pound swordfish, which was subsequently raised, cleaned, and eaten.

tion to their ship." Melville described Moby Dick as a vicious, vindictive, malevolent creature that had already wreaked havoc with the whaling fleet, and taken off Ahab's leg and Captain Boomer's arm.

But despite the overreactions of harpooned whales and the rare instances of battling bulls, sperm whales are for the most part docile, gentle creatures that prefer flight to fight. As Bennett wrote in 1840, "In common with most terrestrial animals which herd together in great numbers, Sperm Whales are naturally timid, and disposed to fly from the remotest appearance of danger: and although many instances occur amongst them of a bold and mischievous disposition, which leads them, when molested, to attack and destroy both boats and man, yet such traits rather belong to the individual." Hal Whitehead (1986) wrote, "During the weeks we spent with sperm whales, the subjects of our research showed themselves to be gentle animals. They are usually shy but occasionally curious in the presence of humans and their boats." Indeed, their armament, consisting of a "battering-ram" nose and a lower jawful of large but not very sharp teeth, is better designed for squid hunting than for battle. Other than armed men in boats, the only known enemies are killer whales, and sometimes large sharks. Off the coast of central California in 1997, Pitman et al. (2001) observed an attack by a pack of some 35 killer whales on nine sperm whales. "The sperm whales appeared largely helpless, their main defensive behavior was the formation of a rosette ('marguerite'—heads together, tails out). When the killer whales were successful in pulling an individual out of the rosette, one or two sperm whales exposed themselves to increased attack by leaving the rosette, flanking the individual, and leading it back into the formation. Despite these efforts, one sperm whale was killed, and the rest were seriously, perhaps mortally wounded."

There are a couple of species of smallish killer whales—smaller than orcas, anyway—that are known as false killer whales (*Pseudorca crassidens*) and pygmy killer whales (*Feresa attenuata*). *Pseudorca* is the larger of the two, reaching a maximum length of 19 feet, and with a skull and teeth that so closely resemble those of its larger cousin that its name actually means "false killer whale." In Galápagos waters, Daniel Palacios and Bruce Mate (1996) were in the water filming sperm whales when they sighted "a group of unidentified large dolphins, porpoising at high speed towards the sperm whales at about 500m from us and the whales. As we approached, we identified these 'dolphins' as a mixed group of 25 false killer whales and 15 bottlenose dolphins." A huge cloud of fecal matter clouded the area where the sperm whales congregated, so the divers were not able to see actual attacks, but they did see small pieces of flesh floating in the murky water. Both species of dolphins (false killer whales are large dolphins) made high-pitched whistles as they darted to and fro, and eventually, the sperm whales raised their flukes vertically above the surface, dived, and were not seen again. Palacios and Mate "believe this interaction was aggressive in nature. The evidence indicating this includes the vigorous approach by the false killer whales, the reaction of the sperm whales to their approach while still at a distance, the formation of the marguerites, the massive discharge of feces by the sperm whales, the strong agitation in the group . . . and the three pieces of flesh we observed."

And then there is the curious case of the pilot whales. Not normally aggressive to other cetaceans, a group of about 30 short-finned pilot whales (*Globicephala macrorhynchus*) were observed in the vicinity of two groups of sperm whales in the north-central Gulf of Mexico. The pilot whales, which are large (up to 20 feet), black, beakless dolphins, "appeared excited, as evidenced by rapid swimming and surfacings, fluke-up dives, and variable movements along the flanks, heads, and flukes of the sperm whales" (Weller et al. 1996). The sperm whales remained close to each other, often huddled together and touching, and formed marguerites numerous times, horizontally and vertically. The pilot whales remained amid the sperm whale group for more than two hours, and at one point, a group of five rough-toothed dolphins (*Steno bredanensis*) joined the party, although they stayed closer to the research vessel than to the whales. Although no actual combat was observed, the pilot whales may have been testing the vulnerability of the sperm whales, or they may have been "engaged in play or that practice of predation with no real intent to harm or kill the sperm whales."

Captured sperm whales are often scarred with round marks that look very much like the marks that would have been left by the suckers on the arms of giant squid. Because it is known that these whales eat squid of all sizes (mostly smaller than *Architeuthis*), it is not an unreasonable assumption that these scars are caused by squid unwilling to be eaten by whales. The suckers of the giant squid (*Architeuthis*) have chitinous "teeth" around the circumference, and the colossal squid (*Mesonychoteuthis*) actually has claws on its tentacles. Some believe that the scars are caused by the teeth of rival males in battles for dominance. Hopkins (1922) described a battle between two bulls that lasted for two

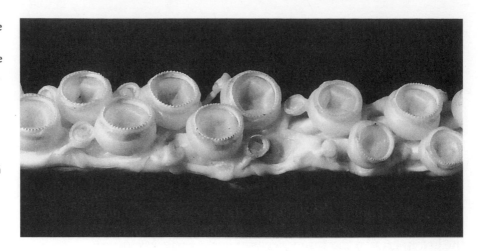

Close-up of the suckers on the tentacles of the giant squid, *Architeuthis*. Each sucker is rimmed with a row of sharp, toothlike projections, which pierce the skin of its prey, often leaving circular scars. (Courtesy of Wolfgang Zeidler, South Australian Museum)

days and resulted in a broken jaw and many missing teeth for the loser, who was killed and captured by the whalers who witnessed the battle. Some investigators, such as McCann (1974), believe that most of the scars on the heads of sperm whales (and beaked whales too) come from the teeth of fighting males, and Caldwell et al. (1966) wrote that "such fights undoubtedly account for the frequent reports that captured male sperm whales have scars or deep scratches on the skin corresponding to tooth marks. Broken teeth found in sperm whales may also be the result of combat." Combat, whether with rival bulls or reluctant squid, is probably the explanation for the scars on the heads of whales, but until somebody actually sees these scratches inflicted, their cause will remain unknown—or at least debatable.

Another unresolved mystery of cetacean behavior is sleep. Most mammals (including us) need sleep, but we are not altogether sure why. Without it, we begin to function badly, and after a certain period of time, our responses deteriorate, and eventually we lose consciousness. (Sleep deprivation in humans is a classic torture technique.) Aristotle pointed out that "all animals which have a blowhole breathe in and out, as they possess lungs. A dolphin has been observed, while asleep, with its snout above water, and snoring in its sleep." Most mammals don't have to worry about breathing while they sleep, but for cetaceans, it can be a serious problem. Exhaling is easy, but whales and dolphins have to think about where they are when they inhale. To avoid a noseful of water instead of air, they must be at the surface. Breathing is therefore a conscious act for cetaceans because they cannot fall into the unconscious or semiconscious state that characterizes sleep in other mammals. Observations of captive dolphins have shown that they sleep differently, with only half their brain at a time. Electrodes hooked up to the dolphin's head measure electricity levels in the brain, and the resulting electroencephalograms (EEGs) demonstrate that in the sleep cycle, half of the dolphin's brain does indeed shut down while the other half is still active. The half of the dolphin's brain that stays awake keeps the animal near the surface, where minor movements of its flippers and flukes periodically raise the blowhole above the surface for breathing.

Although it has been possible to observe bottlenose dolphins, pilot whales, and killer whales in captivity, no such opportunity has ever presented itself for sperm whales. (Physty, the whale I described in chapter 1, was sick, so his behavior—which consisted mostly of lying on his side—was uncharacteristic.) There is no mention of sleep in Whitehead's (2003) comprehensive study of sperm whales, and when other books touch on the subject, it is almost always in reference to captive animals. The repertoire of surface behavior in sperm whales includes logging, where the whale looks not unlike a floating log, the whale resting at the surface with only the dorsal region and the blow-hole exposed, accompanied by little or no forward movement. This makes the whales hard to see at the surface, and it has been suggested that they may in fact be asleep. For the most part, sperm whales hunt their prey at depths where no sunlight reaches, so it matters little if it is day or night. If, however, sperm whales sleep at the surface at night (a completely unfounded supposition), it would explain why nobody has ever identified a sleeping sperm whale.

As far back as 1874, Scammon reported what seemed to be Pacific pilot whales sleeping: "In low latitudes, during perfectly calm weather, it is not infrequent to find a herd of them lying quite still, huddled together promiscuously, making no spout and seemingly taking a rest. Sometimes they assume a perpendicular attitude, with a portion of the head out of the water, as does the sperm whale." Confirmation of the pilot whales' propensity for napping came from Norris and Prescott's 1961 observations of captive animals at Marineland of the Pacific:

> We have repeatedly observed the captive pilot whales sleeping with eyes tightly closed during the day and night. Sleep seems much deeper during nighttime. . . . During the deepest sleep, the whales hang almost immobile in the water with their tails downward at about a 30° angle from the surface. The blowhole and anterior part of the melon are above the surface. . . . The only visible activity is a

slight sculling movement of the tail that serves to keep the sleeping animal on an even keel. The melon, which is composed largely of fatty tissue, may serve as a float allowing these whales to sleep with their blowholes out of water.

For buoyancy in the nose, of course, no creature on earth can approach *Physeter macrocephalus,* so it is not unreasonable to hypothesize similar behavior in the sperm whale. The pilot whale's blowhole is on top of its head, but the unique location of the sperm whale's nostril at the forward end of its nose would make it possible (and easy) for the animal to breathe inconspicuously while hanging vertically in the water. Sperm whales are already known for spy hopping, where the whale's entire head protrudes above the surface as the whale is vertically oriented, often to bring the eyes out of the water for a look around.

One of the most puzzling aspects of cetacean behavior is their occasional predisposition to beach themselves and die in shallow water or on the beach. As Aristotle wrote in *Historia Animalium,* "It is not known for what reason they run themselves aground on dry land; at all events it is said that they do at times, and for no obvious reason." There are recorded instances when they were able to turn around and yet refused to do so; and at other times well-meaning observers have tried to shove or tow them to the safety of the sea, only to have the whales return to the beach and certain death. Many species of odontocetes are given to this curious phenomenon. In a few animals such as the pilot whales, *Globicephala,* and the false killer whale, *Pseudorca crassidens,* the behavior is so common as to be diagnostic. A mass stranding is usually pilot whales, false killer whales, or sperm whales, although other species occasionally strand in numbers. The circumstances and the species vary from stranding to stranding, so it is difficult—if not impossible—to generalize on the reasons or causes of this behavior. It has been assumed that the stranding animals were disoriented, sick, or lost, perhaps because of a dysfunction of their echolocating or navigating systems.

Among the suggested causes of mass strandings are beaches that slope too gradually for efficient reflection of broadcast signals; mass parasitic infestations of the inner ear, which would render the hearing or receiving faculties inoperative; underwater sounds that confuse the animals; unusual weather conditions; predator harassment; and even a death wish, which makes their behavior tantamount to suicide. Any or all of these may contribute to the mass stranding of cetaceans, but none of the explanations seems entirely satisfactory. Some acousticians believe that the sonar capabilities of whales are far too sophisticated for them to become confused by gently sloping beaches, and while parasites certainly can and do infest the inner ear of cetaceans, it seems unlikely that the critical moment of infestation would occur simultaneously in as many as 200 animals.

Part of the school of 48 sperm whales that stranded and died on the beach at Perkins Island, off Tasmania's northwest coast, in January 2009. (Tasmanian Museum)

There are too many underwater sounds nowadays—far more than there were in the days before steam and diesel power, not to mention subsurface drilling operations—and whales stranded long before the industrial revolution. Predators lurking offshore might be responsible for chasing some species into shallow water, but how are we to explain the strandings of killer or sperm whales, two apex predators with virtually no natural enemies? Unusual weather conditions appear to be insignificant in some cases and inapplicable in others. F. G. Wood (1978) of the Naval Oceans Systems Center in San Diego suggested a new hypothesis to explain the phenomenon of mass strandings: "Cetaceans still retain, from their amphibious ancestors, a subcortically induced incentive to seek safety on shore when severely stressed—a 'blind' (but not directionless) emotional response." As Wood himself admitted, such a behavior would be maladaptive; that is, it would be genetically self-destructive and should long ago have been eliminated from the behavioral repertoire of the cetaceans.

Faced with an animal so literally out of its element, Good Samaritans have often tried to refloat the whale, assuming that as a creature of the water, it belonged there and not on dry land. Most efforts of this kind have been marked by the whale's refusal to return to the sea and what has often been interpreted as a repeated suicidal desire to beach itself again, even after it has been towed out to sea. Those who would save the whale by returning it to the water were possibly doing exactly the wrong thing. (After his initial beachings, Physty was in the water for the duration of his "captivity," and after eight days, he swam off strongly under his own power.)

In Farley Mowat's 1972 *A Whale for the Killing*, a fin whale trapped in a pond in Newfoundland was shot hundreds of times by the towns-people, apparently because they could not resist such a huge target. In this case, therefore, there is no question about the whale's injuries—it eventually died from the multiple gunshot wounds—and in extremis, the whale repeatedly beached itself. Mowat, the Canadian author/biologist who chronicled this unfortunate series of events, assumed at first that the injured whale should be in the water, but then realized that perhaps the opposite was true: "The scales were off my eyes and now I saw the truth. She had not grounded by accident, neither had she been beached by the malice of men. She had *deliberately* gone ashore because she was too sick to keep herself afloat any longer. I had misread the evidence, but now it was unmistakable."

It is possible, therefore, to assume that individual whales strand because they are putting themselves in a position where breathing will be easier, or, at least, where they will not drown. As Carleton Ray (1961) put it, "If such diseases involve difficulty in breathing, which seems likely, it may well incline the animals to seek calm, shallow waters and even the fatal pillow the beach affords." The problem of mass stranding is probably connected with this phenomenon, at least insofar as one of the whales or the "leader" is concerned, but there are still too many variables in these situations that we do not understand. The suggested explanations for individual whales on the beach have nothing to do with mass or group strandings. In 1946, 835 false killer whales came ashore at Mar del Plata, Argentina, and all died, but when 29 false killers were found thrashing in the shallows at Loggerhead Key in the Dry Tortugas, as soon as a large male (the leader?) died, the rest all returned to the ocean and swam away (Porter 1977). Because they are so demonstrably gregarious, sperm whales often strand in large numbers. (The only conclusion to be drawn from this is that whales remain together in death as well as in life; no assumptions can be made regarding parasite infestations, mass suicides, injured leaders, or other possibilities.) Sperm whales are by far the largest of the socially organized whales, so a mass stranding is likely to be a momentous event. (Even though some of the balaenopterids are larger, they are not so gregarious and often associate in groups that may be no larger than a pair.) Some of the more noteworthy sperm whale strandings, as recorded in Gilmore (1959), are listed in Table 4.2.

In 1979, Peter Bryant discussed a sperm whale stranding of 59 animals that occurred in 1970, and the largest known sperm whale mass

TABLE 4.2. Sperm Whale Strandings

Year	Site	No. of Whales Stranded
1723	Mouth of the Elbe, Germany	17
1784	Coast of Brittany, France	31
1888	Cape Canaveral, Florida	16
1895?	New Zealand	27
1911	Perkins Island, Tasmania	37
1918?	New Zealand	25
1954	La Paz, Gulf of California	22

As recorded in Gilmore (1959).

stranding, 72 animals in 1974, both in New Zealand. On New Year's Day, 1979, 56 sperm whales were discovered on the beach at Rancho San Bruno, on the coast of the Sea of Cortez in Baja California. (There were 38 males, nine females, and nine animals whose sex could not be determined, because by the time scientists arrived on the scene, Mexican fishermen had already begun to burn the carcasses. The composition of this group raises numerous questions about the school structure and habits of sperm whales because they are not supposed to aggregate in these peculiar proportions, nor are they usually found in the waters of Baja. These waters support a large population of the jumbo squid, *Dosidicus gigas,* which might explain what the whales were doing there, but not why they died there.) On June 16, 1979, 41 sperm whales beached themselves south of the mouth of the Siuslaw River, near Florence, Oregon, and died by the next day. The school, which consisted of 13 males and 28 females, was the largest sperm whale mass stranding in North America. Biologists rushed to the scene because "such mass strandings of sperm whales offer unique opportunities to examine the social structure of entire schools or subgroups" (Rice et al. 1986). Because the teeth of sperm whales lay down a layer of dentin every year, an examination of the teeth revealed that the males ranged in age from 14 to 21 years, and the oldest female was 58. The presence of one or more large, adult males would have classified this aggregation as a bachelor school, but as it was, younger males and sexually mature females suggested that it was a breeding school.

Flukes of a few of the 41 sperm whales that stranded on the beach and died near Florence, Oregon, in June 1979. (American Society of Mammalogists)

In a 1985 summary of the event, Oregon state biologist Bruce Mate, who was at the stranding, wrote,

> Why did the whales die? The whales in Oregon died from heat pros- tration! Whales depend on water to regulate their metabolically produced heat. When the sun eventually came out, their dark skin absorbed more heat from solar radiation and accelerated their inter- nal temperature problem. In reviewing blood and tissue samples later, it was apparent that some of the whales experienced symptoms of heat prostration within six hours of direct sunlight.

For the 41 dead whales on the beach, it was not all that difficult to determine why they died, but the answer to the question of why they had stranded in the first place was not forthcoming.

In his 1989 discussion of the biology of sperm whales, Dale Rice identified no fewer than 13 strandings where all the whales had been males. And a description of a triple all-male stranding that occurred on Sable Island, Nova Scotia, appears in the prologue to Whitehead (2003):

> The group stranding event of three males in January 1997 was observed by researchers working on the island at the time. When first found, at about 09:00, two whales had just stranded, and both

were active, slapping flukes and rolling. A third whale was in the distance offshore and appeared to be swimming roughly parallel to the beach. Eventually this individual moved closer to the beach, appearing to head directly toward the first two, and soon after stranded within 50 m of the others.

Obviously, the third whale stranded because the other two had, but beyond that, there is no way of knowing why the whales behaved the way they did. Was it male bonding? Navigational failure? Communicable disease? Sympathetic suicide? Much more information on the living habits of sperm whales will have to be accumulated before we can begin even to speculate on their dying habits. In January 2010, a dead baby female sperm whale was found on Cannon Beach, northwest of Portland, Oregon. In a local newspaper interview (McCarthy 2010), Debbie Duffield, a cetologist from Portland State University, estimated the age of the whale at "only a few weeks or a month old." Because the dead whale was fresh enough not to have decomposed, a detailed necropsy could be performed. Duffield said, "It gives us a chance to know something more about a species we don't know very much about. It's an insight into a world we don't see very often."

Sometimes it's not the whale that washes up on the beach but rather ambergris, a smelly lump that has come out of the whale and, as with so many elements of sperm whale biology, has mystified people for ages. Here is a detailed description of this mysterious substance, from Daugherty's 1972 *Marine Mammals of California:*

Ambergris is waxy and moist when fresh, dry and brittle when old. The color varies from dull gray through brown to almost black, or may be mottled throughout in alternate layers of light and dark color. There is a characteristic somewhat pleasant earthy odor, intensified by warming in the hand. . . . It floats, even in fresh water. When slowly heated, it commences to soften at about 140°F, and melts between 145° and 150° to a dark, oily liquid. Test it by inserting a heated wire into it; it will melt around the wire forming a dark, opaque liquid. Touched with the finger when partially melted, it is tacky; it adheres and strings. If the wire to which it adheres is re-heated over a flame, it soon emits a white fume with the characteristic odor, and then burns with a luminous flame. It is soluble in absolute alcohol, in ether, in fat, or in volatile oils. It may contain squid beaks.

When a sperm whale stranded, the whole town turned out to see it—and collect the oil. (New Bedord Whaling Museum)

In *The Real Story of the Whaler* (1926), A. Hyatt Verrill gives a long list of what he describes as an "official" record of ambergris "catches" for a period of 73 years. It is too long to reproduce in its entirety, so a few of the more dramatic items from the list are included in Table 4.3. The *Watchman*'s good fortune is corroborated in Starbuck (1878), who states that the ambergris was sold for $10,000, and the *Splendid* was the ship that Frank Bullen (1898) rechristened *Cachalot* in his book. Verrill's list may indeed be correct, but it appears with no supporting documentation. As of this writing (September 2010), gold was selling for $1,293 per ounce.

Although ambergris has been known for millennia, it wasn't until the 18th century that people understood where it came from, and even now, they don't know how it gets there. In *The Whale and His Captors* (1850), the Reverend Cheever describes the mysterious stuff—and even suggests a cure: "The substance called ambergris, and highly prized in perfumery, is obtained from the sperm whale, being formed, it is

TABLE 4.3. Ambergris Catches over 73 Years

Year	Ship	Ambergris Weight (pounds)
1858	Schooner Watchman, Nantucket	800
1866	Bark Sea Fox, New Bedford	150
1870	Bark Elizabeth, Westport	208
1883	Bark Splendid, Dunedin, N.Z.	983

From A. Hyatt Verrill, *The Real Story of the Whaler* (1926).

thought, in that state of the system which calls for a cathartic. A peck of Morrison's or Brandeth's pills, or the homeopathic dose of a pound of calomel or jalap would probably remove obstructions from the creature's viscera."

In 1982, Karl Dannenfeldt reviewed the history of this substance, which he described as "a rare, wax-like pathological growth found in the stomach and intestines of the sperm whale. Its origin is still uncertain, but the condition may be due to an irritation caused by certain indigestible food, especially when the whale has been feeding on cuttlefish (*Sepia officinalis*), a favorite food. The horny beaks or mandibles of the cuttlefish, almost invariably found in ambergris, are indigestible and cause the irritation." In fact, ambergris forms mostly from the beaks of squid, which are also cephalopods, but differ from cuttlefish. *Sepia officinalis,* the common cuttlefish, is found in the Mediterranean and northern European waters, rarely at depths greater than 600 feet, and because sperm whales normally feed at depths far in excess of that, it is unlikely that this species represents a significant part of the sperm whale's diet. (In bygone days, however, squid were generally referred to as *Sepia,* so the confusion is understandable.)

Dannenfeldt's description was taken almost verbatim (including the reference to the cuttlefish) from Bernard Read's 1934 translation of Li Shih-chen's 1597 materia medica, the *Pen Ts'ao Kang Mu.* Read tells us that *Tzu Shao Hua* (ambergris) is obtained "from the oceans of the southwest. . . . In spring when schools of dragons (whales) are about they vomit their saliva which floats on the surface of the water. It is collected by shore natives who sell it for 1,000 cash an ounce. It is also obtained from the bellies of the big fish they cut up. When fresh it is like a fatty gum of a yellowish white color. When dry it forms yellowish black lumps with a fine grain. The old material is purplish black like flying fox dung, shiny and slippery, light in weight, floating on water like pumice-stone and with a rank odor . . . sweet, warming, non-poisonous . . . an aphrodisiac to the male, curing sexual neurasthenia, impotency, gonorrhea, incontinence, given for lack of sexual desire in women."

Before the advent of the sperm whale fishery, ambergris was considered the most unusual (and valuable) product of this most unusual whale. In 1672, John Josselyn described it this way:

> Now you must understand this *Whale* feeds upon *Ambergreece*, as is apparent, finding it in the *Whales* Maw in great quantity, but altered and excrementious: I conceive that *Ambergreece* is no other than a kind of Mushroom growing at the bottom of some Seas; I was once shewed (by a Mariner) a piece of *Ambergreece* having a root to it like that of the land Mushroom, which the *Whale* breaking up, some scape his devouring Paunch, and is afterwards cast upon shore.

This grayish, waxy, peatlike substance is believed to form as an impaction in the intestine of the whale, but this supposition is almost everything we know about ambergris. A whale might vomit up a lump of ambergris in its death flurry, or it might be discovered when the whale is cut open, but it has also been found floating in the open ocean, or washed up on shore, far from any visible whales. According to Clifford Ashley (1926), only "sick" whales are likely to contain this substance, but the relationship between ambergris and the health of the whale is not at all clear. In one peculiar and absolutely unique account, Murphy (1924) found the "characteristic, nodulated facial vibrissae whiskers of a seal" in a lump of ambergris taken from a sperm whale caught off Haiti in 1912. In the original description, he did not venture to guess the species of seal, but later (1933), he wrote that "an animal which could engulf a West Indian monk seal would have no difficulty in taking in Jonah."

This material, pronounced "amber-gree" or "amber-griss" (Dewhurst [1835] and other 19th-century authors spelled it "ambergrease"), was used, according to Melville, "in perfumery, in pastilles, precious candles, hair powders, and pomatum." It is a marbled grayish (or black) waxy substance that is lighter than water and therefore floats. It had been known for centuries from lumps that washed ashore, but it was only in the 18th century, claimed Beale (1835), that about 20 pounds was discovered "in a spermaceti bull-whale." Its primary use was—and still is—as a fixative for perfumes, because it has the property of holding scents. As for its own smell, opinions vary. Ash (1962) wrote that it reminded him of "the scent you smell when you tear up the

This giant boulder of ambergris was extracted from a 50-foot bull sperm whale brought onto the Soviet factory ship *Sovietskaya Rossiya* in 1967. (Courtesy M. V. Ivashin)

moss to uncover the dark soil beneath," but Gaskin (1982) said that "it has the smell of a good cigar."

Ambergris occurs only in sperm whales, and whalers would dream of making their fortunes by finding lumps washed ashore, floating on the surface, or buried in the belly of a captured whale. In 1916, Andrews wrote that "as much as $60,000 worth has been taken from the intestines of a single whale." With the development of synthetic perfume stabilizers, the price has dropped considerably, but in 1972, Nishiwaki wrote that "a recent reaffirmation of its quality has caused a rise in its value. In 1962, ambergris sold at an average of $100 per kilogram." The largest lump ever recorded weighed 983 pounds (442 kilograms), and Robert Clarke (1953) found a "boulder" in the large intestine of a 49-foot bull in the Azores that weighed 926 pounds (416 kilograms). Because the main object of the sperm whale's diet is squid, there are a great number of squid mandibles in the stomachs of a great number of whales, yet ambergris is only rarely encountered. Of course it may be voided or vomited up, but the percentage of whales found to contain this mysterious substance is very small indeed. Whatever it is, it is probably not the direct result of an intestinal irritation. A whale, after all, is a mammal, not an oyster.

Robert Cushman Murphy, author of a 1933 study of the history, mythology, and uses of this material, wrote, "From time immemorial, ambergris has had a fabulous value and, although its ancient uses have with one exception dropped away, it has not, like the bezoar stone or the alchemist's formula, ceased to be prized by the practical moderns." It was incorporated into cosmetics and love potions, and used as a headache remedy and a flavoring for wine. But the "one exception" that Murphy refers to is its use in the fixing of perfume. Even today, when synthetics are available for virtually every commercial chemical function, ambergris is still used—when available—in the manufacture of scents. In 1916, Roy Chapman Andrews wrote that "it is exceedingly valuable, the black ambergris being worth at the present time an ounce, and the gray, which is of superior quality, $20. As much as $60,000 worth has been taken from the intestines of a single whale." Somehow, sperm whales evolved a propensity to respond to an intestinal irritation by producing ambergris, just one of the strange developments in the 50-million-year-long evolutionary process that has brought us the sperm whale.

V

The Social Lives of Sperm Whales

IN 1984, I worked on a screenplay for a documentary movie called *Whales Weep Not*. The title came from a D. H. Lawrence poem of that name, which begins with the line, "They say the sea is cold, but the sea contains the hottest blood of all, and the wildest, the most urgent." The film, shot in and around Sri Lanka, contained the first shots ever taken of blue whales underwater, and it featured the sperm whale research of a newly ordained whale biologist named Hal Whitehead, who had arrived there in a World Wildlife Fund–sponsored sailboat called *Tulip*. The film showed the first people ever to swim with sperm whales in the wild. In his 1990 book *Voyage to the Whales*, Whitehead expresses his dissatisfaction with the filmmakers who had joined them in March 1983: "No trial was more dreaded than the arrival of the film crew. The colorful American crew arrived in Trincomalee on March 23 to make a TV film publicizing the whales, W.W.F., our project and Sri Lanka. It included two experienced underwater cameramen with tales of sharks and James Bond movies, an ex-mercenary, and the fast-talking New York–based producer, bragging and promising."

It was this New York–based producer who had offered me the job of writing the script for the film, which was eventually voiced by Jason Robards Jr. At one point, I was promised a trip to Sri Lanka, but I never made it, and given Whitehead's attitude toward the filmmakers, it's a good thing I didn't. Of the film crew, Whitehead said, "The producer was going to put together his film from two days' work with the blues and one day with the sperms. The filming had proved surprisingly painless for us, and even interesting in its technicalities, yet I could not but feel that the whales were being treated rather superficially. The film

Hal Whitehead at sea. Because he was on watch for whales early in the morning, he is wearing a headlamp. (Jennifer Modigliani)

will show their form and grace, but not the life of the sperm whale, which is lived over scales much longer than the few seconds the cameramen were with them."

Whitehead's book, published in 1990, is the story of a voyage, and because he was then about to begin his research, he raised many more questions about sperm whales than he was able to answer back then:

The lives of large males are almost unstudied, but we can infer a little. . . . They are often solitary, but sometimes they are seen in small groups. They feed much like the females whom we have studied from the *Tulip,* but they probably dive deeper and attack larger squid or big fish. Do they use the "Big Click" during feeding or is it reserved for their encounters with females or other males on the breeding grounds? Perhaps the "Big Click" is also a signal of might. Females desiring mates or competing males might assess the potential of a particular male from some characteristic of his click. The click, whose original purpose was food finding, might then have evolved to emphasize these other characteristics.

Once on the mating grounds, what is the male's strategy? Does he take over a group of females and hold it for as long as he can, be it days or months? Or, as our two short observations suggest, does he interact only briefly with a group—just long enough to see if there are any available females—before moving on to another group?

For *The Book of Whales,* published in 1980 but actually written two years before that, I devoted 22 pages to the sperm whale—more than twice as many pages as any other species—and the bibliography for this species alone included 171 references. Not one of these references included the name of Hal Whitehead. In contrast, take a look at the references in this book: there are quite a few entries under "Whitehead" and "Whitehead et al.," as well as entries that begin with the names Christal, Dufault, Jaquet, Richard, Waters, or Weilgart, all of whom are listed as Whitehead's coauthors. Since the day I was omitted from the list of people who were scheduled to arrive in Sri Lanka to film aboard the *Tulip,* Hal Whitehead has been a veritable factory for sperm whale studies, and as this prodigious output demonstrates, he has learned far more than he expected about the lives of these weird and wonderful creatures.

Like many aspects of the natural history of this species, the social organization of sperm whales is not fully understood. Comparisons with other social animals, such as sea lions or deer, are not particularly useful because the sperm whale seems to defy comparison, being sui generis and a creature of the deep ocean besides. The basic unit, according to Ohsumi (1971), is the *mixed school,* which includes juveniles, immature males and females, and pregnant and lactating females. *Small bachelors* form separate schools that remain isolated from the mixed schools, and the adult bulls join these schools only during the breeding season. The breeding season, which is not absolutely defined, may last for eight months, including all but the summer months in each hemisphere, when the breeding bulls leave the schools and betake themselves to the high latitudes beyond 40°N and 40°S. "Contrary to negative reports," observed Caldwell and Caldwell in 1966, "females do occur in the colder seas north and south of the 40° parallels . . . but they do not appear in these waters in large numbers." For the most part, the adult males frequent polar waters; younger individuals and mature females remain in temperate and tropical waters. Only a small number of bulls, known as schoolmasters, join the mixed schools for breeding, but because there does not seem to be one dominant male per school, the term *harem* does not strictly apply. Bulls that have failed to obtain a place in a mixed school remain out of the breeding pool and are believed to associate loosely with one another, keeping a distance of a mile or more between themselves. At the conclusion of the breeding season, the schoolmasters leave the tropical or temperate habitat of the females and return to the polar latitudes. These so-called lone bulls are not therefore outcasts that have been driven from the breeding aggregations; they leave for colder climates when their breeding duties have been accomplished.*

Not only do male and female sperm whales differ in destinations,

*That only the large bulls migrate to the higher latitudes is demonstrated by the presence of the Antarctic diatom *Cocconeis ceticola* only on males over 40 feet (12.2 meters) long and by the marked differentiation of whale lice on large males and on all other classes. Only the bulls 40 feet or longer are infested with the cyamid species *Cyamis catodontis;* all others show only *Neocyamus physeteris.*

Sexual dimorphism in the sperm whale: males can be twice the size and weight of females. (Richard Ellis)

but they also differ noticeably in their habits. The species is polygamous—or, more accurately, it is polygynous, as males will mate with more than one female—but the old theory that one bull services and protects a herd of females is not accurate. In fact, the social structure of sperm whale society is extremely complicated. Rather than being a male-dominated system, it has been variously described as a "maternal family group" (Ohsumi 1971) or an "extended matricentric family" (Best 1979). When breeding bulls are not in evidence, the schools appear to be led by one or more females, but "there is no evidence that females are dominant to all males (including adult bulls) when these are present in the school" (Best 1979). These designations were used by Ohsumi (1971) on the basis of whale marking, whale sightings, and the unfortunate expedient of "catching all individuals which form the same school." Best's discussion owes much to Ohsumi, but he added a great deal to our attempts at understanding what is undoubtedly one of the most intricate social arrangements in nature.

On the basis of the whalers' tales (and also his readings, particularly of Beale and Bennett), Melville devoted a chapter to "Schools and Schoolmasters," the social structure of the sperm whales' aggregations. "In cavalier attendance upon the school of females," he wrote, "you invariably see a male of full-grown magnitude, but not old; who upon any alarm, evinces his gallantry in the rear and covering the flight of his ladies. . . . The contrast between the Ottoman and his concubines is striking; because while he is always of the largest leviathanic proportions, the ladies, often at full-growth, are not more than one-third the bulk of the average-sized male." The dominant male "keeps a wary eye on his interesting family. Should any unwarrantably pert young Leviathan coming that way, presume to draw confidentially close to one of the ladies, with what prodigious fury the Bashaw assails him, and chases him away!" Melville recognized that the bulls visited the females only during the equatorial feeding season, "having just returned from spending the summer in the Northern seas, and so cheating summer of all unpleasant weariness and warmth." After breeding, the males remove themselves from the harem: "My Lord Whale has no taste for the nursery, however much for the bower; and so being a great traveler, he leaves his anonymous babies all over the world; every baby an exotic." The schools "composing none but young and vigorous males . . . offer a strong contrast to the harem schools. . . . The young males, or forty-barrel-bulls are by far the most pugnacious of all Leviathans, and proverbially the most dangerous to encounter; excepting those wondrous grayheaded, grizzled whales, sometimes met, and these will fight you like grim fiends exasperated by penal gout."

Except for the part where the males betake themselves to the northern seas, Melville's description of the social arrangement of sperm whales is remarkably similar to that of more familiar mammals, where males dominate a group of females, fight off interloping males, and breed with the females in the harem. This arrangement is found in a variety of mammals, including such diverse species as elephant seals, elk, lions, and even chimpanzees, but it does not apply to sperm whales. Would you really expect sperm whales, different in almost every respect from all other mammals, to conform to some sort of behavioral norm? Of course not, and they don't, but there is a surprising similarity to another huge mammal.

In what they called "A Colossal Convergence," Weilgart, Whitehead and Payne (1996) described the remarkable similarities between the sperm whale, the largest toothed whale, and the largest terrestrial mammal, the African elephant (*Loxodonta africana*). Just as sperm whales are able to communicate over long distances, so too can elephants. Females announce their availability for breeding by sending out a series of low-pitched calls, and reproductively receptive males converge from all directions. Elephant calls, far below the range of human hearing, also serve to attract distant family members to the scene of excitement, distress, or separation. The subsonic sounds of elephants were discovered by Katy Payne, who, with her ex-husband Roger, was the first person to record and analyze the songs of humpback whales.

Elephants and sperm whales have unusual but remarkably similar life history parameters, wide-ranging behavior, and ecological success, as well as the largest brains, respectively, on land and in the ocean. Their societies are based on matrilineal groups of about ten related females that often form temporary associations, of a few days or so, with other female groups. After leaving their mother's group when they are about six years old, male sperm whales become increasingly solitary and range to higher latitudes as they grow to about one and a half times the length and three times the weight of females—the most extreme case of sexual size dimorphism among cetaceans. Another striking parallel with elephants is in the delayed age of effective breeding by males: although male elephants and sperm whales become sexually mature during their late teens, they do not seem to take a significant role in breeding until their late twenties. In the case of elephants, this is because younger males do not enter the behaviorally dominant but physiologically demanding state of musth in the prime breeding season. Male sperm whales in the same age range usually remain in productive high-latitude waters, away from the tropical breeding grounds of the females. In these highly sexually dimorphic species, it probably pays younger males to concentrate on growth rather than competing with their much larger elders for the few breeding opportunities presented by slowly reproducing females, who might bear one young every four or five years. Both large male elephants in musth and large male sperm whales on the breeding grounds rove singly between female groups in search of estrus females, usually spending just a short time with each group on any occasion.

The seasonal composition of sperm whale schools is accompanied by migrations, although the movements are as poorly understood as the formation of the breeding units. Unlike the rorquals, which move toward the poles for feeding and then into more temperate waters for breeding and calving, the sperm whales make no such clearly defined movements. Only adult males are found in the higher latitudes, while adult females and juveniles of both sexes are found in the more temperate and tropical regions. Matthews (1938) suggested that males leave the herd while still sexually active and migrate to the high northern or southern latitudes. Later in the year, they will again participate in a general movement toward the equator. The mass movements of sperm whales, known to the old whalers as "making a passage," often involve large numbers of animals, estimated at 200 to 1,000. Boyer (1946) observed "a school of gigantic proportions, maybe one thousand whales." These migratory herds can be differentiated from breeding herds by their numbers and their steady one-directional movement. On the Solander Grounds, south of New Zealand, "as far as the eye could reach," wrote Bullen (1898), "extending all round one half of the horizon, the sea appeared to be alive with spouts—all sperm whales, all bulls of great size."* The sperm whale is an extremely gregarious animal, and it is believed that schools—especially those composed of females—remain together for many years. From the recovery of tags, it has been shown that the schools are stable units, experiencing little or no change in their composition. Best (1979) gave examples of whales that were marked together and were recaptured in the same school five or even ten years later.

After a gestation period of about 15 months, the calf is born into the mother's social unit. The baby is approximately 13 to 14 feet long and weighs about a ton. Opportunities to view the actual birth of a sperm whale are rare indeed, but Gambell and associates (1973) described their observations in South African waters, where they saw an individual "hanging vertically in the water with its head protruding above the surface." (Berzin [1972] quoted another account by Pervushin in which the same vertical attitude was observed on the part of the female, but in this case, a 13-foot calf was seen attached to the female by an umbilical cord.) An hour after their arrival on the scene—they had found the pod by locating a green dye marker that had been dropped from a spotter aircraft—they saw a newborn calf, "about 3.5m long, with strongly wrinkled skin and distinct light pigment streaks on the right flank pos-

*Some are less inclined to accept Bullen's descriptions as factual. In many instances, such as the one in which he described a surface battle between a sperm whale and a giant squid, his report is unique in all the literature.

terior to the dorsal fin. The dorsal fin itself was folded over to the left, and the tail flukes were slightly curled, as in fetal whales." There were five whales in addition to the vertical female, and from the similarity in size, it was assumed that all were females. On other occasions, groups of 24 to 40 whales were seen in the company of a calving female (Gambell et al. 1973). One of the few instances in which a newborn sperm whale was examined (and illustrated) by a scientist occurred in Bermuda (Wheeler 1933), where a 13.2-foot female was captured by fishermen and exhibited alive for two days. It was uniformly black in color, showing white markings around the lips and the umbilicus, and, of course, it had no teeth. In September 1979, the first baby sperm whale—in fact, the first sperm whale of any age until Physty—was held briefly in captivity. A newborn female, later estimated to be less than a week old, stranded on a beach in Rockaway, Oregon, and was transported 300 miles on a flatbed truck to the Seattle Aquarium. There, despite around-the-clock care and attempts at feeding her a mixture of goat's milk, krill, and vitamins, the animal died. She was 12 feet long and weighed 800 pounds.

In most mammals, marine or otherwise, mothers provide some degree of nurturing care to their offspring, beginning with the exclusively mammalian trait of nursing. In their 1966 discussion of caregiving ("epimeletic") behavior, Melba and David Caldwell emphasized attention directed toward young or individuals in distress, which they termed "succorant" behavior. They noted many instances where some members of a school of sperm whales clustered around an injured calf, and in fact, whalers would sometimes harpoon a calf with the idea of drawing other members of the group within killing range. As Bennett observed in 1840 (from the deck of a whaleship), "The females, when attacked, will often endeavor to assist each other, and those that are uninjured will remain for a long time around their harpooned companions." Bennett also observed that when a whale is struck from a boat, "others, many miles distant from the spot, will almost instantaneously express by their actions an apparent consciousness of what has occurred . . . and either make off in alarm or come down to the assistance of the injured companion." As Caldwell and Caldwell (1966) wrote, "Most of the literature on epimeletic behavior in the cosmopolitan sperm whale is found in the literature of the nineteenth century." In the late 20th century, this would change dramatically as humans modified their activities from whale slaughtering to whale observing.

As with all mammals, sperm whale babies are nursed by their mothers. Scammon (1874) described the nursing behavior in this way: "The new-born cub . . . obtains its nourishment from two teats, situated one on each side of the vaginal opening. In giving suck, it is said that the female reclines on her side, when the calf seizes the teat in the corner of its mouth, thereby giving the milk-food immediate passage to the throat." (The use of "it is said" in this description makes it likely that Scammon himself did not witness the actual nursing.) Typically, lactation lasts about two years, "but there is evidence that sperm whales continue to suckle to an apparent age of 13 years and 31 feet" (Best 1979). Mothers and other adult females form groups that are dedicated to the feeding and care of young calves, but the mothers do not stay close to their offspring unless they are actually nursing. Young whales can swim horizontally well enough, but they are unable to dive very deeply, and they cannot accompany their mothers on deep, long-lasting feeding dives. Because young whales are vulnerable to predation by killer whales or sharks, they have to be protected during their mother's dive, so babysitters take over. This arrangement results in nonsynchronous diving, where the females alternate their foraging dives to ensure that there are enough adults at the surface at any one time to protect the calves from predators. "Babysitting in sperm whales," wrote Hal Whitehead in 1996, "seems to be a form of alloparental care [the raising and care of offspring by individuals other than the biological parents.] Its benefit may have been an important factor in the evolution of sociality in female sperm whales."

Then in a 2009 study, Gero, Engelhaupt, Rendell, and Whitehead showed that in addition to babysitting, adult females in a group actually nurse calves that aren't theirs. Gero et al. note that "the matrilineal social organization of the sperm whale functions to provide vigilant allomothers for calves at the surface while mothers make deep divers for food. . . . We examined patterns of adult-infant interactions for 23 sperm whale calves in the Sargasso and Caribbean Seas. . . . For all calves studied in the Caribbean, we found that 1 female provided most of the allocare but did not nurse the calf, whereas in the Sargasso, multiple females provided care for, and nursed the young." It is not known whether an allonursing female, known not to be the mother, was related to the mother or was just part of the group dedicated to caregiving. It is clear, however, that females in groups cooperate to raise their offspring, aptly echoing the African proverb, "It takes a village to raise a child."

The family that nurses together, hunts together, according to Bruce Mate of Oregon State University. With Jorge Urban of the Autonomous University of Baja California Sur, Mate affixed data recorders to three sperm whales feeding on Humboldt squid in the Gulf of California. The researchers found that the hunting whales coordinated their dives so that one made deeper dives, perhaps to locate the prey, or perhaps to herd the squid into a formation that would make it easier for the others to catch them. They observed that the same individual did not always make the deepest dives. The researchers tracked the three-dimensional movements of the whales for 28 days with sensors that recorded the whales' movements at the surface as well as the depth and duration of the dives. In a presentation at the 2010 American Geophysical Union's Ocean Sciences meeting in Portland, Oregon, Mate speculated that this tag-team hunting might show that the whales work in teams to hunt squid, and rotate the toughest assignment (Rojas-Burke 2010).

Sperm whales are weaned at approximately two years of age, and from then on, they must be able to dive to feed themselves. The teeth do not erupt until puberty (around age nine in the females, 25 in the males), so it is evident that teeth are not necessary for feeding. Female sperm whales reach sexual maturity when they are roughly 29 feet (9 meters) long. At this point, growth slows and they produce a calf approximately once every five years. Females are physically mature when they are approximately 30 years old and 35 feet (10.6 meters) long, at which time they stop growing. Along with the size dimorphism, adult male and female sperm whales differ so dramatically in many respects that it is possible to think of them almost as different species. Females are much more gregarious and altruistic than males. Best (1979) proposes two principal functions for sociality in female sperm whales: cooperative foraging and communal care of calves. Young males usually leave the nursery school at an age of four or five years, when they are 25 to 26 feet long, and begin a gradual movement into higher latitudes with increasing age, returning again to breed when they are about 27 years old. Even though males are sexually mature at this time, they often do not actively participate in breeding until their late 20s. For about the first 10 years of life, males are only slightly larger than females, but males continue to exhibit substantial growth until they are well into their 30s. Males reach physical maturity around the age of 50, when they are 52 feet (16 meters) long.

After they are weaned, young whales segregate into juvenile schools composed of young males and young females. At sexual maturity, about 10 years of age for the females and 25 years of age for the males, the females join a mixed or nursery school—the same social unit at different times of the year—whereas the young males form bachelor schools. Most females will form lasting bonds with other females of their family, and on average 12 females and their young will form a family unit. Although females generally stay with the same unit all their lives in and around tropical waters, young males will leave when they are between four and 21 years old and can be found in bachelor schools, comprising males that are about the same age and size. As males get older and larger, they begin to migrate to higher latitudes (toward the poles), and slowly the bachelor schools become smaller, until the largest males end up alone. When the large, sexually mature males are in their late 30s or older, they return to the tropical breeding areas to mate.

One such area is the Galápagos Islands, where Whitehead & Co. have been studying the sperm whales for more than two decades. Because of the great size differential, it is easy to distinguish males from females, so differences in behavior are relatively easy to identify. When there are bulls attendant on a mixed school, it is known as a breeding school, and there may be more than one breeding bull present at a time. These schools vary in size, but Gambell (1972) has estimated one bull per 10 females. On the breeding grounds, males are sometimes seen alone, whereas females never are. Wrote Whitehead (1993): "There are indications that males avoided one another, perhaps by listening for the 'slow click' vocalizations made by males approximately 75% of the time. One incidence of possible aggression between males was observed, and many males possessed parallel scars on their heads, presumably made during aggressive encounters. Males moved between groups of females, spending approximately 8 hours with each group. . . . The evidence is consistent with males' maximizing their expected reproductive success by roving between groups of females." (This strategy is similar to that of African elephants, where males move independently between groups of females.)

The maximum age of sperm whales has been estimated at 32 years (Clarke 1956) to 50 years (Gaskin 1972) to 70 years (Nishiwaki 1972). (In a *National Geographic* article in 1941, Remington Kellogg, then considered America's foremost authority on cetaceans, wrote, "sperm

My son, Timothy, at Sea Life Park in Hawaii around 1982, getting acquainted with a bottlenose dolphin. (Richard Ellis)

whales live for 8 or 9 years," a figure that was picked up by Frank Dufresne who, in *Alaska's Animals and Fishes* [1946], said that the sperm whale "has one of the shortest life spans of any whale, usually not living more than 8 or 9 years.") Undisturbed by whalers, sperm whales can live to the age of 70—perhaps more. Estimations of age have been made from the examination of a cross section of the teeth, particularly the upper teeth, which, being unerupted, are not worn down by normal usage (Nishiwaki et al. 1958). Age can also be estimated by examining a cross section of the lower mandible, where laminations are accumulated at the same rate as in the teeth (Laws 1961; Nishiwaki et al. 1961). Is an older sperm whale a smarter sperm whale?

In April 1980, a symposium was convened in Washington, D.C., under the auspices of the International Whaling Commission to discuss cetacean behavior and intelligence, as well as the ethics of killing cetaceans. The part of the meeting that was not concerned with humane methods of killing whales was devoted to discussions of the intelligence of cetaceans, and it was soon apparent that the participants not only could not agree on what intelligence was, they also could not determine how to find out. Wood and Evans (1980) wrote, "It is a common tendency to think of large brains as indicating 'intelligence,' although there is no universally agreed upon definition of the term." Jerison (1978) avoids the usual difficulties by suggesting that "biological intelligence may be nothing more (or less) than a capacity to construct a perceptual world." Whatever the criterion of intelligence—human or cetacean—if it has anything at all to do with brain size, the bottlenose dolphin (*Tursiops truncatus*) will be found near the top of the scale. Of all known mammals, only the larger baleen whales, the sperm whale, the elephant, and the killer whale have brains that are actually larger than that of an adult bottlenose dolphin, and when various ratios are applied that calibrate brain weight to body weight, the bottlenose comes out at the top. In a paper delivered at the IWC conference, Ridgway (1980) said, "The surface area-to-volume ratio is about one-third larger for those odontocetes than for man (i.e., the cetacean brain is more convoluted)."

He concluded his remarks by saying, "The IWC should encourage scientists and governments in obtaining measures on cetaceans so that we can start to understand why cetacean brains are so large." As Hockett (1978) has written, "Brains are metabolically expensive, and don't get bigger (phylogenetically) unless in some fashion they are more than paying for their upkeep."*

So although the definition of intelligence is a variable, it is usually assumed that brain size is a factor in the equation: whatever intelligence is, an animal with a tiny brain isn't going to have very much of it. The sperm whale, therefore, with the largest brain ever known to exist on earth should be the smartest animal that ever lived, followed by the larger baleen whales (the largest animals ever known), the elephants, large delphinids (killer and pilot whales), bottlenose dolphins, and then *Homo sapiens*. Obviously this linear progression does not work; there must be some sort of a correction for body size in proportion to brain size. In an essay presented at the 1980 IWC Conference on Cetacean Behavior, Intelligence, and the Ethics of Killing Cetaceans, H. J. Jerison said:

> The measure is structural encephalization, as estimated from brain/body relations. . . . Encephalization is measured by the ratio of the volume of the brain to the surface area of the body; it's not the brain/body weight ratio. This measure is relevant for an interesting reason. The volume of the brain (or its weight) reflects the amount of information the brain can handle. This is due to several features of the brain's geometry that are not always fully appreciated. First, the brain consists of nerve cells and other cells that are, on the average,

*Some of the mormyrids, smallish fishes found mostly in African freshwater lakes and streams, turn this equation on its head. They are not thought to be particularly intelligent—whatever that means for fishes—but the cerebellum is greatly enlarged, giving them a brain-to-body-size ratio comparable to that of humans. Some of the mormyrids, known as elephant-nosed fishes (or just elephant fishes) because of their trunklike mouthparts, have weak electrical capabilities that they use to detect prey and perhaps to communicate with one another. The electrolocation functions are believed to be processed in the enlarged cerebellum area of the brain. In the mormyrid *Gnathonemus petersii*, the brain is responsible for 60 percent of the fish's total oxygen consumption, as compared with 20 percent in humans (Nilsson 1996). Curiously, some species of mormyrids without the extended mouthparts are known to home aquarists as "baby whales," even though they are completely fishlike in appearance.

surprisingly similar to one another in different species. Second, the number of these cells that are packed into a brain is directly related to the size of the brain. And, third, the number of connections that nerve cells make with one another also seems to be related to the size of the brain. So brain size is a measure of total information-processing capacity as this would be described in information-theory. The body's surface gets into the picture, because it is related to the expected information-processing capacity for an average animal. By taking the ratio of brain size to body surface we estimate total processing capacity, adjusted for the amount required to handle ordinary body functions for average animals. That is a useful definition of intelligence from a neurobiological perspective.

We therefore have a means by which "neurobiological intelligence" can be calculated, and Jerison concludes,

> We know too little about the nature of the adaptations among whales that require encephalization, but this is certainly due primarily to the technical difficulties in working with completely aquatic animals and to the weaknesses of our scientific institutions when faced with what must be exotic adaptations. The brain is too demanding an organ in the economy of the body for us to assume that such considerable expansion would occur without large amounts of tissue. . . . We can assume that unusual mental processes occur in whales and we should also assume that these will be quite different from those that we know in ourselves.

Gihr and Pilleri (1979) calculated the cephalization ratio on the basis of a "body length–body weight to body weight–brain weight relationship," and in a highly technical study (well suited for the workers of the Brain Anatomy Institute of the University of Bern), they concluded that the species with the highest level of cephalization were "the smaller marine Odontoceti, *Phocoena phocoena, Tursiops truncatus, Grampus griseus, Orcinus orca*, as well as *Inia*, the Amazon dolphin, a member of the Platanistoidea."

Even when we recognize the magnitude of cetacean intelligence—as a function of the relative magnitude of the cetacean brain—we are still a long way from understanding how it is applied to the everyday business of living underwater. We know that a large proportion of the brain is devoted to the analysis and processing of auditory information:

> Fundamental to the echolocation capability of odontocetes is their neuroprocessing of such sounds—the integration, analysis, and interpretation of acoustic information contained in the echoes. . . . Although there is much variation in the size of odontocete brains, both absolutely and by whatever index of cephalization one chooses to use, in general the brains of these animals are characterized, in gross morphology, by large size, with greater width and height than length; great development of the cerebral hemispheres; and intricate convolutions of the cortex. (Wood and Evans 1980)

There have been other suggestions advanced for the development of the large brain in cetaceans. Among the more unusual is that of Fichtelius and Sjölander (1972), who wrote that "the brain of modern man is not as remarkable as we like to believe. As a result of cultural evolution, and particularly as a result of the invention of writing, the greatest part of its capacity lies outside the individual." Because the dolphin has not developed writing—and therefore culture—its communications capabilities have had to develop in a compensatory fashion. Whereas humans can store their collective information in books and computers, dolphins, lacking the hands to build these artifacts, have had to develop the massive brains to store all the knowledge in dolphin history. In his discussion of the anatomical basis of intelligence, Morgane (1974) wrote, "It does seem that the quantity and quality of gray matter (the neocortex) in brains can be taken as a definite index of relative efficiency of those brains in the regulation of behavior."

We have not yet developed a reliable system for measuring the intelligence of cetaceans. There is no question that they are extremely capable animals, with a highly sophisticated auditory sensitivity, but all of the criteria suggested as gauges of their intelligence fail because of our fundamental inability to bridge the gap between those sound-oriented water dwellers and our own land-based, vision-oriented sensibilities. By some still-unrevealed evolutionary path, the dolphins developed their capabilities for a completely aquatic existence, just as bats developed capabilities for a life that is lived in the dark, upside-down, and on the wing. Like bats, dolphins are proper mammals—they breathe air, are warm-blooded, and give birth to live young that they nurse—but they are fishlike in design and habitat. (The only other mammals that spend their entire lives in the water are the sirenians—the manatees and dugongs. All other aquatic mammals—seals, sea lions, walruses, otters—come ashore for some of their activities.) The cetaceans perforce live at the interface of air and water; no matter how sophisticated their aquatic skills, they must surface periodically to breathe.

Water does not transmit light very well, but it is a superb conductor of sound. (Sound travels approximately five times faster in water than it does in air.) Therefore, animals that live in the water must depend more on their auditory senses than on their eyesight. No matter how well an animal sees (and some dolphins see quite well in air; a bottle-nose can leap 15 feet out of the water to grab a fish with its teeth), the very medium in which it lives—no matter how clear—does not permit an emphasis on vision except in short-range situations. It is this situation, not some supernaturally sensitive sensory apparatus, that has been responsible for the cetaceans' dependence on hearing. It is the best thing to use under the circumstances. (The same is true for bats with their ability to echolocate insects in the dark. They use high-frequency sounds to find their prey because neither they nor any other creature can see in total darkness. Some see better in reduced light than others, but an animal that had to depend exclusively on its vision to find its prey in the dark would soon starve.)

Although we do not understand the language of the odontocetes, it now seems clear that it is a well-developed communications system, originated by large-brained creatures whose capabilities are among the most remarkable on earth. They exist in a world where they are almost totally integrated into the medium in which they live. As human divers know, in the enveloping, primeval world of water, which itself constitutes such a large proportion of the animal's corporeal being, gravity as we know it vanishes, and it is replaced by the sublime feeling of weightlessness. In the water, the dolphin is one with its environment in a way that we landlocked humans can only dream about.

Sperm whales have brains six times more massive than human brains. Dolphins' brains are slightly larger than humans', and so are cows' brains. Humans have brains three times more massive than chimpanzees do, even though we share 98 percent of the same genes. The parts of the brain that correlate sight, hearing, smell, touch, motor control, and thinking are the frontal lobes, and whales and dolphins have proportionally larger frontal lobes than we do. This might account for the intelligence of cetaceans: they have been developing their large brains and frontal lobes for approximately 15 million years, as contrasted with our puny one million.

Is the cetaceans' large brain an evolutionary by-product of its echolocation capabilities, or is it a response to its return to a fully aquatic lifestyle? In a 2007 article, Lori Marino, an expert on cetacean neuroanatomy at Emory University in Atlanta, wrote:

> The adaptation of cetaceans to a fully aquatic lifestyle represents one of the most dramatic transformations in mammalian evolutionary history. Two of the most salient features of modern cetaceans are their fully aquatic lifestyle and their large brains. . . . It has been hypothesized that the large, well-developed cetacean brain is a direct product of adaptation to a fully aquatic lifestyle. The current consensus is that the paleontological evidence on brain size evolution in cetaceans is not consistent with this hypothesis. Cetacean brain enlargement took place millions of years after adaptation to a fully aquatic existence. . . . Although echolocation has been suggested as a reason for the high encephalization level in odontocetes, it should be noted that not all aquatic mammals echolocate, and echolocating terrestrial mammals (e.g., bats) are not particularly highly encephalized. Echolocation is not a requirement of a fully aquatic lifestyle and, thus, cannot be considered a sole effect of aquaticism on brain enlargement. These results indicate that the high encephalization level of odontocetes is likely related to their socially complex lifestyle patterns that transcend the influence of an aquatic environment.

The large, convoluted brain of the odontocetes could very easily represent a parallel evolutionary path to that of humans. The effective transmission and reception of information have developed in odontocetes for completely different purposes than in humans because they differ dramatically in everything else, including the very medium in which they live. The development of large brains in odontocetes and humans may be the most remarkable manifestation in all of nature of the evolutionary principle of convergence: the acquisition of the same biological trait in unrelated lineages. (The wings of birds and bats are the classic example of convergent evolution; the wings serve the same function, but they are altogether different in development and structure.)

The entrance hall of the New Bedford Whaling Museum. In this mural, a blue
whale peers down on a group of bottlenose dolphins. I painted it at 30 by 40 inches,
and it was blown up to 30 by 40 feet. (New Bedford Whaling Museum)

Regardless of how it is measured or defined, there would seem to be
no question that the most intelligent of the odontocetes is the familiar
bottlenose dolphin, the species most often seen performing in oceanarium shows. (Some believe that the rough-toothed dolphin [*Steno
bredanensis*] is as smart or smarter, but a sign of their intelligence
might be their unwillingness to tolerate captivity, so very little work has
been done with them.) Echolocation was first observed in captive bottlenoses, and most of the work with captive cetaceans is done with this

species. The famed dolphin named Flipper was a bottlenose, as were the animals in *The Day of the Dolphin*. Reaching a maximum length of 12 feet and a weight of more than 800 pounds, the bottlenose is a powerful animal, capable of prodigious leaps out of the water. Its dives have been measured to 1,500 feet. In the wild, it is usually found within 100 miles of shore, and there are recognizable populations that remain in a particular bay or estuary all their lives. They were exhibited and studied as early as 1913, when a group of five, caught off Cape Hatteras, were placed in a tank at the New York Aquarium. Of all the dolphins, bottlenoses have proven the most adaptable to training. They breed readily, and by now, most oceanarium animals are born and raised in captivity.

John Cunningham Lilly (1915–2001), author of *Man and Dolphin* (1961) and *The Mind of the Dolphin* (1967), was a physician and neurophysiologist who established a research institute on St. Thomas in the Virgin Islands devoted to fostering interspecies communication. In the early 1960s, Dr. Lilly and his coworkers published several papers reporting that bottlenose dolphins could mimic human speech patterns, and he believed it was only a matter of time before dolphins and people would communicate directly. The 1973 movie *Day of the Dolphin*, based on a novel by Robert Merle and loosely on Lilly's research, featured bottlenose dolphins that had learned to speak (actually *squeak*) English, and they were able to save the president of the United States from an assassination attempt. But even though he worked with bottlenoses, Lilly regarded the sperm whale with a kind of reverence that would have embarrassed Herman Melville. Lilly (1967) held that the sperm whale "probably has 'religious' ambitions and successes quite beyond anything we know. His 'transcendental religious' experiences must be quite beyond anything we can experience by any known methods at this time. . . . Perhaps the sperm whale has gone so far into philosophical studies that he sees the Golden Rule as only a special case of a much larger ethic. Compared with us he probably has abilities that are truly godlike." (Given Lilly's avowed dependence on LSD and other mind-expanding drugs, his regard for the sperm whale as a godhead is not altogether surprising.)

Because of their accessibility and amenability, bottlenoses have been extensively studied in captive situations and in the wild. Books by the dozen have been written about every aspect of their lives, and scientists have made careers out of studying them. But sperm whales do not fit comfortably in oceanarium tanks, even if we were capable of capturing one alive. Only Physty has ever been studied in captivity. Further, their behavior in the wild has to be viewed mostly from the deck of a ship—or, rarely, by a diver brave enough to enter their world and film them underwater. For centuries, our knowledge of sperm whales came from whalers (or whalers turned authors); we knew a lot more about how they died than about how they lived.

Of all the cetacean modifications for life in the ocean (for example, development of flippers, blubber instead of hair, migrating nostrils), the giant brain is the most difficult to account for. After all, the other marine predators, such as sharks, large bony fishes, and crocodiles, all highly efficient predators, are not known for their intellect or problem-solving skills. Of course, the argument could be offered that these mammalian interlopers developed large brains in order to compete with their dumber rivals, but this suggests a determinism usually absent from evolutionary theory.

Are dolphins as smart as apes? As humans? In 2001, Diana Reiss and Lori Marino published a groundbreaking study in which they showed that bottlenoses can recognize themselves in a mirror, an ability previously recognized only in great apes, African gray parrots, and humans. At the New York Aquarium at Coney Island (Brooklyn), Reiss and Marino marked captive dolphins with temporary black ink and watched the dolphins investigate the marks, many of which were not visible except with a reflecting surface (mirror) in the tank. The dolphins twisted and turned to see the marks on their bodies. "Collectively," wrote Reiss and Marino, "these findings provide definitive evidence that the two dolphins used the mirror (and other reflective surfaces) to investigate parts of their bodies that were marked. These findings, therefore, offer the first convincing evidence that a non-primate species, the bottlenose dolphin, is capable of MSR [mirror self-recognition]." A couple of years later, scientists (one of whom was Diana Reiss) marked three Asian elephants, and they discovered that these ponderous pachyderms were also capable of MSR. Joshua Plotnik, Frans de Waal, and Diana Reiss (2006) wrote:

> After the recent discovery of MSR in dolphins (*Tursiops truncatus*), elephants thus were the next logical candidate species. Animals that possess MSR typically progress through four stages of behavior when facing a mirror: (i) social responses, (ii) physical inspection (e.g.,

looking behind the mirror), (iii) repetitive mirror-testing behavior, and (iv) realization of seeing themselves. Visible marks and invisible sham-marks were applied to the elephants' heads to test whether they would pass the litmus "mark test" for MSR in which an individual spontaneously uses a mirror to touch an otherwise imperceptible mark on its own body. Here, we report a successful MSR elephant study and report striking parallels in the progression of responses to mirrors among apes, dolphins, and elephants. These parallels suggest convergent cognitive evolution most likely related to complex sociality and cooperation.

To date, no sperm whales have been exposed to MSR experiments. It would be more than a little difficult to mark a wild sperm whale and ask it to look in a mirror, and with the exception of Physty, no sperm whale has ever been maintained in captivity. But the sperm whales' large, convoluted brains, their intricate vocalizations, and their complex social behavior patterns strongly suggest that MSR would not be much of a challenge for *Physeter*. (Remember also that the social structure of sperm whale families closely parallels that of elephants.)

In 2007, sixteen scientists, in disciplines that included paleontology, neuroscience, molecular biology, animal behavior, and cytology, decided that "the time is ripe to present an integrated view of cetacean brains, behavior, and evolution based on the wealth of accumulated and recent data on these topics. Our conclusions support the more generally accepted view that the large brains of cetaceans evolved to support complex cognitive abilities." Marino et al. looked at data collected from captive dolphins (mostly bottlenoses, of course) and field studies of bottlenoses, killer whales, and sperm whales, and concluded:

> Evidence from various domains of research demonstrates that cetacean brains underwent elaboration and reorganization during their evolution with resulting expansion of the neocortex. Cortical evolution, however, proceeded along very different lines than in primates and other large mammals. Despite this divergence, many cetaceans evince some of the most sophisticated cognitive abilities among all mammals and exhibit striking cognitive convergences with primates, including humans. In many ways, it is because of the evolution of similar levels of cognitive complexity via an alternative neuroanatomical path that comparative studies of cetacean brains and primate brains are so interesting. They are examples of convergent

evolution of function largely in response, it appears, to similar societal demands.

At the 2010 American Association for the Advancement of Science (AAAS) Annual Meeting in San Diego, Lori Marino said that bottlenose dolphins "may be Earth's second smartest creature, after humans of course." Dolphins are bigger than humans (and adult male *Tursiops* can be 12 feet long and weigh 800 pounds), so they are expected to have bigger brains—1,600 grams versus 1,300 grams—but they "also have a very complex neocortex, the part of the brain responsible for problem solving, self-awareness, and various other traits we associate with human intelligence" (Grimm 2010). "They are," said Marino, "the second most encephalized beings on the planet." This realization has led researchers and philosophers to suggest that dolphins ought to be treated more like people than like wild animals. Marino further said that "the very traits that make dolphins interesting to study make confining them in captivity unethical." Thomas White (2007), a philosopher at Loyola Marymount University in California, suggested that dolphins aren't merely like people; they may actually *be* people, or at least nonhuman persons. For the most part, we have been defining "people" as those creatures who can read this book, but that may not be inclusive enough. In defining what it means to be a person—awareness of their environment, emotions, personalities, self-control, ethical treatment of others—"dolphins fit the bill," said White.

Measuring the endocasts (the filled braincase) of fossil whales, we find that *Pakicetus* had a brain the size of a walnut and that of *Ambulocetus* wasn't much larger, so we have to figure out how to make the quantum leap from a small-brained protocetacean to *Tursiops,* an animal that many dolphin trainers say is smarter than they are, and then to the animal with the largest brain in history. Although brain size does not directly correlate to intelligence, animals with small brains relative to body size are usually considered to be less intelligent—whatever that means—than those with larger brains. The brain of an adult sperm whale weighs about 20 pounds, as compared to human brains, which weigh about three pounds. Of course, a sperm whale weighs 100,000 pounds to a human's 150, so it ought to have a bigger brain—yet the blue whale, which can weigh 300,000 pounds, has a smaller brain than *Physeter macrocephalus.* Sperm whales may use their large brains for complex communications (among other things), but we need to know why cetaceans such as dolphins and sperm whales developed these enormous brains.

Lori Marino, Daniel McShea, and Mark Uhen (2004) have written that the archaeocetes never developed large brains, so brain size was obviously not connected to their return to the aquatic environment. (Seals, sea lions, and manatees, which also evolved from terrestrial ancestors, never developed particularly large brains.) Marino et al. noted:

Toothed whales (order Odontoceti) are highly encephalized, possessing brains that are significantly larger than expected for their body sizes. In particular, the odontocete superfamily Delphinoidea (dolphins, porpoises, belugas, and narwhals) comprise numerous species with encephalization levels second only to modern humans and greater than all other mammals. Odontocetes have also demonstrated behavioral faculties previously only ascribed to humans, and, to some extent, other great apes.

If big brains are not related to the aquatic environment, what could have been responsible for this exotic evolutionary modification? It cannot be echolocation, because bats echolocate, and they don't have large brains. (The development of flight essentially precludes large brains anyway, because brains are heavy and cannot be carried around by flying mammals that usually do not weigh more than a mouse.) What do humans have that most other mammals do not? Language. Are the underwater clicks, bangs, squeaks, whistles, and chirps of odontocetes the equivalent of human language? So far, these noisy odontocetes have not seen fit to communicate with humans—although it has been said that dolphins have trained human beings to throw them a fish whenever they jump through a hoop—but it is not at all unreasonable to assume that big brains are important when odontocetes communicate with one another.

"Stereotypical calls produced by members of a social group that vary among populations have been called dialects," wrote Dudzinski, Thomas, and Gregg (2009), "and have [now] been described in at least two species of odontocetes." In British Columbia, matrilineal groups of killer whales have repertoires of call types that are unique to each pod. John Ford, curator of marine mammals at the Vancouver Public Aquarium, has been studying killer whales' communications for more than a decade. He reports that the dialects are composed of the whistles and calls the animals use when communicating underwater, quite distinct from the high-energy, sonarlike clicks that the whales emit when echolocating. "Other kinds of sounds," wrote Ford, Ellis, and Balcomb (1994), "mostly whistles and burst-pulsed signals that resemble squeals, squawks, and screams, are used for social communication within and between groups. . . . Each group of whales produces a specific number of these discrete calls, which together form its dialect. The group's dialect is apparently learned by each individual. Probably by mimicking its mother as a calf." Differences between the dialects can be as small as those that distinguish regional dialects of the English language, or as large as those between Japanese and English. In a comprehensive article about culture in whales and dolphins in *Behavioral and Brain Sciences* in 2001, Luke Rendell and Hal Whitehead wrote, "The complex stable and behavioral cultures of sympatric groups of killer whales appear to have no parallel outside humans, and represent an independent evolution of cultural faculties."

Two years later, however, Rendell and Whitehead published "Vocal Clans in Sperm Whales." In 1977, Watkins and Schevill had identified sperm whale "codas," stereotyped sequences of three to 40 broadband clicks usually lasting less than three seconds in total, but in 2003, Rendell and Whitehead categorized these codas into six acoustic "clans" for populations in the South Pacific and the Caribbean: "Clans have ranges that span thousands of kilometers, are sympatric, and contain many thousands of whales and most probably result from cultural transmissions of vocal patterns. . . . We suggest that this is a rare example of sympatric cultural variation on an oceanic scale. Culture may thus be a more important determinant of sperm whale population structure than genes or geography, a finding that has major implications for our understanding of the species' behavioral and population biology." Rendell and Whitehead continue:

> The function of the codas is unknown, and the function, if any, of coda dialects is thus a matter for speculation. However, we do know that members of sperm whale groups will take considerable risks to help group-mates under predatory attack and also provide allomaternal care of calves within groups. We suggest that coda dialect performs a signature function in this context, allowing units to iden-

tify others of the same clan within a highly mobile sperm whale society, and perhaps mediating seemingly altruistic exchanges such as communal defense and allomaternal care. It is important to know whether clan signatures form boundaries to these exchanges; if they do, then sperm whale clans would be the largest cooperative groups known outside humans.

In a 2009 article about whales (mostly the gray whales of Baja) Charles Siebert discussed the unique characteristics of sperm whales:

The sperm whale . . . has been found to live in large, elaborately structured societal groups, or clans, that typically number in the tens of thousands and wander over many thousands of miles of ocean. The whales of a clan are not all related, but within each clan there are smaller, close-knit, matriarchal family units. Young whales are raised within an extended, multitiered network of doting female caregivers, including the mother, aunts and grandmothers, who help in the nurturing of babies and, researchers suspect, in teaching them patterns of movement, hunting techniques and communication skills. "It's like they're living in these massive, multicultural, undersea societies," says Hal Whitehead, a marine biologist at Dalhousie University in Nova Scotia and the world's foremost expert on the sperm whale. "It's sort of strange. Really the closest analogy we have for it would be ourselves."

Killer whales are also highly social animals with intricate family relationships, and they are the sperm whales' only predator. That both incorporate dialects into their vocal repertoire is probably of no significance; killer whales prey on just about everything that swims, including dolphins, porpoises, great whales, seals, sea lions, sharks, fish, and squid. When Weilgart and Whitehead acoustically tracked sperm whales off the Galápagos Islands in 1985, they heard "distinctive loud, ringing clicks, called 'slow clicks,' [that] were highly correlated with the presence of mature male sperm whales." They hypothesized that "slow clicks may be a sign of a mature or maturing male and may inform other sperm whales on the breeding grounds of its competitive ability and maturity." Again in 1992, Weilgart and Whitehead recorded Galápagos sperm whales over several months. They observed that the codas (short, patterned series of clicks) "were found to be temporally very clustered, and could be categorized into 23 fairly distinct types. . . . Codas may function principally as a means of communication, to maintain social cohesion within stable groups of females, following periods of dispersion following foraging." Both male and female sperm whales would appear to have their own "language," used to identify one another and to announce their presence and condition to potential rivals.

Should we be calling this "culture"? What is culture, anyway? Rendell and Whitehead (2001) say that "there is little consensus on this issue; the term *culture* is defined in an array of subtly different ways within the literature." They list no fewer than 15 definitions, but they favor the one proposed by Boyd and Richerson, from a 1985 book entitled *Culture and the Evolutionary Process:* "Culture is information or behavior acquired from conspecifics through some form of social learning." Most of the cultural observations of behavior modification have been of those species most readily observed—captive bottlenose dolphins or wild killer whales—but the study (and recording) of sperm whales in the wild are beginning to reveal complex behavior transmitted within a particular group, or even from generation to generation. Group-specific coda repertoires, group-specific movement patterns, and communal defense methods are ethnographically indicative of culture in sperm whales. The conclusion drawn by Rendell and Whitehead (2001) seems indisputable, says philosopher Michael Allen Fox: "Based on a wealth of empirical data and cautiously worded inferences, *whales and dolphins live and develop within their own unique species-specific, even group-specific cultural contexts.*"

"Culture in Whales and Dolphins" consists of an introductory target essay by Rendell and Whitehead (2001) and open peer commentary by qualified professionals comprising the readership of the journal *Behavioral and Brain Sciences.* There are 39 commentaries, submitted by philosophers, cetologists, sociologists, anthropologists, animal behaviorists, and psychologists. Some agree completely with Rendell and Whitehead; some disagree with some elements of their essay; and some disagree strongly. For example, Patrick Miller of Woods Hole Oceanographic Institution wrote, "Rendell and Whitehead overstate the weak evidence for social learning in cetaceans as a group, including the current evidence for vocal learning in killer whales. Ethnographic techniques exist for genetic explanations of killer whale calling behavior, and additional captive experiments are feasible. Without such tests, descriptions of learning could be considered pseudo-scientific, ad hoc auxiliary assumptions of an untested theory." P. J. B. Slater of the University of St. Andrews, Scotland, wrote, "While cetaceans clearly show social learning in a wide variety of contexts, to label this as 'culture' hides more than it reveals: we need a taxonomy of culture to tease apart the differences rather than hiding them in a catch-all category." Peter Tyack, a whale biologist at Woods Hole, wrote that "Rendell and Whitehead adopt a weak definition of culture to allow low standards of evidence for marine mammals, but they do not adequately rule out genetic factors or individual versus social learning. They then use these low standards to argue that some whales have unique cultures only matched by humans."

Rendell and Whitehead answer every criticism in an authors' response that they call "Cetacean Culture: Still Afloat after the First Naval Engagement of the Culture Wars," but enough questions were raised by the respondents to suggest that culture in cetaceans has not been conclusively demonstrated. Many of the critics were concerned with definitions—particularly of the troublesome catchall term *culture*—but there is no question that the group behavior of bottlenose dolphins, killer whales, and sperm whales, among the most vocal of all odontocetes, is different from that of most other cetaceans, and indeed, different from most other mammals. (The only mammals to whom we unreservedly assign culture—under any and all definitions—are our acculturated selves.). Bottlenoses and killer whales are available for more creative testing because they can be observed in captive situations, but sperm whales, found in only their deep-water, offshore habitat, have shown us only a glimpse of their modus vivendi. They evidently "talk" to each other and segregate into clans with a common dialect, but perhaps Melville had the answer about sperm whale intelligence: "Has the Sperm Whale ever written a book, spoken a speech? No, his great genius is declared in his doing nothing particular to prove it. It is moreover declared in his pyramidal silence."

We now know that sperm whales are anything but silent, and their vocalizations circulate throughout the world's oceans for purposes we don't quite understand. Still, in his 2003 book *Sperm Whales: Social Evolution in the Ocean,* Whitehead wrote, "Sperm whale populations appear to be strongly structured along cultural lines, and their cultural differences seem to have functionally important aspects. . . . In killer whales, it is clear that culture affects important aspects of behavior, including diet, foraging strategies, and social conventions. There are strong indications that this is also true for bottlenose dolphins. As these two are by far the best studied cetaceans, it seems entirely plausible that culture is an important determinant throughout much of Cetacea, and that identity within cultural groups may form an important part of how these animals see themselves."* Sounds a bit like *Homo sapiens,* doesn't it?

"Play," wrote Hal Whitehead, "can be defined as activities that, while having no immediate function, may benefit the animal later in life—for instance, by improving its physical abilities or social skills. The category of play, however, tends to become a garbage-can for any behavior whose function is not obvious." Cetaceans ranging in size from the smallest porpoises to the largest whales will, on occasion, jump out of the water. To the eye of the human beholder, this certainly looks like fun, and as far as we can tell, it serves no useful function. Some species, such as the humpback, are characterized by their breaching behavior, and as Melville wrote, "He is the most gamesome and lighthearted of all whales, making more gay foam and white water generally than any of them." Humpbacks often breach repeatedly, launching their 40-ton bulk from the water and reentering with a prodigious splash, as many as 20 times in succession. (It has been suggested that the loud sounds created by the crash landing are signals to other whales in the vicinity, but for whales such as humpbacks, with an intricate repertoire of songs, grunts, groans, whoops, and warbles, splashing as a communication device seems redundant and unnecessarily labor-intensive.) In their review of sperm whale behavior, David and Melba Caldwell (1966) identified breaching and lobtailing—whacking the flukes on the surface to make a thunderclap of a sound—as "the two types of sperm whale behavior which come under the definition of play; activities performed for the sake of the activity itself . . . rather than for any observable effect upon the environment, and behavior that cannot be shown

*In the 2000 book *Cetacean Societies: Field Studies of Dolphins and Whales,* edited by Janet Mann, Richard Connor, Peter Tyack, and Hal Whitehead, four species were selected to demonstrate the social lives of cetaceans: bottlenose dolphins, killer whales, sperm whales, and humpbacks. In the introduction, the editors wrote, "In general, easy access in coastal waters has been a key factor in determining which species of whales and dolphins have been most closely examined. However, the inclusion of a chapter on the open-ocean, deep-diving sperm whale shows that this need not be the case."

to contribute immediately to any of the recognized internal drives that sustain life."

Many dolphin species are famous for their play behavior, none more so than the bottlenose, whose antics, in and out of captivity, have virtually defined cetacean playfulness. Bottlenoses play with toys in their tanks; chase and harass other occupants; somersault out of the water; mimic their trainers' voices, clickers, and whistles; and, perhaps because of their permanent grin, seem to be enjoying themselves in the company of people. In the wild, dolphins often initiate contact with humans, play with toys and sponges, and do a lot of inspired leaping, but they also ride the bow waves of ships and boats, an activity they obviously cannot pursue in a tank. Some species, such as bottlenose, white-sided, and common dolphins, will pick up a moving vessel from miles away and ride the pressure wave pushed by the ship, apparently for the fun of it. For this joyful exuberance, Melville called them "Huzza Porpoises," and wrote:

Bottlenose dophins are famous for their playful behavior, whether inspired by trainers or self-generated. (Richard Ellis)

> I call him thus, because he always swims in hilarious shoals, which upon the broad sea keep tossing themselves to heaven like caps in a Fourth-of-July crowd. Their appearance is generally hailed with delight by the mariner. Full of fine spirits, they invariably come from the breezy billows to windward. They are the lads that always live before the wind. They are accounted a lucky omen. If you yourself can withstand three cheers at beholding these vivacious fish, then heaven help ye; the spirit of godly gamesomeness is not in ye.

Early observers believed that dolphins riding before fast-moving vessels were somehow moving faster than the vessels, but it was eventually determined that the dolphins were actually utilizing the pressure wave of the vessel for a free ride, and could only move at the speed of the vessel. Norris and Prescott (1961) discussed this "assisted locomotion" in various small cetaceans; they noticed that "none of the animals regularly beat its tail when stationed close to the bow. . . . On several occasions the porpoises turned partly or entirely on their sides and it was then obvious that the flukes were held stationary in relation to the body." In *Whale Hunt,* a chronicle of a whaling voyage from 1849 to 1853 in the ship *Charles W. Morgan,* Nelson Cole Haley tells of cruising off New Zealand in heavy seas, when he spotted "twenty or thirty good sized [sperm] whales tumbling about when the big seas would catch them and almost turn them over. Sometimes one could be seen on the crest of a wave. As it broke, he would shoot down its side with such speed a streak of white could be seen in the wake he made through the water. . . . I have never seen whales at play before or since." Close to shore in many parts of the world, bottlenoses have been observed bodysurfing—riding the curl of an incoming wave—just as humans do. People do this for fun, and it appears obvious that whales and dolphins do too.

Are there other parallels between cetaceans and humans? When Patrick Hof and Estel Van der Gucht (2006) examined the brain of an adult humpback whale and several other cetacean species, they found that the humpback's cerebral cortex, the part of the brain where thought processes take place, was similar in complexity to that of smaller cetaceans such as dolphins. Humpbacks, of course, are the whales that sing, and their complicated songs, which they change yearly, strongly suggest some sort of conscious modification. Rendell and Whitehead (2001) wrote, "While the mechanisms underlying this process are not fully understood, horizontal cultural transmission almost certainly plays an important role in maintaining song homogeneity as there is no conceivable environmental trigger for such a pattern of variation." They also feed on schools of small fishes by releasing bursts of bubbles in sequence as they spiral toward the surface. This "bubble-net feeding"— a technique used by no other whale species—herds the fish to the surface, where the whale lunges into the school with its huge mouth agape and swallows them by the thousands.

Present in the humpback's cerebral cortex are spindle cells, which are also found in the brains of large-brained toothed whales—and humans. Spindle neurons probably first appeared in the common ancestor of hominids about 15 million years ago, since they are observed in great apes and humans, but not in lesser apes and other primates. In cetaceans, they evolved earlier, possibly as early as 30 million years ago. It is possible that they were present in the ancestors of all cetaceans but were retained during their evolution only by those with the largest brains. It may also be that they evolved several times independently in the two cetacean suborders. Part of this process may have taken place at the same time as they appeared in the ancestor of great apes, which would be a rare case of parallel evolution. As Hof and Van der Gucht (2006) observed, "In spite of the relative scarcity of information on many cetacean species, it is important to note in this context that sperm whales, killer whales, and certainly humpback whales, exhibit complex social patterns that included intricate communication skills, coalition-formation, cooperation, cultural transmission and tool usage." Furthermore, "It is thus likely that some of these abilities are related to comparable complexity in brain organization in cetaceans and in hominids." The authors conclude: "Cetacean and primate brains may be considered as evolutionary alternatives in neurobiological complexity and as such, it would be compelling to investigate how many convergent cognitive and behavioral features result from largely dissimilar neocortical organization between the two orders."

Tezio Ogawa (1901–1984) was an eminent anatomist who pioneered studies of cetaceans in Japanese waters. He became professor of anatomy at Tokyo University in 1944 and professor emeritus in 1962. In 1946, he helped to establish the Whales Research Institute and was one of its directors until 1960. Throughout this period, he worked on the anatomy and histology of cetaceans, often sending his students to join the Japanese Antarctic whaling fleet (usually as ships' doctors) to study the anatomy of whales and collect specimens. In a paper presented at a 1962 meeting of the Society of Primate Researchers in Japan, he wrote, "In the world of mammals there are two mountain peaks; one is Mount Homo Sapiens and the other is Mount Cetacea."

VI

Battle of the Giants

IN *Moby-Dick,* as the *Pequod* sails northeast of Java, the harpooner, Daggoo, sees a "strange spectre" from his perch on the main masthead:

> In the distance, a great white mass lazily rose, and rising higher and higher, and disentangling itself from the azure, at last gleamed before our prow like a snow-slide, new slid from the hills. Thus glistening for a moment, as slowly it subsided, and sank. Then once more arose, and silently gleamed.

They lowered the boats and "gazed at the most wondrous phenomenon which the secret seas have hitherto revealed to mankind":

> A vast pulpy mass, furlongs in length and breadth, of a glancing cream-color, lay floating on the water, innumerable long arms radiating from its centre, and curling and twisting like a nest of anacondas, as if blindly to clutch at any hapless object within reach. No perceptible face or front did it have; no conceivable token of either sensation or instinct; but undulated there on the billows, an unearthly, formless, chance-like apparition of life.

The accoutrements of whaling and natural history that Melville weaves into his story are as accurate as any 19th-century biologist's— and a lot better written. Thus, if the whalemen knew anything about giant squid, Melville recorded it in *Moby-Dick.* He wrote: "By some naturalists who have vaguely heard rumors of the mysterious creature, here spoken of, it is included among the class of cuttle-fish, to which, indeed in certain external respects it would seem to belong, but only as the Anak of the tribe."*

The Yankee sperm whalers, plying the world's oceans in pursuit of the cachalot, often saw their prey, in its death throes, regurgitate large pieces of something, and they often hooked one of the pieces to get a closer look. In 1856, Charles Nordhoff (grandfather of the Charles Nordhoff who wrote *Mutiny on the Bounty* with James Norman Hall) wrote a book called *Whaling and Fishing* (1856), in which he described what the whalemen called "squid," but which he believed to be "a monster species of cuttle fish":

> The animal seldom exhibits itself to man; but pieces of the feelers are often seen afloat, on good whaling ground. I have examined such from the boats, and found them to consist of a dirty yellow surface, beneath which appeared a slimy, jelly-like flesh. Of several pieces which we fell in with at various times when in the boats, most had on them portions of the "sucker," or air exchanger with which the common cuttle-fish is furnished, to enable him to hold the prey about which he has slung his snake-like arms. These floating pieces are supposed to have been bitten off or torn by the whales, while feeding on the bottom. Many of those we saw were the circumference of a flour barrel. If this be the size of the arms, of which they probably have hundreds, each furnished with air exhausters the size of a dinner plate, what must be the magnitude of a body which supports such an array?

Of all the well-known predator-prey combinations, none are more intimately connected than the sperm whale and the giant squid. The sperm whale preys on other teuthids, but it is likely that nothing preys on full-grown giant squid except the sperm whale. The giant squid has often been accused of causing the circular scars on the head of the whale, either in an attempt to eat the whale or, more likely, in the struggle not to be eaten. There are those who believe that *Architeuthis* is a sluggish animal, neither powerful nor aggressive, but suppositions like that are never going to interfere with the creature's enduring reputation as a man-eating, ship-grabbing, whale-wrestling monster.

*In the Old Testament (Numbers 13:33), Anak is the father of a race of giants that threaten the Israelites during their exodus from Egypt.

Alleged fight between a sea serpent and a sperm whale, seen from the barque *Pauline*, July 8th 1875
From *The Illustrated London News*, November 20th 1875

In 1875, viewers aboard the bark *Pauline* off Brazil saw what some described as a fight between a sea serpent and a sperm whale. Except for the eye, it is quite easy to see the serpent as the tentacle of a giant squid. (New Bedford Whaling Museum)

It has long been a part of sperm whale lore that it has to dive to prodigious depths to seek out the giant squid, its favorite prey. The size and alleged ferocity of *Architeuthis* provide more than enough ammunition for the creative writer or overimaginative naturalist, and titanic battles between the squid and its archpredator, the sperm whale, will continue to appear.

In *The Whale and His Captors* (1850), the Reverend Henry Cheever, who sailed aboard the whaler *Commodore Preble,* presents an account (not seen by him, but "sworn before a Justice of the Peace in Kennebec, Maine, in 1818") of what appears to have been an encounter between a sperm whale and a giant squid:

> The serpent threw up its tail from twenty-five to thirty feet in a perpendicular direction, striking the whale by it with tremendous blows, rapidly repeated, which were distinctly heard, and very loud, for two or three minutes; they then both disappeared, moving in a south-west direction; but after a few minutes reappeared in-shore of the packet, and about under the sun, the reflection of which was so strong as to prevent their seeing so distinctly as at first, when the serpent's fearful blows with his tail were repeated and clearly heard

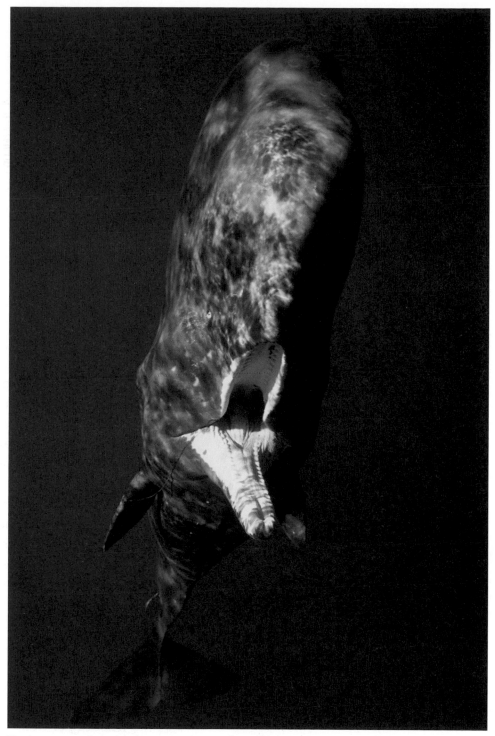

Head-on sperm whale. Coming at you: the most incredible animal on earth. (SeaPix)

Physty on the beach. I put my children in the painting of Physty stranding on the beach, but they didn't come to Fire Island until the next day. (Painting by Richard Ellis)

Richard Ellis (red suit) touching Physty. I swam up to Physty and put my hand on the part of his nose where the popping noises seemed to be coming from. (Photo by Anne Doubilet)

PHYSTY

The True Story of a Young Whale's Rescue

Written and Illustrated by
Richard Ellis

Cover of *Physty*. (Richard Ellis)

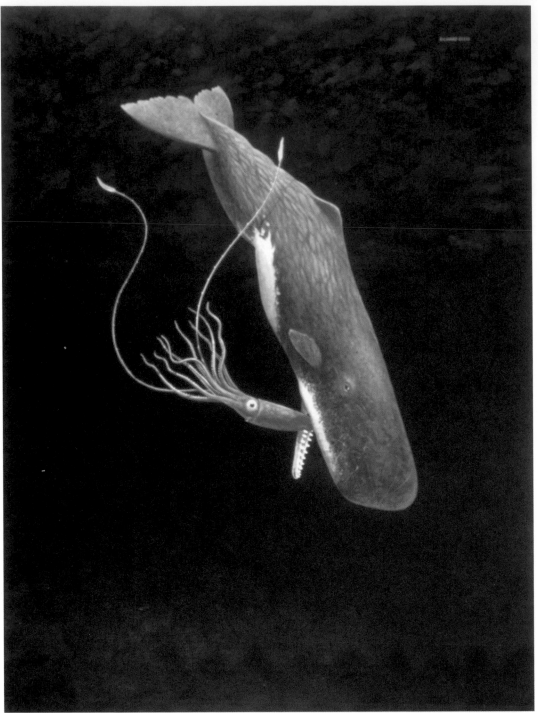

Sperm whale with giant squid. Because I was getting so many requests for an illustration of a sperm whale and a giant squid, I painted this picture, showing that the whale is attacking the squid, not vice versa. (Painting by Richard Ellis)

Pair of sperm whales. Male sperm whales, half again as large as females, take themselves to the high polar latitudes to feed and wait for the breeding season. (Painting by Richard Ellis)

Giant squid. The giant squid has cells in its skin that it uses to change color. In this painting, *Architeuthis* is bright red. (Painting by Richard Ellis)

Sperm whale eating giant squid. The first-ever photograph of a sperm whale feeding on a giant squid. (Photo by Tony Wu)

Giant squid tentacle. The red tentacle is proof positive that it was *Architeuthis*. (Photo by Doug Siefert)

Cover of *Audubon* magazine. (Painting by Richard Ellis)

Sperm whale surfacing. Whales can easily exhale underwater, but they must come to the surface to inhale. (SeaPix)

School of sperm whales. Sperm whale females with a newborn calf. (SeaPix)

Open mouth at the surface. Sperm whales do not breathe through their mouths; they inhale and exhale through the single nostril at the end of their nose. (SeaPix)

Single sperm whale underwater. The great Leviathan rests near the surface. (Photo by Doug Siefert)

Portrait of a sperm whale. (Photo by Doug Siefert)

Diver touching sperm whale. Friendly whale and friendly diver. (Photo by Tony Wu)

Full mural in dining room. The telephone on the floor gives a sense of scale to the sperm whale mural
I painted in the dining room of my Manhattan apartment. (Photo by Richard Ellis)

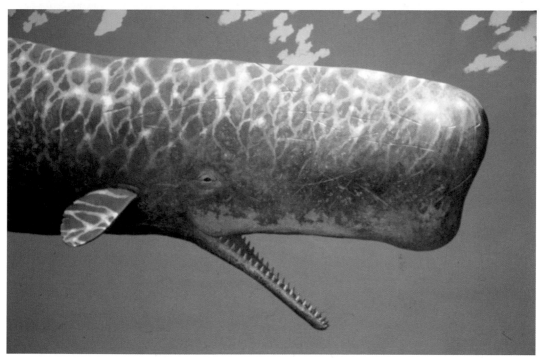

Detail of mural in dining room. A detail of my first mural, painted for the Wehle Museum in Buffalo, New York.
(Photo by Richard Ellis)

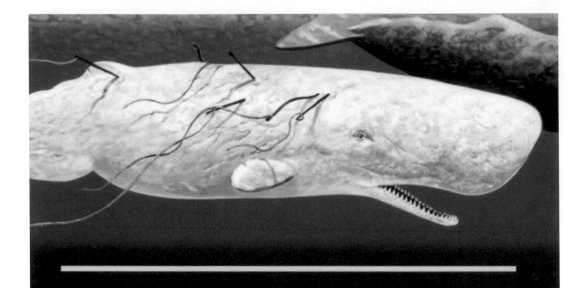

Unpainted to the Last

Moby-Dick and Twentieth-Century American Art

Elizabeth A. Schultz

Cover of *Unpainted to the Last*. (University Press of Kansas)

Richard Ellis at work on the Moby Dick murals. I could only work on two canvases at a time, so I painted in assembly-line fashion, moving the right-hand canvas aside and bringing the new one in at the left. (Photo by Russ Kinne)

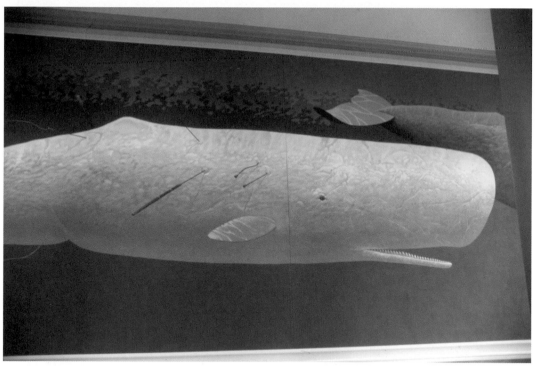

The mural in place. Partially installed in the New Bedford Whaling Museum, Moby Dick shows the harpoons "caught and twisted" in his white flank. (New Bedford Whaling Museum)

Reentry Fifty tons of breaching whale makes a mighty splash upon reentry. (Painting by Richard Ellis)

Sperm whale stranding. "It is not known for what reason they run themselves aground on dry land; at all events it is said that they do at times, and for no obvious reason" (Aristotle). (SeaPix)

Sperm whale family. The bonds among sperm whale family members are strong. (Painting by Richard Ellis)

Sperm whale and triangular squid. Off the Azores in 2009, Australian diver/photographer Wade Hughes snapped this shot of a sperm whale with a large squid (probably *Taningia danae*) in its mouth. (Photo by Wade Hughes)

as before. They again went down for a short time, and then came up to the surface under the packet's larboard quarter, the whale appearing first, and the serpent in pursuit, who was again seen to shoot up his tail as before, which he held out of water for some time, waving it in the air before striking, and at the same time his head fifteen or twenty feet, as if taking a view of the surface of the sea. After being seen in this position a few minutes, the serpent and whale again disappeared, and neither was seen after by any on board. It was Captain West's opinion that the whale was trying to escape, as he spouted but once at a time on coming to the surface, and the last time he appeared he went down before the serpent came up.

Because one cannot actually see the head of *Architeuthis* when the animal is in the water—in fact, a squid doesn't exactly have a head—it seems that both the "head" and the "tail" of this creature were the tentacles of a giant squid, assigned to either end of a sea serpent.

If the battle comes to the surface and there are people to observe it, we might get some idea of what the struggle really looks like. On July 8, 1875, the crew and officers of the bark *Pauline* saw a battle between a sperm whale and *something*. As later reported in the *Illustrated London News*, Captain George Drevar was some 20 miles off Cape San Roque, Brazil, when the *Pauline* came upon "a monstrous sea serpent coiled twice round a large sperm whale." The sea serpent conquers the whale in Drevar's narration, pulling the hapless cetacean below the surface, "where no doubt it was gorged at the serpent's leisure." A week later, Captain Drevar was still in the same latitude, now 80 miles from shore, when he "was astonished to see the same or a similar monster. It was throwing its head and about 40 feet of its body in a horizontal position as it passed onwards by the stern of our vessel." Drevar described the serpent's mouth as "always being open," but with this exception, the creature sounds suspiciously like a giant squid.

Frank Bullen, like Melville a whaleman turned author, had no compunctions about describing a clash between a squid and a whale. (Bullen's narrative, *The Cruise of the "Cachalot"* [1898], is replete with improbably theatrical episodes of courage—usually his own—and wildly unlikely behavior on the part of various animals, and most historians are inclined to dismiss his book as more fiction than fact.) The book contains a vivid description of a battle between the two gigantic predators:

A very large sperm whale was locked in deadly conflict with a cuttle-fish, or squid, almost as large as himself, whose interminable tentacles seemed to enlace the whole of his great body. The head of the whale especially seemed a perfect net-work of writhing arms—naturally, I suppose, for it appeared as if the whale had the tail part of the mollusc in his jaws, and, in a business-like, methodical way, was sawing through it. By the side of the black columnar head of the whale appeared the head of the great squid, as awful an object as one could well imagine even in a fevered dream. Judging as carefully as possible, I estimated it to be at least as large as one of our pipes, which contained three hundred and fifty gallons; but it may have been, and probably was, a good deal larger. The eyes were very remarkable from their size and blackness, which, contrasted with the livid whiteness of the head, made their appearance all the more striking. They were, at least, a foot in diameter, and seen under such conditions, looked decidedly eerie and hobgoblin-like.

Bullen's narrative appears to owe something to Melville's, especially as concerns the coloration of the squid. (In Melville's novel, the living squid is a ghostly white color, and the harpooner, Daggoo, actually mistakes it for Moby Dick, crying out, "There! there again! there she breaches! right ahead! The White Whale, the White Whale!") The amount of salt required to season Bullen's story is evident from his setting: the cuttlefish and the whale staged their epic battle by moonlight.

Because sperm whales have been hunted commercially for almost three centuries, we have had more than ample opportunity to examine their stomach contents. In the early days of the fishery, a great deal of useful information was lost. The blubber was stripped off alongside the ship, and the carcass, along with the stomach contents, was discarded. It was only when the whales were hauled up on the decks of the great factory ships that the stomach contents were spilled out on deck. Whalers observed their quarry, in its death throes, vomiting up great hunks of what could only have been giant squid. As we might expect, Melville discusses this phenomenon in *Moby-Dick:*

For although other species of whales find their food above water and may be seen by man in the act of feeding, the spermaceti whale obtains his food in unknown zones below the surface; and only by inference is it that anyone can tell of what, precisely, that food consists. At times, when closely pursued, he will disgorge what are supposed to be the detached arms of the squid; some of them thus exhibited exceeding twenty or thirty feet in length. They fancy that the monster to which these arms belonged ordinarily clings by them to the bed of the ocean; and that the sperm whale, unlike other species, is supplied with teeth in order to attack and tear it.*

And some 50 years later, in *Denizens of the Deep* (1904), a natural history of sea creatures, Frank Bullen wrote much the same thing:

Every officer, to say nothing of the men, must have known of the very real existence of the great Squid, since scarcely a sperm whale can be killed without first ejecting from his stomach huge fragments of this popularly believed by seamen to be the largest of all God's creatures. Not only so, but in every book which has been written about the sperm whale fishery some allusion to the great Cuttle-fish will surely be found, although it must be admitted that so much superstitiously childish matter is usually mixed up with the facts as to make the latter difficult of belief.

Aboard the fictional *Cachalot,* Bullen espies "great masses of white, semi-transparent-looking substance floating about, of huge size and irregular shape," and asks the mate to tell him what they could be. "When dying," the mate explains, "the cachalot always ejected the contents of his stomach, which were invariably composed of such masses as we saw before us; he believed the stuff to be portions of a big cuttle-fish, bitten off by the whale for the purpose of swallowing." Bullen hooks one of the lumps and draws it alongside:

It was at once evident that it was a massive fragment of cuttle-fish— tentacle or arm—as thick as a stout man's body, and with six or seven sucking discs or *acetabula* on it. These were about as large as a saucer, and on their inner edge were thickly set with hooks and claws all

*For all his vaunted genius and historical accuracy, in this instance, Melville turned out to be wrong about both the squid and the whale. Most squid do not cling to the seabed with their arms, and the teeth of sperm whales, located in only the lower jaw, are probably used only to capture the squid, pincer fashion, not to tear them.

round the rim, sharp as needles, and almost the size and shape of a tiger's.

In *Denizens of the Deep*, Bullen speculated on the relationship between the sperm whale and the giant squid:

> The gigantic Cuttle-fish must be very prolific. He is the principal food, the main support of the sperm whale, and as this vast mammal's numbers are incalculable, and each individual needs, at the very lowest computation, a ton of food to keep him going, the numbers of mollusca upon which he feeds must be proportionate. As to the numbers of sperm whales I may say in passing, that it has several times been my lot to witness an assemblage of cachalots, all of the largest size, covering an area of ocean as far as the eye could reach from the masthead of our ship in every direction. . . . Only to think of the amount of food required for that stupendous host makes my mind reel.

While Bullen's mind reeled, others tried to find out how much food a sperm whale actually needed. In his comprehensive study of the whales captured from the Durban fishery between 1926 and 1931, L. Harrison Matthews (1938) examined the stomachs of 81 sperm whales, and of these, the stomachs of nearly all of them contained the remains of cephalopods, among other things. Most of them were small, averaging about a meter in length, but the "very large cephalopods were represented only by beaks in the stomachs and scars on the skin." There is no question, however, that sperm whales occasionally battle and ingest giant squid: According to Rees and Maul, one was even regurgitated in a state where it still showed signs of life. A whale harpooned off Madeira in 1952 had vomited up a 34-footer that weighed about 330 pounds and writhed on the flensing deck until it expired. Robert Clarke (1955) was present at the whaling station at Porto Pim on the Azorean island of Fayal in 1955 when a giant squid was discovered in the stomach of a 47-foot-long whale. It weighed 405 pounds, and it measured 34 feet, 5 inches from the tip of the tail to the tip of the longest tentacle.

When Victor Scheffer, a respected marine biologist, wrote of an encounter between a sperm whale and a giant squid in *The Year of the Whale* (1969), he had the full weight of science behind him. Nevertheless, he fabricated an encounter that neither he nor anyone else has ever witnessed:

The pressure is now one hundred tons to the square foot; the water is deathly cold and quiet. At a depth of three thousand feet he levels off and begins to search for prey. The sonar device in his great dome is operating at full peak. Within a quarter-hour he reads an attractive series of echoes and he quickly turns to the left, then to the right. Suddenly he smashes into a vague, rubbery, pulsating wall. The acoustic signal indicates the center of the Thing. He swings open his gatelike lower jaw with its sixty teeth, seizes the prey, clamps it securely in his mouth, and shoots for the surface. He has found a half-grown giant squid, thirty feet long, three hundred pounds in weight. The squid writhes in torment and tries to tear at its captor, but its sucking tentacles slide from the smooth, rushing body. When its parrot beak touches the head of the whale it snaps shut and cuts a small clean chunk of black skin and white fibrous tissue. . . . He crushes the squid's central spark of life, its gray tentacles twist and roll obscenely like dismembered snakes. . . . Now at ease, the bull turns the dead beast and leisurely chomps it into bite-sized pieces, each the size of a football, and thrusts them mechanically into his gullet with muscular tongue.

Except for the part where the whale chomps the squid into bite-sized pieces (Victor Scheffer would have known that the whale's jaws are unsuited for chomping things into pieces), this is probably what it looks like when a whale catches a squid. The squid might struggle a little more, and it has more than "sucking tentacles" to grip the skin of the whale: it has sharp-toothed suckers that can dig into the whale's sensitive skin.

Because giant squid and sperm whales are known to inhabit Norwegian waters, a colossal conflict is a distinct possibility there. In *Norges Dyreliv* ("The Animal Life of Norway"), Einar Koefoed relates a tale about the battle of these titans: "A Norwegian whaler once saw a large whale with the body of a giant squid between its teeth. . . . The giant squid had thrown its tentacles, as thick as ropes, around the head of the whale so that it could not open its jaws. Suddenly the squid let go and the whale submerged. After a while it surfaced again with the crushed squid in its mouth."

There are very few contemporary accounts of a battle between a giant squid and a sperm whale, but just before the outbreak of World War II, J. W. Wray was sailing off the Kermadec Islands, north of New Zealand, when he spotted a disturbance in the water. As he tells the story in *South Sea Vagabond,* he sailed over to investigate, "when, to our amazement, a giant tentacle, easily twenty-five feet long, came out of the water, waved around for a second, and crashed back into the water again. A few seconds later our hearts stood still as the fore part of a large whale shot out of the water barely thirty yards away and, encircling its head, its enormous tentacles thrashing the water, was a truly villainous giant octopus. . . . Two more giant tentacles emerged, making a terrific commotion on the water." Although Wray called the animal an octopus, the length of the tentacles makes the case for a giant squid, and the battle with the whale (unidentified, but occurring in an area well-known for sperm whales) certainly suggests the traditional antagonists.

In "Body Scarring on Cetaceans—Odontocetes," a paper published in the *Scientific Reports of the Whales Research Institute of Tokyo* in 1974, Charles McCann proposed several theories about the scars on the heads of various whales. He also included some rather bizarre observations about sperm whales and squid, which, as far as I know, have not appeared elsewhere. He wrote:

> The enormous head of the sperm whale occupies almost one-third of the total length of the animal. It is provided with the cushion of spermaceti, a waxy looking substance already referred to. To specu-

late, this "cushion" functions as a shock-absorber when the animal uses its head as a battering ram in offense and defense. In addition, the array of large teeth serve as weapons of defense and offense, and they function as grasping organs in dealing with the large slithery cephalopods (particularly the Giant Squid) which form a large proportion of the food of Sperm Whales.

In dealing with some of the larger cephalopods, there seems to be a suspicion that the sperm uses its head to pummel the squids after grasping them with its jaws (the fleeing squid, going backwards as is its wont, would be grasped by the tentacles). Evidence of battles with large cephalopods is impressed on the facial region of the Sperms.

The image of a sperm whale grasping a squid by the tentacles and bashing it on the bottom is indeed an unusual one, as is the picture of the whale pummeling the helpless squid with its battering-ram head. We still do not know precisely how the whales catch the squid, but McCann's speculations are too absurd to be taken seriously.

In *The Depths of the Ocean* (1912), written with J. Murray, Johann Hjort made one of the most often quoted—and misquoted—remarks ever made about the giant squid. In 1902 Hjort was aboard the research vessel *Michael Sars* in the North Atlantic when they came across a small giant squid floating dead at the surface north of the Faroes. There was nothing spectacular about this sighting, but then Hjort wrote, "In 1903 in Iceland I had the opportunity of making an interesting observation, showing the gigantic dimensions of these squids." The ship visited the whaling station at Mofjord, where there were two freshly killed whales, a sperm whale and a right whale. Hjort wrote:

Inspecting the cachalot, I saw around its enormous jaw several long parallel stripes, consisting, as closer scrutiny revealed, of great numbers of circular scars or wounds about 27mm in diameter. . . . It occurred to me that these scars must have been left by the suckers of giant squid, and following up this idea I found in the whale's mouth a piece of squid tentacle 17cm in maximum diameter. In the stomach of the whale many squid-beaks of various sizes were found, the largest measuring 9 cm in length, besides some fish bones, and the men who had shot the whale told me that in its death flurry it disgorged the arm of a squid 6 meters long.

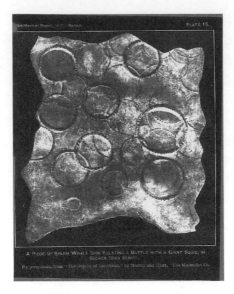

Photograph of sucker marks on the skin of a sperm whale, from Murray and Hjort's 1912 *The Depths of the Ocean*. In the original, the largest sucker mark is about an inch in diameter. (Murray and Hjort, *The Depths of the Ocean*, 1912)

In Hjort's book there is a picture of the "skin of the cachalot with marks from the struggle with *Architeuthis*. Nat. size." The scars that Hjort measured at 27 mm were a little over an inch in diameter, and in the photograph, the largest one measures one inch across. The picture has been reproduced in almost every discussion of sperm whales and squid, but somehow, the diameter of the circular scars has increased to impossible proportions, perhaps through a confusion of the 27 mm scars and the 17 cm–diameter *tentacle,* which equals 6.63 inches.*

Many people assume that if anyone could see a giant squid, it would be Jacques Cousteau, with his thousands of hours exploring the depths of the world's oceans. Indeed, in his *Octopus and Squid: The Soft Intelligence* (written with Philippe Diolé and published in 1973), Cousteau records just such a sighting:

> When I had reached 800 feet, I saw, through a porthole, a very large cephalopod, only a few yards from the minisub, watching the vehicle

*A few of the more egregious examples follow: in the 1977 Time-Life book *Dangerous Sea Creatures,* we read that "an ordinary giant squid of 50 feet leaves teeth-ringed sucker marks measuring between three and four inches across on a whale, but sperm whales have been captured with tentacle marks 18 inches across." In *The Guinness Book of Animal Facts and Feats,* Gerald Wood wrote that scars "measuring up to 5 in. in diameter have been found on the skins of sperm whales captured in the North Atlantic," and Willy Ley—who should have known better—wrote (in 1987's *Exotic Zoology*), "Another claim goes for marks on the skin of such a whale of a sucking disk over 2 feet in diameter."

as it moved slowly past. I could not take my eyes from the mass of flesh, though it seemed not at all disturbed by the presence of the minisub. It was an unearthly sight, at once astonishing and terrifying. Was it sleeping? Or thinking? Or merely watching? I had no idea. It was there, nonetheless, enormous, alive, its huge eyes fixed on me. Then suddenly, it was gone. I did not even see it move, though I am sure that an animal of that size is able to move with extreme rapidity by means of water jets from its funnel. The impression made was one of size and power. I can understand how formidable a giant squid must be.

This may indeed be a description of *Architeuthis*, but it is so vague that one wonders whether it really happened, or whether Cousteau invented it. The location is not given (only that it was "during one of *Calypso*'s expeditions in the Indian Ocean"), nor is the date. All we can deduce is that it occurred before 1973, the publication date of the book. Other than "watching the vehicle as it moved slowly past," what was the animal doing? Did it move with tail or tentacles first? What color was it? For that matter, what color were its huge eyes? (How did an animal with one eye on each side of its cylindrical head look at Cousteau with what he described as "those great unblinking *eyes*"?) And what does the last sentence mean? Because of its "size and power," did Cousteau understand how formidable an opponent a giant squid must be, or, having seen an animal that was a very large squid but not *Architeuthis*, did he wonder what a real giant squid might be like? At no point in the description can you find the words "I saw a giant squid." It is all carefully worded implication, leaving it to the gullible reader to make the connection.

Cousteau cross-references this discussion with "See *The Whale: Mighty Monarch of the Sea*, by Jacques-Yves Cousteau and Philippe Diolé." Published in 1972, the year before the octopus and squid book, *Mighty Monarch* reveals that also "in the Indian Ocean," *Calypso* came across a "large white object" floating on the surface, which the crew retrieved and identified as "a piece of a giant squid's tail. The front part of it is torn, and it is covered with punctures similar to those inflicted by the teeth of a cachalot or a pilot whale."* They also retrieve "a piece

---

*On the page immediately following the description of the squid's tail ("covered with punctures similar to those inflicted by the teeth of a cachalot or a pilot whale") is a glaring example of Cousteau's confused cetology. Even after he has found a piece

of flesh shaped like a saucer, or rather like a plate. It is one of the squid's suction cups. Dr. François measures it and announces that its diameter is 24 inches. This, obviously was a 'small' giant squid. Its body probably measured between 8 and 10 feet—in addition to the large arms, of course." When they attempted to cook the pieces, they found that the tail was "so tough that we could not cut it," and "as for the suction cup, it was too horrible for us to describe. It was as though we had tried to make a meal out of a hunk of soft rubber."

There follows a brief discussion of the "fantastic Karken [*sic*]," in which this sentence appears: "No man has ever seen an *Architeuthis* except as food not yet digested in the stomach of a sperm whale." Did he forget that *he* had seen one—or did he just decide to leave it out of the whale book? (If he hadn't actually seen one, then we can date the reported *Architeuthis* sighting to the time between the preparation of the two books.) In a discussion of eating, or trying to eat, pieces of *Architeuthis*, would he have forgotten that he had already described this "mass of flesh . . . enormous, alive, its huge eyes fixed on me"? And how could a man who claimed to have seen a living giant squid write that the "suction cup" they found measured 24 inches in diameter? (An ordinary dinner plate is 10 inches in diameter. A large garbage can lid is 20 inches in diameter. A New York City manhole cover is 27 inches in diameter.) I suggest that both events are fabrications: there was no sighting of a giant squid, and there was certainly no attempt to cook a 24-inch-diameter suction cup, because there is no such thing as a 24-inch-diameter suction cup.*

According to Clarke (1980), the largest circular scars on the whale's head have come from *Architeuthis*, the squid with the largest suckers. He wrote, "I have not yet seen conclusive evidence to suggest that

---

with tooth marks, he declares, "The cachalot does not pulverize his food, nor does he chew. He does not even really bite. Instead he swallows his food whole, in a gulp."

*It is difficult to decide what to make of the mistakes, misspellings, and misrepresentations in the Cousteau book on cephalopods. If someone gets the facts wrong, the errors might be attributed to poor research, sloppy editing, or, in this case, imperfect translation. But what are we to make of such flagrant distortions? Because they are obviously not accidental, we must assume that the authors included this material intentionally, knowing that there was no possible way to verify it. Cousteau's reputation as a popularizer of marine natural history and conservation should have encouraged more, rather than less, dedication to accuracy.

sucker scars are larger than 3.7 cm [1.44 inches] across." In 1872, the Reverend Harvey measured the suckers of the Bonavista Bay specimen at 2.5 inches in diameter, and in his 1938 monograph on sperm whales, L. Harrison Matthews wrote, "Nearly all male Sperm whales carry scars caused by the suckers and claws of large squids, scars caused by suckers up to 10 cm. in diameter being common. The claw marks take the form of scratches 2–3 m. in length, and appear to be of more frequent occurrence than sucker marks." Matthews's measurements of 10 centimeters—3.9 inches—are so much larger than any other recorded sucker dimensions that one suspects some sort of error, either in measuring or in transcription.

One of the most useful—and reliable—books ever written about squid is Malcolm Clarke's massive "Cephalopoda in the Diet of Sperm Whales of the Southern Hemisphere and Their Bearing on Sperm Whale Biology," published as a *Discovery Report* in 1980. As Clarke wrote in the introduction to this 324-page study, "Whale and squid biology are clearly closely linked, and consideration of both subjects in one paper is necessary to avoid duplication." Since beginning the study in 1962, Clarke (and various colleagues) have examined the stomach contents of 461 sperm whales, collected at the whaling stations of Durban and Donkergat in South Africa; Cheynes Beach in Albany, Western Australia; the island of South Georgia; and the British pelagic factory ships *Southern Harvester* and *Southern Venturer*.

Of the giant squid, he says, "Twenty-three buccal masses and eight large pieces of flesh belonging to *Architeuthis* were collected from whales caught off Durban, Donkergat and Albany." Despite the popular perception that sperm whales feed regularly on giant squid (see Roper's statement above), Clarke's research showed that "*Architeuthis* only comprises 0.26% of the beaks collected." The largest *Architeuthis* beak represented a squid weighing 120 to 180 kilograms (264 to 396 pounds), probably not in the "monster" class. There were no stomachs sampled with the remains of more than one *Architeuthis,* which suggested that the whale was catching solitary animals, or that "their large size may facilitate escape if a shoal* is attacked."

*In British ichthyological terminology, the word *shoal* is used where Americans use *school.* The words come from the same 16th-century Dutch root, *schole,* which meant a group of aquatic animals or floating objects, and have nothing to do with education.

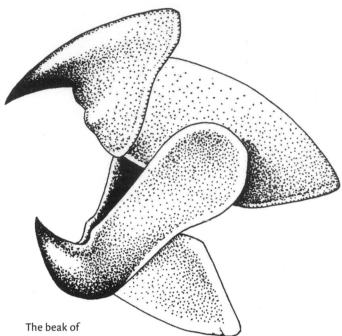

The beak of *Architeuthis*, shown actual size. (Richard Ellis)

As described in chapter 5, sperm whales have the largest brains of any animal that has ever lived. They lead complex social lives, and they are capable of making a great variety of sounds that are undoubtedly part of an intricate communications system. The immediate future holds little promise of interspecies communication, but if any creature can reveal the secret life of *Architeuthis*, it would certainly be *Physeter macrocephalus*, the giant whale that knows enough about the giant squid to find it and (occasionally) capture it in the darkness of an icy black ocean, a mile below the surface.

In the ongoing attempt to capture a giant squid on film or video, several expeditions have been mounted, always to regions where sperm whales can be seen regularly, on the assumption that the whales ought to know where their food might be found. In January 1997, an expedition sponsored by National Geographic Television, the Smithsonian, the New England Aquarium, and several other institutions left for New Zealand in hopes of obtaining the first images of the living giant squid. Under the leadership of Clyde Roper, one of the world's foremost authorities on *Architeuthis*, they headed for Kaikoura Canyon, off the east coast of South Island, a region known for sperm whales. Peter Benchley had published *Beast*, an analogue of *Jaws* but with giant squid as the man-eaters, so he was asked to go along as the writer. When he couldn't go, he called me and asked me to fill in, but I couldn't go either, so Roper wrote up the details of the adventure, which failed to find a giant squid. In 1997, an expedition to the Azores failed again, but they affixed a "crittercam" (a camera that films what the critter sees) to a sperm whale. Although they got some fascinating footage and accompanying sounds of a deep dive from the whale's point of view, the whales—and therefore the scientists, this time including Malcolm Clarke—didn't see the object of the hunt. Again in 1999, Roper and New Zealand squid expert Steve O'Shea led a BBC-TV expedition to Kaikoura, and even though they filmed lots of sperm whales, they didn't find the elusive *Architeuthis.**

Here, then, is *Architeuthis*, the giant squid, long believed to be the archenemy, as well as the principal food item, of the sperm whale. We have seen that it is actually more of a meal than an enemy: the only way a giant squid can win a battle with a sperm whale is to escape, often with the loss of a tentacle or two. Even though sperm whales do eat giant squid, they don't eat that many. Malcolm Clarke analyzed the stomach contents of sperm whales caught in South African, Australian, and Antarctic waters from 1962 to 1978 (his report was published in 1980), and because the indigestible beaks of the squid consumed by the whales can be identified as to species, he was able to use the beaks to identify the types of squid. (In 1986 he published *A Handbook for the Identification of Cephalopod Beaks*.) He found that the most common species consumed by the Southern Hemisphere whales was *Histioteuthis*, a species known as jewel squid because their entire bodies are studded with lights. (This bioluminescence may have something to do with their predominance in the sperm whales' diet, but it may not.) Of all the squid species in the whales' stomach contents, *Histioteuthis*—which is only about two feet long, counting tentacles—made up a whopping 21.3 percent. *Architeuthis* came in at 0.26 percent.

We know now that *Architeuthis* is an inhabitant of New Zealand waters because so many have been caught there in recent years, but until those individuals began turning up (dead) in fishermen's nets, the range of the giant squid was all over the map. In the 1870s and 1880s, a large number washed ashore on Newfoundland beaches, but after that, they appeared as singletons in such diverse locations as Norway, Ireland, the Bahamas, Japan, Massachusetts, and Tasmania. Only in Newfoundland, New Zealand, and Norway were there enough

*The first-ever shots of a living giant squid were obtained in September 2004 by Tsunemi Kubodera, a Japanese squid specialist, who had deployed a robot camera at 900 meters (2,952 feet) off the Ogasawara (Bonin) Islands, south of Japan, a known haunt of sperm whales (Kubodera and Mori 2005). In December 2006, Kubodera was "fishing" for giant squid in the same area, and as he reeled in the camera, he found that a giant squid had grabbed it and was coming to the surface. Because Kubodera's first images were stills, the 2006 video of the bright red squid flailing at the surface became the first-ever film of a living giant squid.

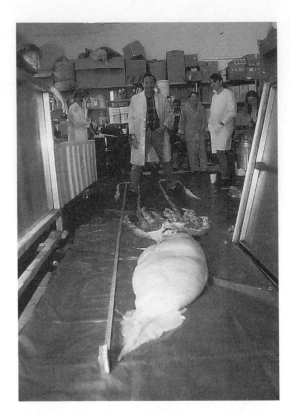

In June 1998, a 27-foot-long giant squid came to the American Museum of Natural History from New Zealand. The author was part of the welcoming committee. (American Museum of Natural History)

stranded squid to identify any sort of pattern, but that sample was far too small to draw any conclusions about where giant squid live. Note that they are mostly cold-water, high-latitude locations—perfect hunting grounds for exiled bull sperm whales in search of a large, ammonium-flavored meal.

VII

"I'll Have the Calamari..."

IF SPERM WHALES feed on large squid, how do they catch them? Unlike many other predator-prey interactions, this one is not at all self-evident. Consider: lions and zebras both live on the African plains. Lion sees zebra, stalks same. Lion chases zebra, catches zebra, feeds on zebra. Both the predator and the prey live in the same place, and they breathe the same stuff in more or less the same way. Now consider bats. They live on insects that they catch on the wing, but the rules are somewhat different. Lions can see the herds of zebras, but the bats have to be able to find and capture the insects in the dark. To accomplish this, bats have developed a highly sophisticated sonar system, where they emit high-pitched sounds that bounce off flying insects. The bats read the returning echoes and adjust their flight to intercept and capture the insects. Although this is a far more complex system than the lion-zebra relationship, it is similar in that both creatures are capable of flying through the air at night. And again, they breathe the same stuff, so neither one is placed at a particular disadvantage by the medium in which they live.

Convergence is the term used in evolutionary biology to refer to the development of similar adaptations in unrelated animals, such as bioluminescence in fireflies and jellyfish, or wings in birds and bats. Bats figure again in an even more surprising convergence: their DNA includes the same protein (prestin) that is found in the DNA of toothed whales, and this protein is expressed in the complex process of echolocation. (Nonecholocating bats and baleen whales do not have prestin in their DNA.) In *Current Biology* (2010), Liu et al. wrote: "The ability of some bats and all toothed whales to produce sonar pulses and pro-

cess the returning echoes for prey detection and orientation (echolocation) is a spectacular example of phenotypic convergence in mammals."
It is more than a little surprising to discover a molecular convergence
in such distantly related groups of mammals, and although the finding does not demonstrate anything more than divergent mammalian
evolution, it does suggest that nature can provide some pretty sophisticated solutions to problems—in this case, finding something to eat.

The sperm whale, like the lion, the zebra, and the bat, is a mammal and breathes air, but its prey does not. But, you will say, seals, sea
lions, and otters are also mammals, and they feed on fish that get their
oxygen without coming to the surface. The seals, etc., chase their prey
visually, and although there are some kinds of seals—Weddell seals, for
example—that are prodigious divers, most of them hunt where they
can see what they are chasing. (Some penguins are also unexpectedly
deep divers.) Sperm whales, however, hunt in a manner that we do not
understand because it has always been hidden from the eyes of investigators by the inaccessibility and opaque darkness of the environment
in which it occurs.

Sperm whales take oxygen into their lungs just the way humans do:
they inhale a breath of fresh air. (Their "spouts" are their exhalations.)
But the hunted cephalopods have gills and obtain their oxygen from
water, and therefore, they do not have to surface to breathe. The whales
have to do their food locating and food catching while holding their
breath, often in the darkness of the depths. Herein lies the greatest
conundrum of the sperm whale's hunting: how can it find and capture
enough food on these deep dives, especially if the prey is a more agile
and faster swimmer—and more critically, if the whale has to abandon
the hunt after a certain period of time because it is running out of air?

How squid are transformed from fast, free-swimming animals into
stomach contents is a mystery. From an apparatus in the head, sperm
whales (and dolphins) can broadcast high-frequency sounds direction-
ally into the water, then read the returning echoes for information on
the identity (and perhaps the condition, speed, texture, and so on) of the
object. The echolocation of dolphins has long been recognized, but the
mechanics whereby the animals actually catch their prey was more
problematical. It is one thing to locate, say, a school of small squid, but
it is quite another to catch enough of them to make a meal. Even when
the echolocating capabilities of odontocetes were understood, there
was a piece missing from the puzzle because the first cetological acous-

"Battle of the Giants" as dramatically interpreted by Canadian artist Glen Loates. It is most unlikely that such an encounter would take place at the surface. (Glen Loates)

ticians simply assumed that the whales found the squid by listening to their echoes, then dashed around, gobbling them up. Upon reflection, this did not appear to be a terribly efficient method of hunting, especially considering the speed and maneuverability of the prey, and also its inherent unwillingness to be eaten. The sperm whale has massive peglike teeth in its lower jaw, so if the whale chased down its prey and snagged it in its jaws, the squid ought to have shown some evidence of having been bitten, but they didn't. One of the first to notice this was the Soviet cetologist A. A. Berzin, who published an exhaustive study of the sperm whale in 1972. After examining the stomach contents of a large number of sperm whales, he wrote:

> The mystery of how this whale feeds deepened in view of the following circumstances. . . . Beale gives an example of a capture of sperm whales in normal condition, one of which was blind while two others had deformed jaws. Up to 10 sperms with badly deformed jaws were recorded in our materials. They were in the same condition as all the other animals and had well-filled stomachs, the contents of which did not differ qualitatively from those of other sperm whales caught the same day. . . . All the above suggests that neither the teeth nor the lower jaw need to participate in obtaining food and in the digestive process.

If the sperm whale does not use its jaws and teeth to capture its food, what does it use? In his study, Berzin reviewed the earlier theories, one of which had been propounded by Thomas Beale, a British surgeon who shipped aboard the whaler *Kent* in 1831–1832. Upon the conclusion of the voyage, he wrote *A Few Observations on the Natural History of the Sperm Whale,* which was published in 1835 and served as one of the primary sources for the cetology chapters in *Moby-Dick.**

In his 1835 essay, Beale mentions several large cephalopods, including the one "discovered by Drs. Banks and Solander, in Captain Cook's first voyage [which] must have measured at least six feet from the end of the tail to the end of the tentacles." "But this last," he wrote, "we must imagine a mere pigmy, when we consider the enormous dimensions of the one spoken of by Dr. Schwediawer, in the Phil. Trans. . . . whose tentaculum or limb measured twenty-seven feet in length." The good doctor's remarks, which were actually presented to the Royal Society by Sir Joseph Banks, include, as a sort of footnote, a story that he heard from someone else, for he says, "One of the gentlemen who was so kind as to communicate to me his observations on this subject, about ten years ago hooked a spermaceti-whale that had in its mouth . . . a large tentaculum of the Sepia Octopodia nearly 27 feet long. This tentaculum did not seem to be entire, one end of it appearing in some measure corroded by digestion, so that, in its natural state it may have been a great deal longer."

Like most sperm whalers, Beale could not imagine how "such a large and unwieldy animal as this whale could ever catch a sufficient quantity of such small animals, if he had to pursue them individually for his food." He suggested that the whale descends to a certain depth, where he "remains in as quiet a state as possible, opening his narrow elongated mouth until the lower jaw hangs down perpendicularly." The jaws and teeth, wrote Beale, "being of a bright glistening white colour . . . seem to be the incitement by which the prey are attracted, and when a

*In C. M. Scammon's 1874 *Marine Mammals of the Northwestern Coast of North America,* Beale's description of the sperm whale's feeding habits appears verbatim as a very long footnote. Scammon also wrote, "the animal's manner of pursuing its prey is not definitely known; but several high authorities maintain, that after descending to the desired depth, it drops its lower jaw nearly to a right angle with the body, thereby exhibiting its polished white teeth, which attract within its reach the swimming food while the creature moves through the ocean's depths; the moment its prey comes in contact with the expanded jaw, it is instantly crushed, and a portion or all is swallowed."

sufficient number are in his mouth, he rapidly closes the jaw and swallows the contents."

Commenting on this theory, Berzin (1972) wrote, "Strange though it may sound, this fantastic hypothesis found many supporters." Scammon (1874) was one, for he paraphrased and illustrated this description, attributing it to "several high authorities." Because he quoted at length from Beale's 1839 account, we may assume that at least one of these "high authorities" was Beale. He also mentioned a whale that was "perfectly blind" and two specimens in which the lower jaw was so deformed "as to render it impossible for the animal to find the jaws useful in catching small fish . . . and yet these whales possessed as much blubber, and were as rich in oil as any of similar size I have seen before or since." It seems unlikely that sperm whales hang motionless in the water, waiting for food to swim into their mouths, but until someone actually observes a sperm whale feeding, this explanation is remotely possible.

In his extensive study called *Whales* (1962), the Dutch cetologist E. J. Slijper suggested a modification of Beale's theory. He wrote: "It is believed that sperm whales do not so much go after this prey as swim about with open mouths, enticing the cuttlefish which seems unable to resist the colorful contrast between the sperm whale's purple tongue and the white gum of the jaws." Slijper knew more about whales than he did about squid, because in the blackness of the depths, the cuttlefish probably could not see the whale's tongue, its gums, or any other part of it. Berzin (1972) noted that "at a depth of some 100 meters, both prey and hunter are invisible to each other. Moreover, squid are known to be much more mobile than sperm whales. . . . even small squid can develop speeds of up to 40 km an hour, and larger specimens still higher speeds, far outdistancing sperms." Obviously, there had to be some device whereby the whale could stop or at least slow down the squid to capture them, and indeed there was. Even earlier than Berzin, another pair of Soviet cetologists, Vladimir Bel'kovich and Alexei Yablokov, had calculated the sound intensity that might be developed in the sperm whale's nose, and in 1963, they collaborated on a short article entitled "The Whale—An Ultrasonic Projector," in which they suggested that the whale might somehow use its nose to project sounds loud enough to stun its prey. Berzin (1972) wrote that by concentrating its sound beam on a selected object, "the animal can create a short-term pressure which must act as an ultrasonic blow capable, even if briefly, of halving, stunning, and paralyzing the object."

In 1983, Ken Norris, of the University of California at Santa Cruz, and Bertel Møhl, of Aarhus University in Denmark, published their hypothesis that odontocetes could indeed debilitate their prey with sound. Other theories, such as Beale's, above, or Clarke's, where the whale maintains a position of neutral buoyancy and waits for a school of squid to swim within range, did not explain the uninjured state of the squid in the sperm whale's stomach, nor did these other theories explain how a large animal like the sperm whale could obtain the required 1,000–2,000 kilograms of food per day that would be required to sustain it. The sonic boom hypothesis not only explains how the cumbersome sperm whale hunts and captures the swift cephalopods, but also answers many other questions that had heretofore been as problematical as the feeding technique. If the sperm whale debilitated or killed its prey with sound, it would go a long way to explaining the unusual construction of the jaws. The whale could use its tooth-studded lower jaw in a pincer fashion to pluck the floating squid out of the water or off the bottom—which then accounts for the lack of tooth marks on the prey. If the squid floated to the bottom, the sperm whale might plow its lower jaw through the sediment to pick them up, which would explain the strange items occasionally found in sperm whale stomachs. In their study entitled "Stones and Other Aliens in the Stomachs of Sperm Whales in the Bering Sea" (1963), Takahisa Nemoto and Keiji Nasu identified stones, sand, crabs, glass buoys, a coconut, and a deep-sea sponge as coming from the stomachs of sperm whales killed by Japanese whalers in the vicinity of the Aleutian Islands. (The largest stone they found weighed just over 3 pounds.) It would also account for the occasional entrapment and drowning of sperm whales in undersea telegraph cables; plowing along the bottom, the whale might accidentally become entangled in a loop of cable, or it might even have mistaken the cable for the tentacle of a giant squid. In a 1957 study entitled "Whales Entangled in Deep Sea Cables," oceanographer Bruce Heezen listed 14 instances of sperm whales trapped and drowned in cables, and wrote, "It is possible that the whales attack tangled masses of slack cable mistaking them for items of food." The deepest recorded entanglement was 620 fathoms, or 3,720 feet.

From the examination of the stomach contents spilled on the decks of whaling ships, we know that the whales eat lots of squid, and by analyzing the number of squid beaks found in the stomachs of captured sperm whales, we can get an idea of the quantity. Finds of 5,000 to 7,000 beaks per whale are not uncommon, and Berzin (1972) mentions one Soviet scientist who found 28,000 beaks in the stomach of a single whale, indicating a feeding frenzy in which 14,000 individual squid were consumed. (Squid have upper and lower mandibles, so the total number of beaks must be halved.) In his 1977 study, Clarke estimated the amount of food required to feed the enormous world population of sperm whales; recognizing the difficulties of estimating whale populations, he wrote, "Estimates of the whale population are, unhappily, notoriously questionable, but a 1973 estimate placed this at 1¼ million." Using a mean weight of 15 tons for males and five tons for females, Clarke (1977) calculated the total weight of the world's sperm whales at 10 million tons, which would require *100 million tons of squid per year* (my astonished italics). Clarke estimated that the 1.25 million sperm whales alive in 1979 eat 100 million tons of squid per year, "more than the 60–70 million tons for the total world annual catch of fish, and [the figure] probably approaches the total biomass of mankind." (More recently, the world catch of fish has reached some 90 million tons, closer yet to the cephalopod figure suggested by Clarke.)

It would appear that such consumption would require a dense concentration of squid—and indeed, squid may be the most numerous large animals in the ocean. Although his evidence is largely circumstantial (or nonexistent), Ivan Sanderson, in *Follow the Whale* (1956), discusses the numbers of squid required to feed the world's population of sperm whales:

> Most people don't even know what a squid is, yet these animals probably make up a greater aggregate of pure animal matter on this earth than any other two kinds of living creatures put together. They exist in countless millions of apparently endless masses in every ocean and sea in the world, and almost three-quarters of this planet is covered by oceans and seas which are on the average nearly two and a half miles deep. Throughout this vast volume of liquid there are probably more squids than anything else.

Malcolm Clarke, a British scientist who specializes in sperm whales and squid, commented on the complex interaction of the two in a 1977 study:

Man's awareness of the existence of large squid came, not from what he caught in his nets, but from monsters floating dead or moribund at the sea surface and from the tales of whalers who had seen, with unbelieving eyes, whales vomit complete or dismembered kraken of immense proportions. Such doubtful tales hardened into drawings and recorded measurements over one century ago, and ever since, man has tried desperately to catch by net and line, these will-o'-the-wisps of the sea. Though our nets have become larger and larger and faster and faster, very little progress and most of that in the last decade, has been made towards catching any deep sea squids greater than half a meter or so in length. In a century, many tantalising glimpses of the deep sea squids have come from strandings on the coast and from the stomachs of toothed whales, particularly the commercially exploited Sperm Whale.

Because the biology of squid is so poorly known (Malcolm Clarke [1966] attributed our lack of knowledge of the giant squid "to our complete inability to catch these active animals in the open ocean"), one of the best tools for learning about them is the examination of sperm whale stomach contents. Clarke (1997) suggested that the stomachs of sperm whales represent our only viable method of estimating squid populations. If samples are not collected from sperm whale stomachs, we will have no idea of the number of squid in the ocean. "If this is not done," he wrote, "we may, at some future date, find it impossible to make even the broadest estimate of some squid populations which have great importance in the sea but are inaccessible to us." We not only can ascertain the numbers of squid consumed, but also can learn a great deal about the squid themselves. Indeed, the whales perform an important service for those who would study squid; no other way of collecting is nearly as productive. Squid can often evade nets and trawls, but they seem to be less successful at avoiding the powerful, deep-diving whale that feeds on them. One of the disadvantages of this method, however, is that the specimens are often partially digested by the time the teuthologists get to look at them.

Even though the giant squid is not the predominant prey item of the sperm whale, it stands to reason that larger prey items will provide more nourishment than smaller ones, and it behooves the predator to structure its hunting effort to maximize its caloric intake. If it takes approximately the same amount of energy to capture a smaller squid as a larger one, it makes sense to assume that the whale would take a giant squid over a little one. It may have to do with light, or sound, or the movement of water, but in the depths of the world's oceans, as the giant whales dive in pursuit of the giant squid, a balance between predator and prey has to be achieved. In order to prevent extinction of the entire race, the squid have somehow to be aware of the presence of the whales, and some of them have to escape.

The teeth of sperm whales are probably not used for feeding, or if they are, they serve only a secondary function. Of a whale with a deformed lower jaw, Spaul (1964) wrote, "This condition would conform with the limited and possible non-essential function of the teeth in the sperm whale." The revelation also comes from the examination of the toothless jaws of animals that are no longer nursing. (The presence of milk in the stomach of a captured whale indicates that it has not yet been weaned.) Young animals seem to feed quite satisfactorily, and animals without teeth or with lower jaws so twisted that they could not possibly use their teeth to feed are usually found in good health as a result of normal feeding activity. It has been suggested (Caldwell et al. 1966) that the teeth of sperm whales are a secondary sexual characteristic more than a feeding mechanism and may be used by males primarily for fighting.

Teeth or no teeth, sperm whales feed on a variety of food organisms, predominantly squid, and very few of the latter show tooth marks. Although they do feed on the giant squid, this cephalopod is by no means the dominant food item in their diet. Robert Clarke (1955) examined the stomach of a 47-foot bull sperm whale in the Azores that contained a squid measuring 34.5 feet (10.49 meters) from the tip of the mantle to the tip of the longest tentacle. The squid weighed over 400 pounds (181.4 kilograms). It has been suggested that sperm whales often regurgitate food items when they are harpooned, which may account for the absence of some of the larger squid in the stomachs of harpooned whales. On the whaling brig *Daisy*, Robert Cushman Murphy (1947 saw a dying sperm whale "belching up squids . . . barrelful after barrelful of the tentacled creatures."

According to Malcolm Clarke (1966), the giant squid can reach a maximum length of 60 feet (18.3 meters) and a weight of almost a ton. After examining the aforementioned 34-foot specimen, Robert Clarke (1955) wrote, "I view with less reserve such a traveler's tale as Bullen's, telling of a struggle he watched at night between a sperm whale and a giant squid at the surface of the sea." (Not everyone found Bullen's story easy to accept. Ommanney [1971] wrote, "Like a good deal of that fine book of travel and adventure, it should surely be taken with a grain of salt.") Even though most of the squid consumed by sperm whales are smaller than the monster described by Bullen, the giant squid has been a part of the literature on sperm whales for so long that it is almost impossible to separate fact from fantasy. In 1856, Nordhoff wrote,

> Whalemen believe them ["the sperm whale cuttle-fish"] to be much larger than the largest whale, even exceeding in size the hull of a large vessel, and those who pretend to have been favored with a sight of the body describe it as a huge, shapeless, jelly-like mass, of a dirty yellow, and having on all sides of it long arms or feelers, precisely like the common rock squid. . . . Many of those we saw were of the circumference of a flour barrel. If this be the size of the arms, of which they have probably hundreds, each furnished with air exhausters the size of a dinner plate, what must be the magnitude of the body which supports such an array?

"The stomachs of sperm whales taken off California and British Columbia," wrote Dale Rice in 1978, "contain mostly the large squid *Moroteuthis robustus*. I have measured squid specimens that were up to four feet five inches from the tip of the tail to the anterior edge of the mantle and eleven feet four inches to the tip of the tentacles." Like *Architeuthis*, this species has substituted ammonium ions for the sodium ions in its muscle tissue, which accounts for its low density and its bitter taste. (Sperm whales don't seem to mind a little taste of ammonia; some of their favorite foods, such as *Architeuthis* and *Moroteuthis*, reek of it, but as sperm whales have neither taste buds nor a sense of smell, a little ammonia isn't going to affect their eating habits.) In fact, the ammonium content (but not the actual taste) of these squid may be one of the factors that contribute to their popularity in the sperm whales' diet. The ammoniacal squid are neutrally buoyant and can therefore remain motionless in the water column. The stationary behavior of the squid may somehow be related to the hunting techniques of the whales; it has to be easier to catch something that is standing still than something that is darting hither and yon.

During the 1963–1964 sperm whaling season in New Zealand, 133 carcasses were examined by cetologist David Gaskin (1967). He found that squid of the genus *Moroteuthis* (probably *M. ingens*, a large South-

Moroteuthis robustus, the so-called Pacific giant squid, can reach a length of 20 feet. (Richard Ellis)

ern Ocean form that reaches a mantle length of six feet) made up 74.8 percent by weight of the fresh squid species in the stomachs. As they crossed the flensing deck at the Tory Channel Whaling Station, Gaskin and Martin Cawthorn (1967) noticed that the squid "were glowing a bluish white, and that the light was visible for many yards." Because the luminescence "came away in the hands" and was therefore not produced by specific light-emitting organs, they realized that it was a "property of the mucous covering of the squid or . . . unicellular organisms held in suspension." There are many deep-sea creatures—siphonophores, echinoderms, jellyfish, and many fish species—that can produce a luminescent mucous, but "generalized luminescence of the kind recorded by the present author [Gaskin 1967] does not seem to have been recorded previously for squid in available literature."*

Gaskin's article was written some time before the development of the various theories about sperm whale feeding techniques, such as the whale emitting sonic blasts to stun the squid or, more recently, that the whale might use sound to light up the depths so it can see what it is hunting. He probably knew about Beale's idea that the squid might be attracted to the white lower jaw of the whale when he wrote,

the author has always considered it unlikely that such a large animal as a full grown male sperm whale, weighing perhaps 50 tons, could conceivably chase separately every squid eaten. Considerable energy must be expended swimming down to depths of perhaps 500 fathoms, without taking into account actually chasing squid, which

*According to Nesis (1982), "The skin of cephalopods is thin but is very complicated in nature. The upper layer—the epidermis—is formed by a single layer cylindrical epithelium with numerous mucous cells. The mucous makes the body of the cephalopod slimy, which makes movement easier in the water. Under the epidermis lies connective tissue containing muscular tissue, chromatophores, and iridocytes."

Originally touted as the first photograph ever taken of a living giant squid (*Architeuthis*), this is actually a picture of a dying *Moroteuthis* in the shallow waters off Japan. (Poppe and Goto, *European Seashells*, 1993)

are fast-moving animals. The long narrow shape of the lower jaw of the sperm whale does not seem compatible with the capture of fast-moving prey at speeds of several knots. The jaw has little lateral movement.

Given the speed of the squid and the whale's difficulty in capturing them, Gaskin suggests that the idea of "active feeding" might have to be dropped and a "plausible method of passive feeding suggested":

> The luminescent mucous described above could easily be transferred to the lining of the sperm whale's mouth and act as a lure to attract more squid. If the mucous collected on or between the mandibular teeth the regular spacings might at a distance give the appearance of a fish or other animal with light organs along its flanks. Predatory fish or squid could be attracted to the light, and the sperm whale could lie almost motionless in the water and swallow the animals as they came to its mouth. The whale might have to capture a few squid actively before enough mucous accumulated.

It is an ingenious construct, but one that might be a little difficult to prove. Besides, it presupposes that sperm whales throughout the world's oceans feed only on mucous-covered bioluminescent squid. What happens when the "passive" whale encounters a squid species that has no mucous?

To demonstrate that the whale's method of capturing its food is still undetermined, a 2002 article by Kirk Fristrup and Richard Harbison was entitled "How Do Sperm Whales Catch Squids?" After reviewing the earlier suggestions by Beale, Gaskin, Norris, and others, the authors offer two more hypotheses that are based on the idea that the whales might hunt by sight. Yes, it's pretty dark down there, but the whales might look upward and see the prey species silhouetted against the light of the surface far above them. (That a sperm whale probably cannot see upward while swimming in its normal configuration is neatly explained by suggesting that the whales might have to swim upside-down.) Hypothesis 1: "Sperm whales use vision to locate and capture prey at mesopelagic depths." Hypothesis 2: "Sperm whales use stimulated bioluminescence to lure prey into their mouths." Reminiscent of Beale's suggestion that the squid might be lured in by the whiteness of the whale's lower jaw and teeth, Fristrup and Harbison say that because it is known that squid are attracted to bright lights (commercial squid fishing uses batteries of lights to attract the squid to the hooks), the squid "might deliberately stimulate [dinoflagellate] bioluminescence in order to provoke attacks by squid. This could be accomplished by moving the jaw, perhaps while swimming, creating a zone of turbulence around the mouth. This display would be further accentuated and reflected by the white lining of the mouth."

Another theory about how sperm whales locate their food suggests that the whales might provide their own light. If the whales' sound bursts cause deep-water dinoflagellates to light up, their bioluminescence might provide enough light for the hunting whale to see—and therefore catch—its prey in what would otherwise be total darkness.* This also does not explain the lack of tooth marks on the captured squid, nor does it account for the squid's ability to outswim and outmaneuver a slow swimmer like a sperm whale. This startling theory appeared without attribution in the National Geographic Society's 1996 *Whales, Dolphins and Porpoises*. Because *Architeuthis* doesn't light up, this idea, however far-fetched, might be used to explain how sperm whales can find the great squid in the depths.

*Even less is known about how squid catch their prey in the depths, but a study published by Fleisher and Case in 1995 shows that some squid hunt by the light generated by their stimulation of light-emitting zooplankton. In the laboratory, *Sepia officinalis* (the common cuttlefish) and *Euprymna scolopes* (a small squid from Hawaiian waters) were observed to capture nonbioluminescent prey much more frequently and efficiently when the water was illuminated by the movement-stimulated *Pyrocistis fusiformis*, a bioluminescent dinoflagellate.

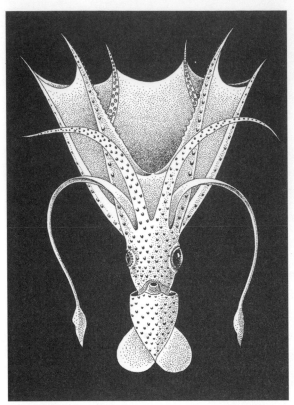

The umbrella squid Histio-teuthis bonnelli is the most common food of sperm whales in some regions. Thousands of beaks have been found in the stomach of a single whale. (Richard Ellis)

It is difficult to envision a feeding sperm whale generating a cloud of bioluminescence and attracting squid to swim into its mouth, but the idea of using light to attract prey in the darkness of the depths is a standard operating procedure for many species of deep-water fishes. Some species are equipped with bioluminescent lures that are used to attract prey, and others with lights that might be used to illuminate the immediate vicinity so that the fish can see in what would otherwise be total darkness. (The squid *Histio-teuthis*—a favorite prey item of sperm whales—has photophores around one of its eyes that act as a searchlight.) Some bioluminescent fishes use lights to attract potential mates, but these lights might also serve a negative function in attracting predators. One of the most unusual demonstrations of bioluminescence has been observed in *Searsia,* a little black bathypelagic fish that can discharge a luminous secretion into the water from a subcutaneous gland just behind its head. (Members of the family are commonly known as tube-shoulders.) *Searsia* appears to be the only fish that bioluminesces outside its body, but the little squid *Heteroteuthis dispar* is also known to emit luminescent ink clouds, and is the only cephalopod known to do so.*

Unlike most other squid, *Taningia danae* does not have two long feeding tentacles, but it does have something more surprising: on the ends of two of its arms are gigantic yellow photophores, the largest light-producing organs in any known animal. These lemon-colored (and lemon-

*In 2009, using remote-controlled cameras at depths up to 5,900 feet off the California coast, researchers from the Scripps Institute of Oceanography, Woods Hole, the Monterey Bay Aquarium, and the University of Göteborg discovered an entirely new genus of swimming worms that can release tiny green "b ombs" of luminescence that burst into light for many seconds and then fade, allowing the three-inch-long worms to escape from predators. The researchers (Osborn et al.) named the species *Swima bombividris,* or "green-bomb swimmer."

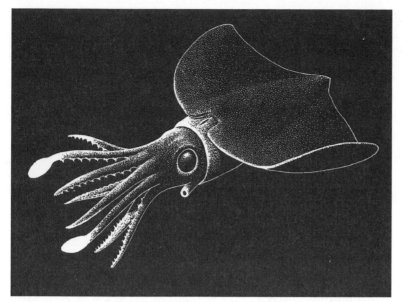

Instead of long feeding tentacles, *Taningia danae* has gigantic yellow photophores on the ends of two of its arms, the largest light-producing organs in any known animal. These lemon-colored (and lemon-sized) photophores can be flashed at will because they are equipped with a black, eyelidlike membrane that can be opened or closed. In addition, this species has claws—likened to those of a cat—on the suckers of its arms. (Richard Ellis)

sized) photophores can be flashed at will because they are equipped with a black, eyelidlike membrane that can be opened or closed. In addition, this species has claws—likened to those of a cat—on the suckers of its arms. The Smithsonian's specimen was dumped into the hold of the Georges Bank fishing boat *Defender* as Captain George Dow emptied his nets. He was 200 miles southeast of Portland, Maine, when his engineer came up to the bridge and asked, "Ever seen a seven-foot squid?"

Not much is known of the biology of *Taningia*, but sperm whales must have some of the answers because most known specimens have been taken from the stomachs of captured whales. (According to a 1993 review by Roper and Vecchione, other specimens have come from the stomachs of sharks, lancetfishes, tunas, wandering albatrosses, and elephant seals.) In 1959, the stomach contents of a whale were examined by Malcolm Clarke at the Canaçal whaling station on the island of Madeira. There were 4,000 beaks, 28 partially digested squid, and "a perfectly intact specimen" of the squid then known as *Cucioteuthis unguiculata,* but now called *Taningia danae.* Clarke (1962) wrote that it was "easily recognized by the large, gelatinous body, the broad fins extending almost to the front of the mantle, the absence of long tentacles (there being only eight arms), and the presence of strong hooks on the arms." The mantle was measured at 140 cm (4.55 feet).

From the records collected by Roper and Vecchione (1993), it appears that *Taningia* has an almost worldwide distribution, having been collected—largely from the stomachs of sperm whales—in the western North Atlantic, off Bermuda, Hawaii, South Georgia, South Africa, Japan, Australia, New Zealand, the Azores—in short, almost everywhere that sperm whales were hunted.* They wrote, "With the addition of the material in this paper, the geographical distribution of *Taningia danae* can be described as truly cosmopolitan with the exception of the polar regions. It occurs in all major ocean basins, in central waters, near oceanic islands, near continental slopes. It occurs in warm, temperate, and sub-boreal waters." A species that is found all over the world, and in such a variety of habitats, is a rarity among cephalopods.

The digestive juices of sperm whales are strong, and the remains of their food items are often corroded beyond recognition. Squid beaks, however, are composed of a tough chitinous material, and they resist digestion much more effectively than the animal's soft parts. A branch of teuthology, emphasized and practiced by Malcolm Clarke, involves the use of squid beaks to identify the species. It is fairly easy to identify the beak of *Architeuthis;* it can be six inches long. That sperm whales do indeed capture and eat giant squid can be seen in Fiscus and Rice's (1974) examination of the stomach contents of sperm whales that were collected off the coast of California from 1959 to 1970: 12 of the 552 whales examined had mandibles of giant squid in their stomachs. But, wrote the authors, "sperm whales may eat *Architeuthis* more often than our records indicate. . . . *Architeuthis* mandibles could be overlooked among remains of *Moroteuthis robustus,* another giant species (although smaller than *Architeuthis*) that is the predominant food of sperm whales off California."

Most of the squid consumed by sperm whales are less than three feet in total length. M. R. Clarke (1962) examined a bull sperm whale in Madeira that had over 4,000 squid mandibles in its stomach. Ninety-five percent of these came from squid that weighed less than 2.2 pounds (1 kilogram); 2.8 percent were from squid between 2.2 and

*In 1980, three specimens, one "in almost perfect condition," were found floating offshore in South Australian waters. When Wolfgang Zeidler (1981) of the South Australian Museum described them, he wrote, "Nearly all known specimens of *Taningia* have been collected from sperm whale stomachs, and it is unusual to encounter them floating at the surface. It is possible that they were regurgitated by sperm whales, and this may be the case for the specimen lacking a head, but the other two were in relatively good condition, and the fishermen estimated that they had died only recently."

22.0 pounds (1 and 10 kilograms); and 2.2 percent came from squid that were estimated to weigh more than 22 pounds (10 kilograms). Compared with other animals that inhabit the oceans in substantial numbers, very little is known about squid. For example, almost all the existing data on the giant squid are derived from stranded specimens or from whole or partial specimens recovered from the stomachs of captured sperm whales. (It was not until 2005 that the first one was seen alive.) In a discussion of the systematics and ecology of oceanic squid, M. R. Clarke (1966) pointed out "that if it were not for the fact that large populations of sperm whales feed mainly on such squid as *Architeuthis, Taningia, Lepidoteuthis, Histio-*

teuthis, and others, these would be considered extremely rare." Other species considered important food of sperm whales are *Dosidicus, Moroteuthis, Gonatus,* and *Calliteuthis.* (Only the generic names are listed here; M. R. Clarke commented [1966], "While a large number of species have been described, revisionists have been struggling to clarify the taxonomic tangle for over half a century.") In other words, many squid species do not have common names, and the scientific names have not always been codified successfully.

The colossal squid, *Mesony-choteuthis hamiltoni,* caught at a depth of 2,500 feet by a Soviet trawler in the Antarctic in 1981. (Alexander Remeslo)

Now *Mesonychoteuthis,* another monster cephalopod, enters the race for the title of biggest squid of all, and therefore the world's largest invertebrate. Though it is known from only a few specimens, current estimates put its maximum length at around 14 meters (46 feet), based on analysis of smaller and immature specimens. Mature adults have not yet been examined, but they probably get even larger. The body of *Mesonychoteuthis* can be as large as or larger than that of *Archi-teuthis,* but its tentacles are shorter. As described by Roper, Sweeney, and Nauen in 1984, *Mesonychoteuthis hamiltoni* is "a very large species" that figures prominently in the diet of Antarctic sperm whales. According to G. C. Robson's 1925 description of the type specimen (based on fragments of two specimens collected from the stomachs of sperm whales taken off the South Shetland Islands), the longest arm was 46.3 inches (118 centimeters) long, and its "hand" was equipped with a series of swivel-based hooks that could be rotated in any direction. The name can be translated as "middle-hooked squid" and refers to the location of the double row of hooks on the middle of each arm, between the basal and terminal ringed suckers. A mature *Mesonychoteuthis* has never been seen, either dead or alive, and almost all of our information comes from the examination of dead and often semi-digested specimens. Its tentacles are comparatively short and thick, and its tail fins are broad, muscular, and heart-shaped. *Mesonychoteuthis* is a cranchid squid, characterized by the fusion of the mantle to the head at one ventral and two dorsal points, and photophores on the ventral surface of the prominent, sometimes protruding, eyes. Holy calamari! The eyes of *Mesonychoteuthis* light up!

In early 2003, a specimen of *Mesonychoteuthis* was retrieved virtually intact from the surface of the waters off Macquarie Island, about halfway between Tasmania and Antarctica. The carcass was examined by Steve O'Shea, of New Zealand's Auckland University of Technology, who commented in a BBC report (Griggs 2003), "Now we know that it is moving right through the water column, right up to the very surface and it grows to a spectacular size. . . . We can say that it attains a size larger than the giant squid. We've got something that's even larger, and not just larger but an order of magnitude meaner." O'Shea opined that this species has one of the largest beaks of any squid and also has unique swiveling hooks on the clubs at the ends of its tentacles, which, he said, "allows it to attack fish as large as the Patagonian toothfish and probably to also attempt to maul sperm whales." The specimen has a mantle length of 8.2 feet (2.5 meters), a larger mantle than any *Architeuthis* that O'Shea has ever seen, and he said that this specimen was immature: "It's only half to two-thirds grown, so it grows up to four metres in mantle length." The squid researchers are calling *Mesonychoteuthis hamiltoni* the "colossal squid." "This animal, armed as it is with hooks and beak, not only is colossal in size but is going to be a phenomenal predator and something you are not going to want to meet in the water," said Kat Bolstad, research associate at Auckland University of Technology.

You probably wouldn't want to meet a thousand-pound squid in the water under any circumstances, but it might be less dangerous than some of the other large teuthids, such as *Dosidicus* or even *Architeuthis*. When Rui Rosa and Brad Seibel (2010) compared the metabolic rate of *Mesonychoteuthis* with that of other large squid, they opined that the

At a length of 40+ feet, the colossal squid is one of the world's largest invertebrates and a sometime prey species for sperm whales. (Richard Ellis)

colossal squid—which they identify as "the world's largest invertebrate"—uses "a slow pace of life linked with very low prey requirements (only 0.03 kg of prey per day)," and they argued that "the colossal squid is not a voracious predator capable of high-speed predator-prey interactions. It is, rather, an ambush or sit-and-float predator that uses the hooks on its arms and tentacles to ensnare prey that unwittingly approach." As with *Architeuthis*, the interactions of *Mesonychoteuthis* with sperm whales are likely to be rare, and not so much a battle of the giants as a struggle by the squid to avoid being swallowed by the deep-diving whale. The huge eyes of *Mesonychoteuthis*—at 11 inches across, the largest eyes of any living creature—are probably helpful for spotting the fishes it hopes to entangle in its hook-studded tentacles, but they also might be particularly useful for spotting hungry sperm whales intent upon dive-bombing them. (Rosa and Seibel call this "predator detection and avoidance.") The colossal squid's low metabolic rate appears to be a function of the deep, dark, cold Antarctic waters in which it lives.

However it is accomplished, sperm whales can locate and capture enough prey items to sustain themselves. But there may be a price to be paid for all this deep diving. When Michael Moore and Greg Early (2004) examined the bones of adult sperm whales, they found substantial evidence of bone death (osteonecrosis), with more erosion and pitting in the larger whales. They concluded:

It therefore appears that sperm whales may be neither anatomically nor physiologically immune to the effects of deep diving. This opens the question of decompression issues constraining surfacing behavior and implies that they and probably other cetaceans may be open to acute embolic injury if forced to surface rapidly. The recent description of acute decompression-like sickness in beaked whales exposed to military sonar may therefore reflect acute nitrogen embolism resulting when decompression sickness avoidance behavior, such as the dive traces previously described in sperm whales, is overridden by extended surfacing.

A study published in 2009 by Hooker, Baird, and Fahlman suggested that beaked whales that surface too quickly from their deepwater feeding dives might get the bends. By means of data gathered from three species (Cuvier's beaked whale, Blainville's, and the northern bottlenose), the researchers observed that under the pressure of a deep dive, the nitrogen in the blood of the whales can remain in the bloodstream as bubbles, and if not allowed to dissolve, it can obstruct blood vessels, causing serious cramps and pain; but if the bubbles obstruct the blood flow to the heart or the brain, they can cause death. (Human divers who have been at depth for any amount of time have to decompress, either by waiting at certain stations for a prescribed amount of time or, in an emergency, by entering a decompression chamber.) For many years, it was believed that deep-diving whales such as sperms and bottlenoses could not get the bends because they have relatively small lungs, and at depths of 300 feet or more, their lungs collapsed completely, forcing the oxygen into the muscle tissue or into rigid respiratory dead spaces.

With the exception of the large and medium-size bulls, the rest of the sperm whale population remains year-round in tropical and temperate waters. We do not know what causes the expulsion of immature males from the mixed schools, but we do know that colder water holds more dissolved oxygen, which is what fish and squid breathe. By no small coincidence, the area of the world ocean north of 40°N and south of 40°S is believed to be the area of greatest abundance of the cephalopods on which sperm whales feed (Kondrakov in Berzin 1972). What is known about the ranges of most squid species comes mostly from commercial squid fishers in places like the Antarctic, eastern Canada, California, Peru, Chile, Argentina, and, not surprisingly, Japan. As Boyle and Rodhouse noted in *Cephalopods: Ecology and Fisheries* (2005), "Clarke (1980) has estimated that over 320 million tonnes of squid are consumed annually by sperm whales"—a figure 300 times larger than the total cephalopod catch by humans, and more than three times the total global fisheries catch by all commercial fishers. They wrote, "By far the greatest consumption of a commercially exploited cephalopod by a predator is *Dosidicus gigas* [the Humboldt squid], fed on by sperm whales. R. Clarke et al. (1988) estimated that when sperm whales were at, or below, the level of maximum sustainable yield between 1959 and 1961, the minimum consumption of this squid was 6.7 million tonnes per year and could have been as high as 20.1 million tonnes."

In their detailed analysis of the feeding habits of sperm whales off the coasts of Chile and Peru (in the Humboldt Current), Clarke, Aguayo, and Paliza (1968) came to the conclusion that the whales "feed practically exclusively on *Dosidicus gigas.*" They examined the stomach contents of 2,403 sperm whales and found that 99.4 percent of the remains were from *Dosidicus.* Such a result is at odds with most other studies because dependence on a single species has not been demonstrated in any other populations. Sperm whales are much more general in their eating habits, and with this questionable exception, they have never been shown to eat only one species of squid. In his commentary on the findings of Clarke et al., Hal Whitehead (2003) wrote that "the beaks of smaller squid found by Malcolm Clarke and colleagues (1976) in the Chilean and Peruvian whales . . . actually came from the prey of *D. gigas,* not of the sperm whales—the 'prey-of-prey phenomenon.'"

In a 1959 article he called "Hunting Sea Monsters," Gilbert Voss said that he did not consider the giant squid *Architeuthis* to be the powerful monster of marine mythology: "The funnel is flabby, the valve is weak and the locking cartilages are mere shallow grooves and ridges. Even the fins are but flimsy narrow bordering flaps, of little or no use as a means of propulsion. Most important of all, the great paired dorsal nerve axons, which control the movements of the body wall or mantle, are missing." He further speculated that "*Architeuthis* is not even the largest of the squid, that there might be a giant squid that could grow to a length of 50 to 60 feet, [and that] they in turn could prey upon and search out the sperm whale. They would be the most powerful fighting machines the marine world has ever produced, and there is no reason to believe they cannot exist." Voss's candidate for the squid world's most "powerful fighting machine" was *Dosidicus,* the Humboldt squid, which

Although better known as a bird painter, Francis Lee Jaques did this illustration of a sperm whale hunting large squid, probably the Humboldt squid, *Dosidicus gigas*. (American Museum of Natural History)

he described as "rulers of their domain, of quite a different nature from *Architeuthis*." "Their bullet-shaped bodies," he wrote, "are heavy and strong, with powerful jet funnels and large fins. Their arms and tentacles are massive and strong, and with their beaks they can bite oars and boat hooks in two and eat giant tuna to the bone in minutes."

It is not the largest of the squid, but *Dosidicus* is by far the largest species caught commercially. It can have a mantle length of four feet (1.2 meters) and a total length, with tentacles, that may reach 10 feet. In the Humboldt Current, Peruvians and Chileans fish for the squid they call *jibia gigante;* and in the Gulf of California, between mainland Mexico and the Baja Peninsula, there is another large fishery for

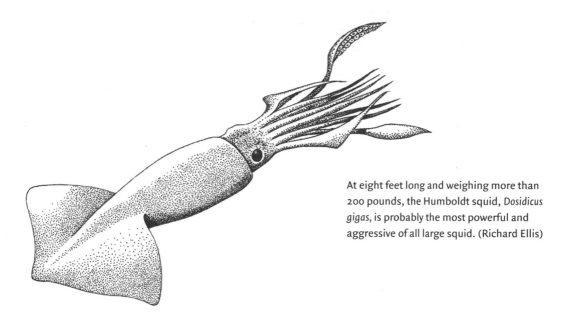

At eight feet long and weighing more than 200 pounds, the Humboldt squid, *Dosidicus gigas*, is probably the most powerful and aggressive of all large squid. (Richard Ellis)

this species. Known in Baja as *calamar gigantes* or *diablos rojas* ("red devils"), *Dosidicus* is a top predator, preyed on only by sperm whales, which probably cannot catch these fast-moving cephalopods but must debilitate them with focused sound blasts. There are only a few squid species larger than *Dosidicus*, and all of them serve as menu items for sperm whales.*

Within its North Pacific range, *Moroteuthis robustus* is known as a favorite food of sperm whales. We know this because the same area of the North Pacific was the favorite hunting ground of Japanese and Russian sperm whalers in the mid-20th century, and they had more than ample opportunity to examine the stomach contents of the whales they brought aboard their factory ships. In a study of the squid retrieved from the stomachs of sperm whales of the Bering Sea and the Gulf of Alaska, Okutani and Nemoto (1964) found that *Moroteuthis* was the favorite food of sperm whales in those waters. They wrote, "Because of its huge size, reddish variegation and rippled skin, this species is easily

*In February 2002, a New York City seafood restaurant called Esca claimed to have giant squid on the menu. Notes on the menu claimed that this was the fabled *Architeuthis,* the submarine attacker from *Twenty Thousand Leagues Under the Sea,* the 60-foot-long monster with eyes as big as automobile hubcaps, and so on. I went to the restaurant and ordered a giant squid appetizer, which comprised a small grilled piece with a drizzle of extra-virgin olive oil. The lack of ammonia convinced me that it was not *Architeuthis,* and when I asked the chef where it had originally come from, he admitted that it had come from Mexico. It was, of course, a filet of *Dosidicus.*

discriminated at the field observation on board the factory ship. The Pacific giant squid is believed to live near the bottom at depths that range from 200 to 600 meters (650 to 2,000 feet), although Hochberg and Fields (1980) report that it is 'occasionally seen swimming at the surface or stranded at the surf line.' "The stomachs of sperm whales taken off California and British Columbia," wrote Dale Rice in 1978, "contain mostly the large squid *Moroteuthis robustus*. I have measured squid specimens that were up to four feet five inches from the tip of the tail to anterior edge of the mantle and eleven feet four inches to the tip of the tentacles."

Wherever sperm whales are found, they find squid to eat. They also eat various kinds of fish, crustaceans, jellyfish, and other marine creatures, but it is clear from studies of stomach contents that in most areas of the world the principal prey of sperm whales are cephalopods (Kawakami 1980). The newly famous colossal squid (*Mesonychoteuthis*) is known only from Antarctic waters, and because of its size, it might be sought by the bull sperm whales that cruise the Southern Ocean while waiting for the siren call of receptive females to the north. Other Antarctic species identified in Malcolm Clarke's 1980 report "Cephalopoda in the Diet of Sperm Whales of the Southern Hemisphere and Their Bearing on Sperm Whale Biology" were *Kondokovia longimana, Moroteuthis knipovitchi,* and *Gonatus antarcticus.*

From the earliest days of industrial whaling, people have been more than a little curious about what sperm whales ate (and how they ate it), so wherever whaling fleets convened in pursuit of sperm whales, the stomach contents of the whales were of more than passing interest. Naturally, different regions produced different results. In the North Pacific, the whales favored *Moroteuthis;* in the North Atlantic (the Azores and Iceland), the Octopoteuthidae (eight-armed squid) predominated in the whales' diet; and in New Zealand waters, the favorite food of the whales—making up a resounding 89.7 percent of stomach contents—is the family Onychoteuthidae, the so-called hooked squid, characterized by hooks on the tentacular clubs. (The Antarctic species *Kondokovia* and *Moroteuthis* are onychoteuthids.) The lives of squid and the lives of sperm whales are so intricately intermingled that it is virtually impossible to write about the life history of one part of this equation without including a detailed discussion of the other.

Sperm whales are also known to eat various fishes, including barracuda, albacore, sharks, skates, and rays. Spotters in airplanes for the

Union Whaling Company of Durban, South Africa, reported a school that "appeared to have a large school of fish ahead, which they followed. Flock of forty to fifty birds diving constantly just ahead" (Gambell 1968). There is a record of a sperm whale with an 8.2-foot (2.5-meter) basking shark in its stomach (Clarke 1956), and Norman and Fraser (1938) mentioned "an authenticated report of a whale in whose stomach a 10-foot shark was found intact." Some of these prey species, such as barracuda and albacore, are swift midwater inhabitants, whereas the rays and skates are bottom dwellers. Andrews (1916) found the remains of "several enormous spiny lobsters" in the stomach of a sperm whale. The habits of the larger squid are poorly known, but most squid are powerful swimmers, capable of quick, darting motions. (Malcolm Clarke [1966] attributed our lack of knowledge of the giant squid "to our complete inability to catch these active animals in the open ocean.") We cannot catch the giant squid, but the sperm whale can, and it can also catch a variety of other food organisms. The sperm whale does not appear to be a particularly agile swimmer—certainly not in the class of an albacore.

Skates and rays are bottom dwellers, and because the sperm whale is known to be an inhabitant of deep waters, one must assume that the whale was somehow feeding on the bottom at significant depths. We have some idea of how deep this species can descend from an unusual source: transoceanic cable companies. Heezen (1957) recorded 14 instances of sperm whales entangled in deep-sea cables, six of which were in about 500 fathoms (3,000 feet), and one, among the deepest dives known for any mammal, was at 620 fathoms (3,720 feet or 1,135 meters).* In most cases in which the whale is identifiable, the cable had been wrapped around the lower jaw of the whale, indicating "that sperm whales often swim along the sea floor [and] become entangled while swimming with their lower jaw plowing through the sediment in search of food."

In 1969, Malcolm Clarke was observing sperm whales from the spotter plane of the Union Whaling Company of Durban. Two whales were sighted; they dived; and when they surfaced, they were shot by the catcher boats. When one of these whales, which had been submerged for 82 minutes, was cut open, its stomach was found to contain two small sharks of the genus *Scymnodon,* known bottom dwellers. "It is difficult to avoid the conclusion that these whales probably dived to the bottom at a far greater depth than we have hitherto thought likely," he wrote in 1976. The depth of the ocean at the location where they dived is in excess of 1,746 fathoms, or 10,476 feet. A dive of almost two miles

---

*Nishiwaki (1972) mentioned a "similar incident . . . on a cable between Lisbon and Malaga which had been set at a depth of 2,200 m [7,216 feet]," but gave no reference to corroborate this figure. The source may have been Heezen's remark about a one-pound lump of "amorphous fish tissue, highly odoriferous," that was hauled up from a depth of 1,200 fathoms. There is no indication in Heezen's report that this material was from a whale, and the captain who hauled up the cable believed it was from an octopus.

surely represents one of the most spectacular accomplishments of an already amazing animal.

Although they obviously prefer squid, sperm whales will eat various other objects, some intentionally and some accidentally. It is reasonable to assume that sperm whales also feed in what might be considered a "normal" manner, chasing down and eating fish and squid at various depths. Although this explains neither the presence of stones and other nonfood items in the stomach nor a well-fed whale with a nonfunctioning lower jaw, it probably accounts for the greater proportion of the eating habits of most individuals. However it is accomplished, sperm whales eat a lot of food. Sergeant (1969) estimated that they consume about 3 percent of their total body weight daily; for a whale of 30 tons, this would amount to some 1,800 pounds (810 kilograms) of food per day.

From the debilitating sonic booms to the pincerlike jaws and the myoglobin in the muscles, it is obvious that sperm whales were designed to hunt and capture free-swimming squid in the depths of the ocean. Still, this search-and-destroy method requires a lot of energy, and if an easier way of feeding became available, it is likely that clever *Physeter* would take advantage of it.

Assume that you're a hungry whale, hanging around the Gulf of Alaska, and your echolocating clicks identify a lot of potential food items, apparently immobile, at a depth of around 350 feet. This sure beats diving to 1,000 feet and chasing a lot of squid around (never mind bone damage from those deep dives), so you drop down to investigate. Hanging on a line, just waiting to be plucked and eaten, are sablefish (*Anoplopoma fimbria*), also known as black cod, the object of a major commercial longline fishery in the area. When their lines are hauled in, the fishermen find missing body parts, crushed tissue, blunt tooth marks, shredded bodies, and lips remaining on hooks, an indication that sperm whales have been feeding on the hooked fish. (Also, sperm whales are often seen surfacing in the vicinity of the longline boats.)

The principal investigator of this phenomenon is Jan Straley, an assistant professor of biology at the University of Alaska Southeast in Sitka. In a 2004 radio interview, she said, "Sperm whales are cueing in on some aspect of fishing behavior and then waiting for these fish to be brought up to the surface on longline gear, and then, at some point in the water column the whales take the fish off the longline gear, off the hooks. . . . We don't know how they're taking fish off the gear.

The distinctive shape of sperm whale flukes strongly suggested to Jan Straley that it was this species that was plucking sablefish from Alaska longlines. (SEASWAP © NOAA scientific research authorization permit number 1700-473 issued to J. Straley)

It's amazing to think that they could even do it when you look at their head. They have that great big head and that tiny lower jaw." But at the November 2009 Sitka Whalefest conference, held in Alaska, Straley showed a video shot in 2006 that contained images of a whale at a depth of about 350 feet mouthing a longline with black cod attached. Although the action was not filmed, the whale managed to separate the fish from the line. If it actually ate the fish, it was out of camera range.

The same opportunistic technique has been observed in the Southern Ocean fishery for Patagonian toothfish (aka Chilean sea bass) according to Ashford et al. (1996): "Interestingly," they wrote, "no partially eaten fish were recovered on the lines during hauls with which sperm whales were associated, although hooks were missing. If the sperm whales were removing fish, they must be doing so by plucking, rather than biting. Such behavior would be consistent with the anatomy and size of the mouth of the species and its tendency to consume prey intact." Both the toothfish and sablefish fisheries are enormous, and occasional sperm whale depredation is not having a marked effect on the overall catch. It is conceivable, however, that sperm whales, with their sophisticated socialization and communication skills, could let the other whales know about the sablefish buffet. The technique has spread to the sperm whales of Vancouver, who are plucking halibut off longlines, and killer whales already know about it, having depredated "at least 20% of the bottom longline sets [and] in Price William Sound, an estimated 25% of the total catch is lost to killer whales" (Yano and Dahlheim 1995).

Some of the rarely seen beaked whales—such as Cuvier's beaked whale and Sowerby's beaked whale—have increasingly been found dead with stomachs filled with plastic bags. The plastic bags are eaten by the whales and become lodged in their stomach and intestines, which at best interferes with normal food intake and at worst completely blocks the digestive system and can lead to death. We cannot know whether sperm whales mistake foreign items for food, but we know that the oceans are littered with plastic garbage, from the surface to the depths. Beaked whale species in particular are highly susceptible to swallowing plastic bags because to the whales, these objects likely strongly resemble squid, their usual target prey. Other species of large whales, which take large mouthfuls of water during feeding, also take in plastic bags by accident and are also at risk. In August 2000, a 26-foot Bryde's whale stranded close to central Cairns in north Queensland and died on the beach. A necropsy found that the whale's stomach was tightly packed with plastic. The baleen whale had swallowed supermarket bags, food packaging, three large sheets of plastic, and fragments of garbage bags. At Tomales Point near Point Reyes, California (north of San Francisco), in August 2008, a 51-foot-long sperm whale washed up dead on the beach, and the necropsy, performed by veterinarian Frances Gulland of the Marine Mammal Center in Sausalito, revealed nearly 450 pounds of fishing net, mesh, braided rope, plastic bags, and a plastic comb. Floating plastic rubbish can too easily be swallowed, and because it cannot be digested or passed, it stays in the whale. It is likely that many whales are killed by consuming floating garbage, but most of them probably die at sea and are never examined.

If the garbage bags don't get you, the pollutants will. Roger Payne, who with his wife, Katy, first studied the right whales of Patagonia and first described and recorded the songs of the humpback, set out a circumnavigation of the world specifically to test the blubber of sperm whales for the presence of persistent organic pollutants (POPs). For almost six years (September 7, 1999, to August 17, 2005), aboard the 91-foot-long (28-meter) floating laboratory *Odyssey*, Payne and his crew of 12 scientists and educators sampled everything from oceanic debris and jellyfish to albatrosses and hammerhead sharks. The primary mission of the voyage was described on the Ocean Alliance Web site (http://oceanalliance.org/voyage/voyage_science.html) as follows:

Today, there are major concerns about the ubiquitous accumulation of persistent organic chemicals in man and animals. Concerns include issues of immunosuppression, neurological function, reproduction and cancer. There are many data that point to possible problems at local and regional levels, but there is a lack of globally integrated data that allow a consistent appraisal of exposure and risk in a manner that covers entire ocean basins. We propose to measure concentrations of polynuclear aromatic and halogenated aromatic hydrocarbons (PAHs and HAHs) in marine mammals, especially great whales from each of the major ocean basins. We will also look for changes in macromolecular biomarkers of exposure including foreign compounds bound to DNA (adducts) and cytochrome P4501A induction. These data will allow a better understanding of the risks associated with chronic exposure to important toxicants.

In a 2005 article in *Science*, Dan Ferber wrote, "Early results are in from the first-ever global survey of toxic contaminants in marine mammals—and they're not pretty. Sperm whales across the Pacific, even in midocean areas thought to be pristine, are accumulating humanmade chemicals called persistent organic pollutants (POPs). DDT was the most common pollutant, followed by polychlorinated biphenyls." The results and experiences of the voyage of the *Odyssey* were broadcast on PBS and streamed regularly to schools and Web sites around the world. Speaking of contaminants on the PBS program *Living on the Earth* three years after the voyage had ended, Payne said:

Many of the substances that we're concerned with are fantastically insoluble in water. They are highly soluble in fat. So what happens is that they end up in the ocean water, anywhere in parts per trillion or parts per quadrillion, so hugely diluted. But as soon as they get into fats they end up in very, very high concentrations because the fats can hold lots of them. The trouble is the animals don't have any way of dealing with them so they store them, and then they get passed on when that animal gets eaten by some other animal. So here on your plate is a pound of swordfish. It took a million pounds of diatoms to create that one pound—not the whole fish—just that one pound. A million pounds is 500 tons, so it took 50 ten-ton truckloads of diatoms to make that one pound of swordfish. You take all those trucks and you park them along a row; it's about ten blocks long, and to the end of that row you attach your liver, and with it you detoxify that entire line of trucks. And that's what you do when you eat a pound of swordfish. And then maybe tomorrow you have another pound. I adore swordfish. I would do anything to have a piece of swordfish except eat it.

Just after midnight on March 24, 1989, the 987-foot-long super-tanker *Exxon Valdez* ran aground, dumping 11 million gallons of crude oil into Alaska's Prince William Sound, before the *Deepwater Horizon* blowout in 2010, the largest oil spill ever to occur in the United States. Joseph Hazelwood, captain of the *Exxon Valdez,* had handed control of the ship to third mate Gregory Cousins when the ship rammed Bligh Reef, tearing a gaping hole in the cargo tanks. Winds and shifting tides spread the oil over 10,000 square miles along the Alaska Peninsula, qualifying it as one of the worst ecological disasters in history. The oil moved for 1,500 miles along the coastline of Alaska, contaminating portions of the Kenai Peninsula, lower Cook Inlet, and the Kodiak Islands. High winds blew the oil slick onto the shore, creating havoc with living creatures. The actual total will never be known, but it has been estimated that at least 100,000 seabirds died, along with 5,000 sea otters; 150 bald eagles; hundreds of seals, sea lions, whales, dolphins, and porpoises; and countless fishes. The devastation of the spawning grounds of pink salmon, black cod, and herring destroyed the livelihood of dozens of fishing communities.

On April 20, 2010, the BP *Deepwater Horizon* offshore drilling rig blew up in the Gulf of Mexico, killing 11 men and spewing oil into the Gulf of Mexico at the rate of at least 700,000 barrels a day, putting it in the running for largest oil spill in history. Much of the oil from the damaged well 5,000 feet down has floated to the surface, but underwater plumes, one of which has been estimated at 10 miles long, three miles wide, and 300 feet thick, have remained below the surface. The effect of these subsurface rivers of oil on marine life has not been studied—this is a brand-new, totally unexpected phenomenon—but however it plays out, it will not be good. The combination of oil spills and wildlife has virtually defined ecological disaster.

The Gulf of Mexico is home to an estimated population of 1,600 to 1,700 sperm whales, said Dr. Randall R. Reeves, a biologist with Okapi Wildlife Associates, Quebec, Canada, and an advisory curator of the New Bedford Whaling Museum. He noted that sperm whales "move through the water column to great depths, and they spend long periods at the surface 'catching their breath.' Exposure to oil is inevitable, and it is very hard to imagine that such exposure will not be harmful to both the whales directly and to their prey." Whales and dolphins breathe air at the surface, and trying to inhale through an oil slick will transport the gooey oil directly into the lungs of the whale. Petroleum in the lungs is toxic. All the fish, squid, crabs, shrimp, and other midwater marine life that breathe water through gills will also be poisoned by this midwater floating death trap. Dead, oil-soaked sea turtles and bottlenose dolphins have already been found washed ashore in Louisiana. Much of the oil will also end up trapped in big eddies—like the infamous Pacific Garbage Patch or the Sargasso Sea—which is where sea turtles and other ocean life congregate. Most of those mortalities will never make their way to shore to be counted, so we really have no idea of the extent of the damage to marine life. Below 175 feet, the velocity of this current drops off dramatically, giving those oil particles more time to clump and sink to the bottom or wash ashore. What will happen when the oil hits those communities that live on the sea floor, such as *Lophelia* coral, remains unknown. There has never been a substantial oil leak a mile down.

On September 19, 2010, BP effectively killed the Macondo Well five months after the April 20 explosion. It will be decades before we know the effect of this drilling disaster on the Gulf of Mexico—and the wildlife. Even though the gushing well is now plugged, this catastrophe will not abate. The oil now polluting the Gulf of Mexico cannot be removed, and it has therefore become a new and deadly component of the once rich ecosystem that was the Gulf of Mexico. This is not a natural disaster, like Hurricane Katrina in 2005. It is an accident caused by deregulation, human carelessness, and disregard for the environment. Its effects on marine life will be felt for centuries. On June 16, 2010, almost two months after the *Deepwater Horizon* blowout, a dead subadult sperm whale was spotted floating in the Gulf of Mexico, some 70 miles south of the spill site. Scientists are planning to analyze samples from the floating carcass to determine the cause of death. In a development possibly related to the BP oil spill, a newspaper report on June 23, 2010, stated that "two pygmy sperm whales died after beaching themselves on Florida's East Coast."

From 2001 to 2005, Bruce Mate of Oregon State University, one of the country's foremost whale researchers, led a study of the sperm whales of the Gulf of Mexico. They tagged 57 individual whales, making this the most intensively studied population of sperm whales in the world. After the explosion of the *Deepwater Horizon* rig, the National Oceanic and Atmospheric Administration launched several studies to learn the effects of the oil on the region's resident sperm whale population, with Mate and Christopher Clark (an acoustician at Cornell University) as lead investigators. The sperm whales' high metabolic rate means they have to eat all the time, and they will be diving regularly through the different layers of oil. (Of course, they have to inhale at the surface, and trying to breathe through a layer of oil might be problematic.) The researchers will lower recording devices to the bottom, at depths ranging from 200 to 1,500 meters, to ascertain where the whales go, and what they do when they get there. How will this undersea gusher affect the whales? Will they be able to avoid the oil in the water? Will they be able to feed through the oil?

VIII

Making Contact

HUMAN INVOLVEMENT with the sperm whale goes back at least as far as biblical times, when Jonah was swallowed by a "great fish." Of the larger whales, only the sperm has a gullet large enough to accommodate a person, and the great fish is usually said to have been a whale. There are those, however, who believe the great fish of the biblical account was a shark. In his *Life of Sharks,* Paul Budker not only discounts the likelihood that a whale was the swallower, but actually identifies the shark by species. Basing his arguments on the writings of the French naturalist Guillaume Rondelet (1507–1566), Budker suggests "that the impossibility of passing a man down the narrow throat of a whale led Rondelet to search for a marine animal capable of swallowing such a large prey and bringing it up whole later on. *Carcharodon,* the white shark, was not a bad choice." Further on, he concludes that "one should, therefore, substitute 'shark' for 'whale' in the story of Jonah, and even, for the sake of complete accuracy, *Carcharodon carcharias.*" In his original description of the great white shark, published in his *Systema Naturae* of 1758, Linnaeus too said that the great white is an enormous fish, and probably the kind that swallowed Jonah.

There is no doubt that a sperm whale is capable of swallowing an object the size of a grown man; there are reports of whales vomiting up pieces of squid "half the size of a whaleboat" after having been harpooned (Ashley 1926), so the whale ought to have no trouble with a puny *Homo sapiens.* There have been many tales of whalers swallowed by whales—most of them apocryphal—but perhaps the most persistent is the story recounted by Edgerton V. Davis, in which a young sealer fell into the sea off St. John's, Newfoundland, and was swallowed up

Many people got their first view of a sperm whale from the 1940 Disney cartoon *Pinocchio*. The bulbous, big-lipped creature at the bottom of the poster could not have been more incorrect. Monstro the whale is a malevolent creature that swallows everybody, sneezes them out, and, breathing fire like a dragon, chases them all over the ocean. (New Bedford Whaling Museum)

by "a huge sperm whale," which was then killed. When the whale was opened, the man was found, "badly crushed . . . and partially digested." Scheffer (1969) was skeptical of this story, claiming it strange that Davis waited half a century to tell his tale: the event was supposed to have happened in 1893 or 1894, but was not published until 1946. In *Whales and Whaling* (1959), Paul Budker reprinted Davis's story in full and found it an "accurate scientific account."*

It was in 1577 that the first engraving of a stranded whale appeared in print. By the turn of the 17th century, more whales had stranded on European coasts, and with the heightened interest in popular science, more engravings appeared. Perhaps because the whales preferred the coasts of the Netherlands, or perhaps because the Dutch had a particular interest in stranded whales, the majority of the early illustrations

*In the 1940 Disney cartoon *Pinocchio*, Monstro the whale—a gigantic amalgam of sperm and baleen whales, with ventral pleats and plenty of teeth—swallows Gepetto the puppeteer (and Gepetto's ship, cat, and goldfish) and then Pinocchio, who escape by building a fire inside the cavernous belly of the whale, causing Monstro to sneeze them out. His mouth aflame, the terrifying Monstro chases Gepetto and Pinocchio until he smashes into the rocks and they wash up on the beach. Disney's whale probably inspired more fear of whales (and misinformation about their dangerous propensities) than any whale since Moby Dick.

Three Beached Whales, a 1577 engraving by the Dutch artist Jan Weirix. Although these are obviously supposed to be sperm whales, they have been drawn with two blowholes. There are at least nine more whales in the surf waiting to strand. (New Bedford Whaling Museum)

of whales were the work of Low Country artists. In these elaborately detailed drawings, the good burghers of Holland are often seen perched on the carcass, standing around in fashionable attire, or occasionally carrying off a bucket of what was probably whale oil.*

The North Sea coast of Holland would appear to be one of those places (others of note are in New Zealand and Cape Cod) where whales

---

*Sperm whales, thought to be extinct in the Mediterranean because so many were caught and drowned in drift nets, have returned to the waters of southern Italy and Sicily. As if to prove it, nine whales beached themselves on an Italian beach at Puglia in December 2009. Seven of them died on the beach, but two managed to return to deeper water and were not seen again. A necropsy performed on December 18, 2009, by Giuseppe Nascetti of Tuscia University at Foggia revealed that the whales had died from ingesting plastic bags, tin cans, and pieces of rope, adding another to the grim list of anthropogenic explanations for cetacean strandings.

When a sperm whale beached itself at Katwijk in Holland in 1598, Hendrik Goltzius drew it for posterity. (New Bedford Whaling Museum)

seem to strand. From 1531 to approximately 1690, some 40 whales of assorted species beached themselves on these shingled coasts. Most of them seem to have been sperm whales. With its huge head, its mouthful of ivory teeth, and—in what appear to be a majority of the cases—its male genitalia prominently exposed, the dead whale must have been a wonder of wonders to the Dutchmen who came to view these beached monsters. It would be another half century before the whalers of Rotterdam and Delft would head for the icy seas of Spitsbergen, where they would hunt a totally different creature, the Greenland right whale.

One of the best documented of these extraterrestrial aliens was a 54-foot bull sperm whale (*potvisch* to the Dutch) that was discovered floundering helplessly in the shallows of Berckhey in February 1598. When the whale died, its carcass was sold for the oil, but its fame lies more in its portrayal than in its products. The Berckhey whale was drawn by the artist Hendrik Goltzius, and this drawing has appeared in countless versions, often accompanied by a descriptive text that marvels at its leviathanic dimensions. In later years, more whales would strand on these beaches and be immortalized by Dutch engravers, but the Goltzius illustration, repeatedly reproduced for 200 years, has probably been used more often than any other cetacean depiction before or since.

First seen thrashing feebly on the rocky foreshore between Beadnall and Seahouses (Northumberland) in January 2010, a 30-foot-long sperm whale stranded and died there. The half-grown male died of unknown causes, but its stomach was empty, and it was assumed that the whale became separated from a pod while hunting for food and starved. The coast guard had to guard the carcass to keep locals from removing the teeth, which they believed to be valuable. Although the northwestern quadrant of the North Sea is somewhat unfamiliar territory for sperm whales, to the south the shores of Denmark, Germany, the Netherlands, and Belgium have seen stranded cachalots for at least four centuries. Indeed, these events are so common that in 1995, a symposium was convened at Koksijde, Belgium (Jacques and Lambertsen 1997), to discuss sperm whale strandings in the part of the southeastern North Sea known as the Wadden Sea (*Waddenzee* in Dutch, *Wattenmeer* in German). This shallow region is composed of tidal flats and wetlands—the perfect environment for sperm whale strandings.

By the early 18th century, the first sperm whale had been killed by a Nantucket whaling captain named Christopher Hussey. This event did not set off an immediate and intensive quest for sperm whales because right whales were still plentiful in New England waters, and the hunters would not have the means or the incentive to pursue the sperm for another 40 years. Around 1750, two innovations vastly broadened the opportunities available to American whalemen: the development of onboard tryworks and the invention of the spermaceti candle by a Newporter, Jacob Rodriguez Rivera (Kugler 1976). Earlier whaling efforts, such as the Dutch and English hunting of the Greenland right whale, involved killing the whales and then towing them to a shore station for processing. (The Spitsbergen village of Smeerenberg, established by the Dutch in the 17th century, was centered on the tryworks, where the blubber of the whales was rendered into oil.) With the development

PURE SPERM
SEWING MACHINE OIL

MANUFACTURED BY

WILLIAM F. NYE.

NEW BEDFORD, MASS.

Sperm oil was the all-purpose lubricant of the 19th century. (New Bedford Whaling Museum)

of tryworks on board the whaleships, the crew could now process the animals wherever they caught them and thus remain at sea until their holds were filled. At first the spermaceti oil was combined with the oil from the blubber and known collectively as sperm oil, but when Rivera invented the smokeless spermaceti candle, the two types of oil were separated. A large bull might have upwards of a ton, or 10 barrels, of spermaceti oil in the case, and the blubber, rendered down, might yield as much as 100, but such large amounts were rare.

At least since Aristotle, men have known that whales were mammals. They breathe air in and out of lungs, they give birth to live young, they nurse them with mother's milk, and they maintain constant body temperature. At some time in their lives, all mammals—even whales and dolphins—have hair. Some fish, on the other hand, give birth to live young, but most species lay eggs. Fish breathe by extracting dissolved oxygen from the water by means of gills, and although most fish have scales, no fish has fur or feathers. A fish moves its tail from side to side; a whale or dolphin moves its tail up and down. Most fishes adopt the temperature of the water they live in, a characteristic that classifies them as cold-blooded. It should therefore be relatively easy to differentiate a whale from a fish, even though they are largely the same shape, but for some reason, the activity of hunting whales was known as a *fishery*. Perhaps it was because fish live in the water, and most water-based hunting targets one kind of a fish or another. In any event, many books about whaling refer to a particular fishery, such as the Greenland fishery, the American fishery, and even the sperm whale fishery. In chapter 41 of *Moby-Dick*, simply titled "Moby Dick," Melville wrote,

> Yet as of late the Sperm Whale fishery has been marked by various and not infrequent instances of great ferocity, cunning and malice in the monster attacked; therefore it was, that those who by accident ignorantly gave battle to Moby Dick; such hunters perhaps, for the most part were content to ascribe the particular terror he bred, as it were, to the perils of the Sperm Whale fishery at large, then to the individual cause.

The first of the New England whalers hunted right whales and probably humpbacks, and it was only a fortuitous accident that spared the last of these baleen whales. It is said, in what may be an apocryphal story, that around 1712, a Nantucketer named Christopher Hussey was blown offshore while hunting right whales and came upon a school of large, bluff-headed whales with forward-angled spouts. Hussey managed to harpoon one and towed it back to Nantucket. It was examined by the curious populace and seen to have peglike ivory teeth where the right whale had baleen plates, as well as a vast reservoir of clear amber oil in its nose. This event, whether it happened exactly this way or not, marks the beginning of the sperm whale fishery, and although the early phases were conducted offshore in the North Atlantic, the whalers soon realized that the sperm whales were to be found in greater numbers elsewhere. The stalwarts of Nantucket, New Bedford, Sag Harbor, Mystic, and numerous smaller ports outfitted their ships for three- or four-year voyages to the ends of the earth in the hunt for the great sperm whale, the animal Herman Melville called "the largest inhabitant of the globe; the most formidable of all whales to encounter; the most majestic in aspect; and lastly, by far the most valuable in commerce."

The Nantucketers dominated the American fishery for about 50 years, but then another group rose to challenge their supremacy. A little Massachusetts village on the Acushnet River achieved a primacy that would soon eclipse Nantucket and every other whaling port in the New World. Around 1760, New Bedford began to send whaling ships to the south. By 1770 the brigs *Patience* and *No Duty on Tea* had crossed the Atlantic, and later, *Rebecca* became the first of the American whalers to double Cape Horn with a full cargo of sperm oil obtained in the Pacific. "Thus began," wrote Scammon in 1874, "the commercial enterprise at New Bedford . . . which has since become, and still is, the whaling metropolis of the world." In 1774 the entire colonial fleet consisted of 360 vessels employing some 9,000 men, and of this total, Nantucket and New Bedford maintained the lion's share.

They cruised the waters of the world, from the poles to the equator, slaughtering every whale they could find, until their holds were filled with the thick oil, and their clothes, decks, and rigging reeked of the smoky smell of the trypot fires. The ships returned to their home ports and unloaded. As soon as they could, they made for the whaling grounds again. It appeared that only a war between nations could interrupt the war between man and whales.

In 1812, many Nantucket whalers en route to gain the sanctuary of their home ports fell to British privateers, and the men and booty were impressed into the service of the British. But despite the privations of war, the Nantucketers, knowing no other business, continued to outfit whaleships for their globe-circling voyages. As they worked the grounds off the coasts of western South America—among the most productive sperm whaling grounds in the world—they now found themselves threatened by a new enemy: Peruvian privateers. Many ships were detained in Talcahuano, Chile, where they had stopped to reprovision before sailing home with their cargo of oil. The Honorable Joel R. Poinsett, who had been sent by the American government to see that the ships were protected, somehow got to lead a troop of 400 Peruvian militiamen against the pirates and, winning the day, gained the release of the ships. Meanwhile, off the Massachusetts coast, British privateers were harassing Nantucket whalers, and they applied for some sort of arrangement whereby their fisheries might be protected. No such protection was forthcoming, and as Starbuck (1878) has written, "the people found the history of their sufferings during the Revolution repeating itself with a distressing pertinacity and fidelity, and they bade fair to perish of starvation and cold."

By 1815, with the treaty of Ghent signaling the cessation of hostilities, the Nantucketers hastened to reestablish their only industry, and within a year, the wharves were again stacked with greasy casks. The fleet had been reduced to 23 vessels by the end of the war; by 1820 there were 72 ships, brigs, schooners, and sloops flying the flags of

Nantucket owners. Although the British tried to assert themselves before 1812, their efforts to break their former colony's monopoly on the whale fishery did not survive the War of 1812. During the war, the American whaling industry was seriously damaged by the British practice of impressing crews and capturing ships. In response, the United States frigate *Essex* roamed the South Pacific recapturing ships that had been taken by British privateers, and virtually swept the seas of British whaleships. (This *Essex* was not the whaleship that became one of the most renowned of all the whaling fleet, distinguished by being stove and sunk by a whale.) By the time the war ended in 1815, the British fleet was in ruins. The beleaguered American whaling industry was resurrected, and entered a period of unmatched prosperity, based partly on the growth of population and economic activity at home, and partly on the needs of the British market for American oils. It was the beginning of the industrial revolution, and the new machinery needed lubricants. By 1833 the American whaling navy numbered 392 ships and more than 10,000 sailors. In another decade, both figures would double.

Despite Britannia's glorious maritime history, it could not compete with the dogged tenacity of the Yankee whalers, and the oceans belonged to the Americans. Under Captain Joseph Allen of Nantucket, the ship *Maro* discovered the rich Japan Grounds in 1820, and in two years' time, there were 30 whalers working there. Fast on their transoms came the whalers, who would discover that sperm whales also enjoyed the warm tropical waters of various islands in the South Pacific. The coasts of Zanzibar, the islands of the Seychelles, the icy coasts of Kamchatka, and even the mouth of the Red Sea were explored by the far-ranging whalers. It was obvious that the Yankees regarded the whole world as their private whaling preserve. By 1846, the year that is generally held to have been the high point of the New England whale fishery, there were a total of 735 ships, brigs, schooners, and barks flying the flags of American owners, with an aggregate weight of 233,189 tons and a book value of over $21 million.

The desire to participate in the whale-oil bonanza was not restricted to New England, and even though we tend to associate whaling with New Bedford and Nantucket, many other towns sent out whaling ships. In Massachusetts, there were Salem, Gloucester, Marblehead, Provincetown, and Edgartown on Martha's Vineyard; in Connecticut, New London, Stonington, and Mystic; and on Long Island (New York), Sag

Wanderer, the last of the square-rigged whalers, set sail from New Bedford on August 27, 1924, and ran aground the next day on Cuttyhunk Island, off the Massachusetts coast. (New Bedford Whaling Museum)

Harbor, Amagansett, East Hampton, and Southampton. Even places not normally associated with whaling joined in. In 1834, the whaleship *Ceres* was sold to a Wilmington concern and set out on Delaware's first whaling voyage. The three-year trip was a failure; it returned home with its holds only 40 percent full. Another ship, the *Lacy Ann,* set out in 1837 for the Pacific, and was successful enough for the Wilmington Company to pay its only dividend in 1840. North Carolina is another state where whales were occasionally taken, but the catch consisted primarily of right whales, which occasionally ventured close enough to shore for the whalers to set out after them in small boats. This fishery began around 1667 with the first settlements, but by the Revolution, the whales were gone and the Carolina fishery ended.

Except for enforced slowdowns during the Revolution and the War of 1812, the American sperm whale fishery flourished for about a century, roughly from 1750 to 1850. It reached its zenith in the decade 1840–1849, when over 126,000 barrels of sperm oil reached American ports (Harmer 1928).* The fishery was in decline by 1850, when, for the

*One barrel equals between 30 and 33 gallons. Scoresby (1820) reckoned 252 gallons to the ton, or almost eight barrels. Estimates for the average sperm whale ran between 25 and 40 barrels per whale, although some whales exceeded this by three or even four times.

first time, sperm oil imports fell below 100,000 barrels. The decline is attributable to the opening of the rich North Pacific and western Arctic whaling grounds for bowheads, as well as the increased expenses of outfitting whaling ships. It is often said that the sperm whale fishery dropped off because of the decrease in the number of whales. (Matthews [1938] wrote that "the decline could only have been due to over fishing.") Other historians, such as Harmer (1928), have attributed the demise of the fishery to the introduction of petroleum. Although the 1859 discovery of oil in Pennsylvania certainly contributed to the extinction of the American sperm whale industry, it occurred some 20 years after the fishery began its precipitous decline.

In *Moby-Dick* a French whaler appears as the *Bouton de Rose,* which Stubb translates as "Rose Bud." Melville tells us that French whalers were known as "Crappoes," from the not particularly complimentary *crapaud,* or "toad." They are, he wrote, "but poor devils in the Fishery; sometimes lowering their boats for breakers, mistaking them for sperm whale spouts; yes, and sometimes sailing from their port with their hold full of boxes of tallow candles, and cases of snuffers, foreseeing that the oil they get won't be enough to dip the Captain's wick into." (Stubb talks the French captain into cutting loose his whales because they are rotten, and as one floats away, the *Pequod's* mate spears a savory lump of ambergris, "worth a gold guinea an ounce to any druggist.") Although they sent ships out until 1868, du Pasquier (1982) believes that "French whalers were no longer an important factor in whaling after 1850."

The earliest entries into the Greenland whale fishery—the Dutch and the British—sailed north, and their enterprise was therefore known as the northern fishery. The hunt for sperm whales, begun (maybe) by a Nantucketer, was pursued all over the world, but largely in the South Pacific and the South Atlantic, and was therefore the southern fishery. The Dutch, whose economy in the 17th century revolved largely around baleen and whale oil, had turned the Greenland fishery into a great success story (until the whales ran out) but did not fare so well in the southern fishery. Following a long tradition of northern whaling, the burghers of the Netherlands made a feeble attempt to join the sperm whale sweepstakes in the Southern Ocean. Encouraged by the endorsement of King Willem I, the Dutch made an attempt to hunt sperm whales in the South Atlantic and the South Pacific, but for a whaling nation, they met with remarkably little success. Their first vessel was the New England–built *Logan,* under the command of Reuben

Coffin, a member of one of the most respected families in Nantucket whaling history. From 1826 to 1830, the *Logan* plied the seas in search of whales, but evidently it found very few, and those it found it could not catch because of the inexperience of the Dutch crew. In 1832 the *Eersteling* ("Firstling") went to sea under the command of Captain H. F. Horneman, who seems to have had no experience whatsoever in the hunting of whales. The results of this venture were almost a foregone conclusion, exacerbated by the discord between the Dutch sailors and the British who had been brought aboard to advise. The *Prosperina* sailed from Rotterdam in 1836, and in two years, it found and killed 19 whales. The firm of Reelfs Brothers sent out the *Anna & Louisa* to the South Seas in 1833, and based on the scantiest reports of good fortune, they bought the old *Prosperina,* refitted it, and named it *Zaidpool* ("South Pole"). As with every other Dutch sperm whaling voyage, these too failed to repay the initial investment. It appeared that whatever whale oil was going to be used in Holland would be delivered there by Yankee whalers.

By the end of the 19th century, almost all the known stocks of right whales and bowheads had been severely depleted. The only whales consistently hunted were the sperm whales, and these were getting harder and harder to find. As with so many aspects of 19th-century life, the industrial revolution intervened in the history of whaling, and individual initiative was overtaken by technological advances. Steam whalers replaced the square riggers, and other oils began to seep into the lighting and lubricating industries.

As early as 1830 an illuminating oil known as camphene was being distilled from turpentine. That it was ill-smelling and volatile did not so much militate against its use as encourage the search for other substances. Some of the more astute New Bedford merchants had guessed at what was coming with the discovery of petroleum, and in 1859, Messrs. Howland, Taber, Delano, Wood, and Hicks (names heretofore indelibly associated with the whaling industry) had erected a factory for the distillation of petroleum. Cottonseed, linseed, and palmseed oils were also being used in soap making, rope dressing, and leather tanning. Whale oil, which had defined the industry and been responsible for the making of history, fortunes, and literature—and the unmaking of thousands of whales—was becoming an anachronism. The age of petroleum was about to begin.

After peaking around the middle of the 1800s, the New England whale fishery began a decline from which it never recovered. When gold was discovered at Sutter's Mill in California in 1848, whaleships experienced the wholesale defection of their crews—and sometimes their captains as well—as soon as they docked in San Francisco. The Civil War saw the sinking of many whaleships and the scuttling of others. Confederate raiders wreaked havoc on the Yankee whaling fleet, and the unforgiving ice of the Arctic crushed 33 whalers in the autumn of 1871. In 1857, the powerful New Bedford fleet had consisted of 324 vessels, but 50 years later, it was reduced to 19 ships and barks, 12 schooners, and a brig. The balance of power had shifted to San Francisco, but there too the handwriting was on the wall, and only feeble attempts were made to revive the economically moribund American whaling industry. (Various whale species would be killed from California land stations during the 20th century, but neither the number of whales killed nor the effects on the local economy were significant.) The cumulative slaughter of various whale species, particularly the sperm whale, obviously contributed to the decline of the industry, as the whalers had to spend more and more time at sea for a steadily declining return.

As collecting the oil became more cost- and time-intensive, the introduction of petroleum as a cheaper substitute for whale oil drove the price even further down. From a high of $1.77 per gallon in 1855, sperm oil had fallen to 40 cents in 1896. Investors were unwilling to spend money on this unremunerative industry. Greasy casks lay rotting on the wharves, and the masts of the whalers rocked in the swell of the harbor, some of them never to sail again. One owner scuttled his vessel in the Acushnet River, and another volunteered a whaleship to be burned as a Fourth of July spectacle.

A rise in the price of whalebone revived California whaling. Steam whalers headed for the Arctic to harvest the bowheads, but this industry too was affected by forward-marching industrial technology, and just as cheaper petroleum substitutes were taking the place of whale oil, spring steel was replacing whalebone for corset stays. The New Bedford whalers stubbornly persevered and continued to send ships and men to sea. However, by the turn of the 20th century, American pelagic whaling was just about over. The last of the square-rigged Yankee whalers to set out in pursuit of whales was the bark *Wanderer,* which departed from New Bedford on August 25, 1924. It encountered a fierce northeasterly gale, ran aground, and was wrecked on the rocks at Cuttyhunk Island the next day.

GRAND BALL GIVEN BY THE WHALES IN HONOR OF THE DISCOVERY OF THE OIL WELLS IN PENNSYLVANIA.

The capture of the first sperm whale off Nantucket in 1712 may be apocryphal, but there is no doubt that Nantucketers were in full cry after the cachalot by the middle of the 18th century. (By 1748, Boston newspapers were running advertisements for "Sperma Ceti.") The fishery was in decline a little over a century later. Fortunes were made and lost; cities rose and fell; men lived and died, all for the magical substance in the head of the whale. The enduring legacy of the sperm whale fishery will not be the houses of New Bedford and Nantucket; it will not be the salvaged harpoons and the scratched whale teeth that remind us of the whalemen and their victims; it will not be the logbooks and diaries that told of their pleasures and hardships; it will not even be—if an absence can be a monument—the massive destruction of the whales themselves. Rather, the legacy of the sperm whale fishery will be found only in the pages of Melville's epic narrative, the most powerful parable ever written of the eternal conflict between man and beast. *Ars longa, vita brevis.*

Throughout the 19th century, square-rigged whaleships from New England plied the seas in search of the great cachalot, but in the great scheme of things, they actually didn't kill all that many, largely because chasing down a single whale from a rowboat and then throwing a spear at it may be picaresque and dangerous (and the stuff of great literature), but it is hardly an efficient way of reducing whale populations. In a 1935 study, Charles Haskins Townsend analyzed the logbooks of 744

American vessels that carried out 1,665 whaling voyages from 1792 to 1913 and accounted for the death of 36,908 whales, an average of 305 whales per year. His time frame includes the start of the sperm whale fishery, when there were few ships engaged in it, and also the end of the fishery, when mechanized whalers had forsaken sperm whales for the blue and fin whales of the Antarctic, but even at the height of the fishery in the mid-19th century, there was never a year in which 10,000 sperm whales were killed. (Other nations were engaged in sperm whaling, of course, but none approached the number of whales killed by the doughty Yankees.)

Were sperm whale populations decimated by 19th-century longboat whaling? Not nearly as much as they were during the 1960s, when Soviet and Japanese catcher boat fleets worked the North Pacific south of the Aleutians, harvesting sperm whales with techniques and in numbers that would have astonished Captain Ahab. Cannons mounted on the catcher boats shot eight-foot-long exploding grenade harpoons into the whales. Then the whalers towed the dead whales to 500-foot-long factory ships, where they were winched aboard through a slipway in the stern and processed on deck. Whereas the Yankee fishery could account for 36,000 whales in a 121-year period, the Soviet and Japanese killed almost that many in the North Pacific in 1968 and 1969 alone.* Under pressure from antiwhaling governments and conservation organizations, the International Whaling Commission continued to reduce sperm whale quotas for Japan and the USSR, until 1980, when the number fell to zero. It was probably the first time in 200 years that commercial whalers were not killing sperm whales. Despite the first two centuries of slaughter, the sperm whale was not then considered endangered, but it is considered endangered now.

Despite decades of optimistic estimates, the global sperm whale population suffered heavy losses from mechanized whaling. Estimates of the total sperm whale population during the onslaughts of the North Pacific Soviet and Japanese whaling fleets were based on the traditional "catch per unit effort" methods, which only estimated the difficulty in

*In a 1980 International Whaling Commission report, Japanese sperm whale expert Seiji Ohsumi provided the totals for sperm whaling in the North Pacific from 1910 to 1966. The numbers are broken down into "Japan, coastal" (total for those years, 78,220); "Kuril Islands" (28,399); "Japan, Pelagic" (48,232); and "USSR, Pelagic (106,654)." The grand total of sperm whales killed in the North Pacific during this 56-year period is 261,505.

finding and killing whales from one year to the next. "The estimates of 1.5 to 2 million animals," wrote Hal Whitehead of Dalhousie University in Nova Scotia, in a paper published at the 2002 meeting of the International Whaling Commission, "have no valid scientific basis." Before the advent of sperm whaling in the mid-18th century, Whitehead suggests that there were about 1,110,000 animals, but on the basis of mark-recapture techniques, acoustic censuses, and visual censuses, he estimates that there are now only approximately 360,000 sperm whales left in the world's oceans. The revised estimate now means— among other things—that 20th-century whalers, grossly mistaken in their estimates of the numbers of sperm whales, were actually hunting a declining species, and the remaining whalers, like the Japanese, who continue to argue for a quota of sperm whales, will be targeting a depleted population.

IX

How to Catch a Whale

A VOYAGE ABOARD a New England whaler was not a luxury cruise. The voyages were often less than romantic and the weather less than benign. There were indeed fresh breezes, tropical sun, and occasionally vast herds of cachalots, but there was also the tedium of years of sailing (the record seems to be the 11-year voyage made by the ship *Nile*, out of New London, from 1858 to 1869), as well as gales, blizzards, typhoons, hurricanes, mountainous seas, and howling winds. The crew's quarters were stinking holes; their food was cheap, coarse, and maddeningly monotonous; the work itself was dirty and dangerous. In the 19th century, the hierarchy of officers and men, so important to the successful operation of a whaling vessel, was rigidly observed, and nowhere was the distinction more evident than in their respective living quarters. The captain lived in relative luxury; the ship's officers had smaller cabins; the boat steerers, the cooper, and the steward occupied the steerage, an irregular compartment fitted with plain bunks. The crew was in the forward section just below the main deck, which followed the shape of the ship: it went from a fairly wide cross section to a narrow, cramped, triangular warren, where the ship's timbers formed the walls and the pounding of the waves formed the ambiance. The lower portion of the foremast often kept the occupants of the fo'c'sle company, reducing their limited space even further, and the only light that entered this literal and figurative rat hole came from the hatchway cut in the deck for the purpose of giving access to the ladder that allowed the men to climb in and out of their quarters. When the weather turned foul, the hatch was closed, and there was no light but stubby candles, and no ventilation whatsoever. The number of men who occupied this wretched space often exceeded 20.

The fo'c'sle of a whaler, as shown in the 1922 film *Down to the Sea in Ships*. (New Bedford Whaling Museum)

No whaleman was ever paid a wage, except in unusual circumstances. If, for instance, a full ship had to take on additional hands on the way home, their share of the profits would be zero (since they had not participated in the whaling), and they were paid a monthly wage. Ordinarily, each man, from the captain to the cabin boy, received a percentage of the profits—called a lay—at the end of the voyage.

The distribution differed from vessel to vessel. Larger ships could carry more oil, and therefore the profits to the crew were likely to be proportionately higher—but while a successful voyage could be better for the captain and the officers, it meant precious little to the foremast hands. (On an unsuccessful voyage, where the profits were low or non-existent, the crew might receive nothing at all.) The captain might earn 1/8 or 1/10 of the net proceeds, while a mate could earn 1/15 and a harpooner 1/90. Ordinary seamen could hope at best for 1/150, and there are instances in the records when a green hand signed aboard for 1/350. What did this mean in terms of actual money? On board the *Addison*, first mate Ebenezer Nickerson, whose lay was 1/18, earned $845. Robert Baxter, the second mate with 1/35, earned $554.83, and a boat steerer named Narcisco Manuel, with 1/90, got $376.56. Compare these figures with those of the crew: John Martin, at 1/175, earned a total of $31.95, and Francis Finley got $92.08. During six consecutive voyages totaling 1,128 days at sea from 1845 to 1868, the average lay per voyage on the Salem whaler *James Maury* was $321.21, or about 26 cents a day. This compared unfavorably with wages then being paid to unskilled laborers ashore (an average of 90 cents a day), but landlub-

Casks of oil line the New Bedford wharf. (New Bedford Whaling Museum)

bers did not get to visit exotic Pacific islands where they might be eaten by cannibals or risk their lives fighting gigantic whales.

The whaleman's food and bunk space were generously provided without charge, but throughout the voyage, he was docked for various items that he had to buy from the ship's stores. Additional items of clothing, tobacco, knives, needles, and even thread were charged to each man's account, and if he required spending money in a port of call, this too was deducted from the final reckoning. This was a period when the master's voice was law, and if a man needed a new shirt or a pair of boots, he could "either pay up or go naked." Although most of the whalemen signed aboard voluntarily, they usually did not know of the dangers and hardships that lay ahead of them, and the profit sharing that at the outset sounded so attractive often deteriorated into an enforced and dangerous risk sharing.

Among the more unusual charges assessed to a whaleman was the cost of desertion. If a man jumped ship, his account included the cost of recapturing him, an expense that was obviously nullified if he remained at large. On the other hand, there were captains who rewarded the lookouts with bonuses for the sighting of whales. This exercise was glorified in *Moby-Dick,* where Ahab nails a gold doubloon to the mainmast and exhorts his crew, "Whosoever of ye raises me a white-headed whale with a wrinkled brow and a crooked jaw; whosoever of ye raises me that white-headed whale, with three holes punctured in his starboard fluke—look ye, whosoever of ye raises me that same white whale, he shall have this gold ounce, my boys!" The "Spanish ounce" that was offered to the crew was a 16-dollar gold piece.

On a three- or four-year voyage, a man might earn $100, but the items billed to him often exceeded this amount, so many hands returned to port not only with no spending money, but in debt. The only thing to do to work off this indebtedness was to sign on for another voyage, thus starting the insidious process all over again. If and when they made it back to port, the whalemen were set upon by all sorts of land sharks, eager to assist them in disposing of their wages by enticing them into taverns, brothels, and other iniquitous dens where they could make up for the pleasures they had been denied for the past several years.

The system of wages aboard a whaler was obviously not conducive to enthusiasm or hard work. In response to the brutal discipline often administered by the captain, there was bound to be apathy, indifference, and suspicion on the part of the foremast hands. There was also a profound class distinction between the officers and the men. Despite the abuses, hardships, and low earnings that characterized the industry, however, the labor supply was somehow adequate to meet its needs. As Hohman (1928) has written, "The steady stream of men pouring into the forecastles proved sufficient to counteract the continuous labor leakage caused by death, illness, incapacity, discharge and desertion." It was possible (although uncommon) for a dedicated seaman to work his way up through the ranks, and there are instances when a green hand, or even a cabin boy, raised his lay from 1/150 to 1/15, and after perhaps 20 years at sea (in four- or five-year increments), a man might command a whaling vessel.

The *Benjamin Tucker*, a New Bedford whaler, brought back 73,707 gallons of whale oil, 5,348 gallons of sperm oil, and 30,012 pounds of whalebone in a voyage that ended in 1851. At the prevailing prices—43 cents a gallon for whale oil, $1.25 a gallon for sperm oil, and 31 cents a pound for bone—the gross value of this cargo was $47,682.73. From this, $2,362.73 was variously deducted, leaving a net of $45,320.00 to be distributed. But before the profits were divided, the owners took a substantial percentage off the top to compensate for their initial outlay—and also because these flinty New Englanders were not in the business for the thrill of the chase. In general, the owners took between 60 and 70 percent of the profits. On the 1805–1807 cruise of the *Lion*, the various oils yielded a total of $37,661.02. Of this, $24,252.74 went directly to the owners, leaving $13,408.28 to be divided among the captain and the crew for two years of work.

Of course, profits from the whaling industry were not restricted to the owners. They had to repair, refit, and reprovision their ships, which provided work and income for the shipwrights, chandlers, coopers, rope makers, carpenters, and blacksmiths, and ready markets for the farmers and greengrocers. The entire township of New Bedford benefited from the outfitting and victualing of the armada of ships that annually departed it wharves, loaded with food, clothing, and supplies, most of which were bought from local merchants.

The captain had his own cabin, with a proper bunk, a washstand, a table, and perhaps even a sofa and some extra chairs. The captain's quarters of the whaleship *Florida* "opened off the after cabin on the starboard side and extended nearly to the end of the forward cabin. A small room and a toilet room were aft of the stateroom. A large swinging bed was in the captain's cabin instead of the usual fixed berth" (Williams 1964). The gimballed bed was a special innovation designed by Captain Thomas Williams because he planned to bring Mrs. Williams along.

Occasionally a captain took his wife, and even more infrequently, he took his entire family. Captain Williams, of the ship *Florida* out of New Bedford, was accompanied by his wife for a voyage that lasted from September 1858 to October 1861. During the voyage, Eliza Azelia Williams gave birth to two children, who spent the first years of their lives at sea. She also kept a detailed journal of her adventures, which allows us a most unusual perspective of life aboard a whaleship. The voyage commenced on September 7, 1858, in New Bedford, and on January 12 of the next year, Mrs. Williams gave birth to a baby boy, whom they named William. (William's arrival might help to explain her seasickness early in the voyage, when she wrote, "it remains rugged and I remain Sea sick. I call it a gale, but my Husband laughs at me and tells me I have not seen a gale yet. If this is not one I know I do not want to see one.") On August 5, 1859, off the rugged coasts of Sakhalin in the Okhotsk Sea, the *Florida* spoke the *Eliza F. Mason*, and Mrs. Williams visited another "lady ship," where the captain had brought "a Lady Companion, and a little Girl that they brought from the Bay of Islands, New Zealand." On February 27, 1860, Mrs. Williams wrote, "We have had an addition to the Florida's Crew in the form of our little daughter."

United States maritime law decreed that a logbook be maintained by the mate or the first officer. (The term *logbook* originated with the practice of casting a log overboard affixed to the ship by a knotted line. The speed at which the line played out—measured in knots—determined the speed of the ship, and the daily records were originally kept in a book reserved for that purpose. Later, *logbook* came to mean the book used for the keeping of all the ship's records.) For the most part, logbooks and journals were kept by the masters. Although rarely educated in the classical sense, some of these men could read and write passably well, and their records have given us an enduring picture of life aboard a whaleship. Even though the maintenance of a logbook was mandatory, it obviously served the whalers particularly well because the appearance of whales at a known latitude and longitude in one season might enable the whalers to predict their reappearance at the same location the following year and thereby avoid aimless wandering.

The more mundane entries consisted of the ship's position, the number of whales caught, and illness and injury aboard ship, but additional dramatic possibilities were vast. Whaling historian Stuart Sherman (1965), in his introduction to the catalog of the logbook collection of Paul Nicholson, listed "castaways, mutinies, desertions, floggings, women stowaways, drunkenness, illicit shore leave experiences, scurvy, fever, collisions, fire at sea, stove boats, drownings, hurricanes, earthquakes, tidal waves, shipwrecks, ships struck by lightning, men falling from the masthead, hostile natives, barratry, brutal skippers, escape from Confederate raiders, hard luck voyages and ships crushed by ice." That is not to say that all logbooks read like *Moby-Dick;* dramatic events occurred only infrequently, and most of the daily entries—when the ship was not engaged in killing whales—consisted of remarks on the weather, wind direction, location, and whatever else the keeper of the logbook deemed pertinent.

It is not surprising that few of the foremast hands kept records; their quarters were not conducive to literary pursuits, and besides, many of them couldn't write. Francis Olmstead (1936) could. Of the literary aspirations of his fo'c'sle companions, he wrote,

The forecastle of the *North America* is much larger than those of most ships of her tonnage, and is scrubbed out regularly every morning. There is a table and a lamp, so that the men have conveniences for reading and writing if they choose to avail themselves of them; and many of them are practicing writing every day or learning how to write. . . . When not otherwise occupied, they draw books from the library in the cabin and read; or if they do not know how, get someone to teach them. We have a good library on board, consisting of about two hundred volumes.

J. Ross Browne, a journalist who shipped aboard the New Bedford whaler *Bruce* in 1842, kept a journal of his experiences that was published in 1846, with major revisions, as *Etchings of a Whaling Cruise*. Browne wanted to do for whaling what Richard Henry Dana had done for merchant sailing in 1840—that is, exaggerate the problems so that necessary changes would be implemented. Although his account may contain a certain amount of propaganda in the form of negative commentary, he was aboard a whaler for more than a year, and because he is regarded as a reporter and not a writer of fiction, much of the material contained in his book can be taken as fact. Here is Browne's description of the place in which he lived:

The forecastle was black and slimy with filth, very small, and hot as an oven. It was filled with a compound of foul air, smoke, sea-chests, soap-kegs, greasy pans, tainted meat, Portuguese ruffians and sea-sick Americans. . . . In wet weather, when most of the hands were below, cursing, smoking, singing and spinning yarns, it was a perfect Bedlam. Think of three or four Portuguese, a couple of Irishmen, and five or six tough Americans, in a hole about sixteen feet wide, and as many perhaps, from the bulkheads to the fore-peak; so low that a full-grown person could not stand upright in it, and so wedged with rubbish as to leave scarcely room for a foothold. It contained twelve small berths, and with fourteen chests in the little area around the ladder, seldom admitted of being cleaned. In warm weather it was insufferably close. It would seem like an exaggeration to say, that I have seen Kentucky pig-sties not half so filthy, and in every respect preferable to this miserable hole; such, however, is the fact.

Rats were more numerous on whaleships than on any other vessels, probably because of the profusion of blood and oil that soaked the decks, despite the regular scrubbings. They were more than any ship's cat could cope with, and then as now, there was nothing that could cope with cockroaches. They were endemic aboard the whalers, and for many seamen, the roaches were a more predominant aspect of a whaling voyage than whales. Francis Olmstead (1936) wrote that they made "a noise like a flush of quails among the dry leaves of the forest." He added, "They are extremely voracious, and destroy almost everything they can find: their teeth are so sharp, the sailors say, that they will eat the edge off a razor."

In *Nimrod of the Sea* (1874), William Davis describes roaches as serving a useful purpose: "His chief recommendation is his insane pursuit of the flea," but then he goes on,

> It is a horrible experience to awaken at night, in a climate so warm that a finger-ring is the utmost cover you can endure, with the wretched sensation of an army of cockroaches climbing up both legs in search of some Spanish unfortunate! It reminds me of how many times I have placed my tin plate in the overhead nettings of the forecastle, with a liberal lump of duff reserved from dinner, and on taking it down at supper, have found it scraped clean by the same guerrillas. They leave no food alone, and have a nasty odor, which hot water will scarcely remove. But one becomes philosophical at sea in matters of food.

The crew's rations aboard a whaleship ranged from bad to disgusting, but, as Browne (1846) says, "a good appetite makes almost any kind of food palatable." He describes the usual fare on board the *Bruce* (which he has, for culinary and other reasons, renamed the *Styx*): "I had seen the time when my fastidious taste revolted at a piece of good wholesome bread without butter, and many a time I had lost a meal by discovering a fly on my plate. I was now glad enough to get a hard biscuit and a piece of greasy pork; and it did not at all affect my appetite to see the mangled bodies of divers well-fed cockroaches in my molasses; indeed, I sometimes thought they gave it a rich flavor." Fresh vegetables were taken on at the outset of a voyage and were often picked up when the vessel put in for provisions, but unless they were used quickly, they rotted. (By Browne's time, the causes of scurvy were known, but if the vegetables were used up and the ship was cruising somewhere off the

Aleutian Islands, there was not much anyone could do to prevent the dread disease.) Because of their inability to store much water—and to prevent it from spoiling—the whalers hardly ever drank it. Scammon (1874) tells the story of one captain, who, to preserve the dwindling water supply, had the drinking cup hung from the royal masthead, requiring any man who wanted a drink to climb all the way up after the cup. They drank longlick, a mixture of tea, coffee, and molasses, and if the cook was imaginative, he prepared something known as lobscouse (or simply "scouse"), which was a hash made of hard biscuits that had been soaked in the greasy water left over after boiling the salted meat. The mainstay of the whaler's diet was salted meat, which was supposed to be pork or beef, but was occasionally horse. In *Omoo*, Melville described the meat on board a whaleship:

> When opened, the barrels of pork looked as if preserved in iron rust, and diffused an odor like a stale ragout. The beef was worse yet; a mahogany-colored fibrous substance, so tough and tasteless, that I almost believed the cook's story of a horse's hoof with the shoe on having been fished up out of the pickle of one of the casks.

Because the everyday food was so often inedible—Nordhoff (1856) described the duff made by a certain cook as "that potent breeder of heartburns, indigestion, and dyspepsia . . . the very acme of indigestibility," and Ben-Ezra Ely (1849) wrote, "no swine that gleans the gutters ever subsisted on viler meat and bread than did our crew"—the opportunity to eat something fresh was a blessing. The cook prepared seabirds, whatever fish they could catch, turtles, dolphins (off the African coast, Nordhoff describes the harpooning and subsequent eating of a hippopotamus). Because they were engaged in the capture of 50- or 60-ton mammals whose carcasses they would otherwise leave for the sharks, they often ate the meat of the whales. On the eating of various parts of the whale, usually during the trying out, Browne (1846) wrote,

> About the middle of the watch they get up the bread kid [a kid was a wooden tub] and, after dipping a few biscuits in salt water, heave them into a strainer, and boil them in oil. It is difficult to form any idea of the luxury of this delicious mode of cooking on a long nightwatch. Sometimes, when on friendly terms with the steward,

they make fritters of the brains of the whale mixed with flour and cook them in the oil. These are considered a most sumptuous delicacy. Certain portions of the whale's flesh are also eaten with relish, though, to my thinking not a very great luxury being coarse and strong.

It was a different world above decks. On December 28, 1856, the crew of the New Bedford whaler *Addison* caught a porpoise, and Mary Chipman Lawrence, the captain's wife, wrote in her journal, "The meat looks very much like beef. The oil is contained in the skin, which they will boil out tomorrow. Had some of the meat fried for dinner and some made into sausage cakes for supper. They are as nice as pork sausages." If a further demonstration of the disparity between the fare of the men and that of the officers is required, here is Mrs. Lawrence's description of Christmas dinner for that same year: "roast chickens, stuffed potatoes, turnips, onions, stewed cranberries, pickled beets and cucumbers, and a plum duff. For tea I had a tin of preserved grape opened and cut a loaf of fruitcake."

Unlike their British counterparts, American whalers rarely carried any sort of medical man. It commonly fell to the captain to cope with whatever illness or accident befell his crew, and given the master's experience, it was considerably safer to remain healthy. For internal maladies, whaleships were often equipped with medicine chests, which contained various potions and a manual for their dispensation. (Stories were told of masters who, having run out of medicament Number 12, simply administered equal amounts of Numbers 5 and 7.)

Physical injuries were not uncommon, considering the number of sharp-edged tools, whistling whale lines, and hostile natives—not to mention shipboard arguments between men who were almost always armed with knives. Here again, the master served in the role of surgeon, with the same amount of training as he had as apothecary. In *Nimrod of the Sea*, Davis (1874) tells the gory tale of a whaleman who was yanked from his boat by a kinked line and dragged

In the war between whales and whalers, there were occasional instances when nobody won. (New Bedford Whaling Museum)

some 125 fathoms from the boat. When he was finally picked up, "it was found that a portion of the hand, including four fingers, had been torn away, and the foot sawed through at the ankle, leaving only the great tendon and the heel suspended to the lacerated stump." Equipped with "his carving knife, carpenter's saw and a fish-hook," the captain "amputated the leg and dressed the hand as best he could."

As whaling voyages increased in distance and duration, it became expedient to enlarge the ships. In the early days of the fishery, around 1820, the ships averaged around 280 tons burthen, but within two decades, 400-ton vessels were not uncommon. The move toward bigger whaleships contributed to the decline of Nantucket whaling because there was a prominent sandbar across the harbor, and only the smaller, shallower-draft ships could enter. New Bedford, with its excellent harbor facilities, took up the slack.

Whaleships differed from merchantmen of the time in that they usually carried less sail. More canvas meant more men aloft, and the whalers needed as many hands as possible for the boats. One further characteristic of the whaler was the presence of masthead hoops, in which the lookouts stood during the daylight hours to watch for whales. Square-rigged ships, which gave their name to an era of sailing, ran powerfully before the wind, but they were not particularly handy in headwinds or crosswinds. The whalers did not have to perform any smart sailing maneuvers, nor did they have to sail with great speed. All they had to do was get from one location to another and then lower the boats after the whales. Because of the determined, plodding nature of their craft, the masters rarely sailed at night, preferring instead to furl their sails and wait till dawn before continuing.

It was during the heyday of New England whaling, from about 1830 to 1860, that the fabulous clipper ships reached the zenith of sailing-ship design, with their graceful lines, sharply raked bows, and opulence of canvas. In marked contrast to these oceangoing greyhounds, the whalers were sturdy, bluff-bowed, flat-bottomed sailers, designed more for durability and storage than for speed. For example, the *Lagoda* sailed for 50 years, and the all-time record holder, the *Charles W. Morgan,* sailed for more than 80 years and earned over a million dollars for its owners. The *Lagoda* was copied at half-scale for the New Bedford Whaling Museum, and the *Morgan,* the last of its kind, is now the proud centerpiece of Mystic Seaport in Connecticut.

A typical whaler was 100 to 150 feet long and was especially broad

The bark *Canton* in 1906. Even though they were criticized for their sturdy, utilitarian design, the sight of a whaleship under full sail was enough to stir the heart. (New Bedford Whaling Museum)

in the beam to accommodate the fixtures of whaling: heavy brick try-works on deck, iron cauldrons, cooling tanks, davits for the boats, and, of course, the space required to perform the trying out of the whale. Ordinary seamen, whose voyages did not take four or five years, belittled the whaleships as "built by the mile and cut off in lengths as you want 'em." They were usually painted black, and they traditionally had mock gun ports painted along the sides, supposedly as a deterrent to pirates or hostile savages.

The naval historian Albert Cook Church (1938) wrote: "Whaleships differed materially from any other type of merchant ship or clipper in model and equipment, and in fact, both sides of a whaleship differed from each other above the waterline." The larger ships were equipped with four boats, one on the starboard quarter and three on the port (also known as the "larboard") side. This allowed the cutting stages, which were always to starboard, to be lowered without interference from the davits.

All the requisite equipment would be carefully stowed aboard the whaleboats, from the line, which was carefully coiled in a tub so it could be let out rapidly, to the knife that might be required to cut it if a man got his leg entangled. In addition to the six adult men who would be required to man the boat, Scammon (1874) lists the contents of a fully equipped whaleboat:

> One mast and one yard, one to three sails, five pulling oars, one steering oar, five paddles, three rowlocks, five harpoons, one or two line-tubs, three hand lances, three shortwarps, one boat-spade, three lance-warps, one boat-warp, one boat-hatchet, two boat knives, one boat-waif, one boatcompass, one boat-hook, one drag, one grapnel, one boat-anchor, one sweeping-line, lead, buoy, etc., one boat-keg, one boat-bucket, one piggin, one lantern-keg (containing flint, steel, box of tinder, lantern, candles, bread, tobacco, and pipes), one boat-crotch, one tub-oar crotch, half a dozen chock pins, a roll of canvas, a paper of tacks, two nippers, to which may be added a bomb-gun and four bomb-lances; in all, forty-eight articles, and at least eighty-two pieces.

When a whale or a group of whales was sighted, the lookout shouted, "She blows!" or "Blows!" and when the captain had ascertained "where away," the boats were lowered and the chase began. All the boats might or might not be lowered, depending on the number of whales sighted. If only a single whale was seen, the captain might designate one boat to chase it. The starboard boat was reserved for the captain (or the fourth mate, if the captain chose to stay aboard ship during the hunt); the larboard, waist, and bow boats were for the first, second, and third mates, respectively. Each boat contained a regular crew, consisting of four oarsmen, a headsman, and the boat steerer/harpooner. As Beale (1835) described the strike, "The boatsteerer also at this time pulls the bow oar, but when on the whale he ceases rowing, quits the oar, and strikes the harpoon into the animals, the line attached to which runs between the men to the after part of the boat, and after passing two or three times round the loggerhead is continuous with the coil lying in the bottom of the boat." The boats were double-enders; in case they got turned around in the frenzy of the hunt, they would be able to maneuver, and they were among the most graceful and utilitarian boats ever designed.

The lowering of the boats often took place as the ship was underway; the captain did not come about for the comfort or convenience of his crews. Often in high seas, the graceful whaleboats took off after the whales with the men pulling while facing the stern; the boat steerer was the only man who could see the whales. When they had come within range, the boat steerer exhorted the bow oarsman, who was also the harpooner, to rise up, turn around, brace himself by placing his knee in a notch created for that purpose, and "dart his iron." The earliest harpoons had simple fluted arrowhead-shaped heads, but as the fishery developed, more sophisticated designs were introduced. While the two-flued iron pierced the blubber effectively, its razor edges would occasionally pull out as smoothly as they went in. This led to the introduction of the single-flued toggle iron, which held much better. Harpooners and blacksmiths had plenty of time on board the whalers and in port to work on harpoon design, and all sorts of elaborate heads with toggles, barbs, and swivels were tried. The most successful of these designs was the double-barbed Temple iron, invented in 1848 by New Bedford blacksmith Lewis Temple. A graceful, practical device, the Temple iron consisted of a pointed head that was held in the forward position by a wooden shear pin that broke off when withdrawal forces were applied. This rotated the head 90 degrees in the flesh of the whale, forming a T shape that would not pull out because the flattened surfaces were pulling against the meat or blubber. The iron was fastened to the shaft of the harpoon by a line that was bent to the heavy manila line. The line, which Melville calls the "magical, sometimes horrible whale-line," was originally fashioned of hemp, but this was later superseded by manila rope, which was stronger and more elastic. "Hemp is a dusky, dark fellow," Melville wrote, "a sort of Indian, but Manilla is as a golden haired Circassian to behold."

Even though tradition demanded that the harpoon and the lance be thrown separately, some creative whalemen tried to design an iron that would fasten to and kill the whale simultaneously. A Scottish toxicologist named Robert Christson invented a poison-headed harpoon, equipped with glass cylinders containing prussic acid, one drop of which is lethal enough to kill a man. There is no evidence that prussic acid harpoons were used in the American fishery, but they were carried on some vessels. The likelihood is that the American harpooners thought that they had enough problems killing the whale without worrying about killing themselves.

If the iron was well placed—the ideal spot was in the flank, forward of the hump—the boat was fast to the whale, and the injured animal took off. Sometimes the whale sounded, taking out the line at such speed that the line smoked as it ran out, and the loggerhead had to be doused with water to keep it from bursting into flame. More often, the whale swam at the surface, towing the boat through the waves at a violent clip. Sperm whales are prodigious divers, and no boat could hold enough line for a dive that could be measured in miles. If the whale sounded, another 200-fathom line might be bent to the first, and then another. Eventually, the wounded whale had to surface to breathe.

The lance, also known as the killing iron, was plunged into the "life" of the whale—that is, a vital artery, the lungs, or the heart. The killing iron consisted of a wooden shaft like that of the harpoon, with a leaf-shaped, scalpel-sharp head. It was not thrown but rather stabbed repeatedly into the body of the whale. Melville describes the death throes of a whale:

> The red tide now poured from all sides of the monster like brooks down a hill. His tormented body rolled not in brine but in blood, which bubbled and seethed in furlongs behind in their wake. The slanting sun playing upon this crimson pond in the sea sent back its reflection into every face, so that they all glowed to each other like red men. . . . Stubb slowly churned his long sharp lance into the fish and kept it there, carefully churning and churning, as if cautiously seeking to find some gold watch that the whale might have swallowed, and which he was fearful of breaking ere he could hook it out.

The victory did not always go to the whalers. Sperm whales are immensely powerful creatures, and they do not take kindly to being stabbed with spears. The most frequent problem occurred when the whale took it into its 20-pound brain to retaliate. A fragile, 30-foot whaleboat was no match for an enraged, wounded, 60-ton whale, and the harpooned animal might rise up from the depths and grab the boat in its massive jaws, splintering it into so many matchsticks. Both ends of a wounded whale are lethal. The triangular flukes, which might measure 20 feet across, could function as a formidable weapon, crashing

In this 1835 print, *Capturing a Sperm Whale*, by William Page, the discrepancy in size between the whale and his captors is clearly evident. (New Bedford Whaling Museum)

down on the whaleboat and dumping the men into the sea. Other perils faced the whalemen. The whistling line might take a turn around a leg or an arm, surgically severing it, or yank the man into the water. Even if the boat was not destroyed, it might be upended and its occupants dumped into the ocean. Many of the sailors could not swim, so such a plunge often spelled death.

Beale (1835) claimed that the larger whales are "sometimes known to turn upon their persecutors with unbounded fury, destroying everything that meets them in their course, sometimes by the powerful blows of their flukes, and sometimes attacking with jaw and head." Whales that are "fastened to" would sometimes thrash about wildly, biting at anything within range (Scammon 1874). There seems to be some question regarding the mobility of the sperm whale's lower jaw. Gaskin (1972) said that "the long narrow lower jaw has virtually no lateral movement, so that the sperm whale is capable of little more than a straight up and down biting movement," but Ashley (1926) warned that "his jaw is exceedingly mobile; a boat fifteen feet away at either side is in imminent danger from a rolling sperm whale's jaw." Attacking whales often turned on their backs and swam upside down to bring the lower jaw into play above the surface, a behavior that Ashley referred to as "jawing back."* Even when the whale sounded, the whalers were in danger, for the sperm whale is capable of such deep dives that it might run out all the line. If the line was not cut quickly, the whale might take the boat below.

The whalemen's inability to predict the migratory patterns of their quarry often resulted in prolonged searches, with few or no sightings to show for it. In her 1989 discussion of whalemen's perceptions of their business, Mary Malloy wrote,

> In the middle of the nineteenth century, many journal keepers lamented the long time between whales. The ship *Acushnet* (on whose previous voyage Melville had served) left its home port on 16 July 1845 and had caught only one whale by 1 July 1846. George H. Folger, mate of the *Para* under Captain Daniel F. Worth, reported a similar dismal tale of his long voyage of 1867–71. They took their first whales on 23 April 1868, and then went until 3 December before their next successful strike. On 20 February 1869, they killed a massive 107-barrel bull sperm whale, but were unlucky for the next nine months.

When they actually confronted the whales, the human beings—the world's deadliest predators—usually won the battle and then faced the problem of bringing whale and ship together. If the conquering whaleboat was downwind of the ship, it was a relatively simple matter to sail the ship to the carcass, but if less propitious conditions prevailed, the exhausted whalemen might have to tow the whale back to the ship, often for miles. And then, after an exhausting chase and a laborious haul with a 50-ton deadweight in tow, the real work began. What had been a free-swimming, powerful sea mammal was effectively reduced to a disparate assortment of its parts, the reduction accomplished by literally tearing it apart and boiling the pieces. As Malloy (1989) wrote, "Manning the tryworks was the worst and most degrading part of the whaleman's job—analogous to the butcher's job ashore, but more dan-

---

*Starbuck (1878) wrote, "If the right whale had the habit of 'jawing back' as the sperm whale has, it would be next to impossible to secure him . . . on the tip of the upper jaw there is a spot . . . seemingly as sensitive in feeling as the antenna of an insect. However swiftly a right whale may be advancing, a slight prick on this point will arrest his forward motion at once."

Sometimes sperm whales responded negatively to being chased down and stabbed with a spear. (New Bedford Whaling Museum)

gerous. Captain Robert Brown of the *Emerald* reported on 30 December 1840 that crewman Joseph Jackson fell into a pot of boiling whale oil and lived for twelve painful hours before his death. And it was not unknown for trypots to explode as they did aboard the *Bowditch* on 29 October 1837."

In the Yankee whale fishery, the process of removing the whale from its outer integuments was known as cutting in, and the rendering of the blubber into oil was known as trying out. In the English fishery, these operations were known, respectively, as flensing and making off. As with virtually every other aspect of New England whaling, the cutting-in process was described better by Melville than by anybody else. In *Moby-Dick*, there is one chapter devoted to the actual process and several more to the by-products, including the "blanket," the "funeral," and the "sphinx"—the last referring to the head of the whale after the body and blubber have been separated from it.

The whale was made fast to the ship by lashing heavy chains through its head and around its flukes. The first part of the whale to be brought aboard was the lower jaw, ripped from the head and laid aside to be dealt with later. Then the whale was decapitated, and if it was a small one, the head was brought aboard. But the head of a large whale, often one-third of its 60-foot, 60-ton body, could not be brought on deck (Melville wrote that "even by the immense tackles of the whaler, this were as vain a thing as to attempt weighing a Dutch barn in jeweler's scales") and had to be processed in the water. The "head matter"

The 1922 Hollywood movie *Down to the Sea in Ships* included footage of actual sperm whaling. In this scene, the whalers are towing a dead whale (lower left) back to the ship for processing. (New Bedford Whaling Museum)

was saved for last, however, because the carcass of the whale alongside the ship was threatening to the ship by its weight, and the longer it remained unprocessed, the longer the sharks could wreak havoc on the very outer layer of blubber that was of so much interest to the whalers.

By the use of a complicated series of tackles—described by Melville as "ponderous things comprising a cluster of blocks generally painted green, and which no single man can possibly lift"—the cutting stages were lowered, and the process of removing the blubber commenced. Sitting or standing on the lowered cutting stages, men with razor-sharp cutting spades began to slice into the whale's rubbery outer covering. A massive iron hook was inserted in the first piece to come off, and this was hoisted high into the air while the men on the scaffold sliced the blubber. The whale was rotated in the water, and its blubber "stripped off front the body precisely as an orange is sometimes stripped by spiralizing it." The power for this peeling and dismemberment came from the strong backs of the whalemen, who turned the windlass located forward of the foremast.

As the thick spiral of blubber was peeled from the whale, it was cut into sections approximately fifteen feet long and a ton in weight (the "blanket pieces"). These were dropped through a hatch into the blubber room, where they were stored until the carcass of the whale was completely stripped. (With the removal of the blubber and the head, the remainder of the carcass was left for the sharks.) Workers in the dark, bloody blubber room further reduced the blanket pieces to smaller, more manageable "horse pieces," which were then sliced into "bible leaves," with cuts almost to the skin making them resemble the splayed pages of a thick-leaved book. (It was believed that the opening of the blubber into "pages" made the oil more accessible.) The bible leaves were then forked back up through the forehatch to the men who would place them in the trypots.

Although the trypot fires were usually started with wood, the unmelted skin of the whale made a wonderful fuel, and the whale was therefore cooked in a fire of its own kindling. As the oil was separated from the blubber, it was carefully ladled into a copper cooling tank, where it rested before being casked. Aside from the obvious danger of a fire spreading, the process was—like almost every aspect of whaling—hard, messy, and dirty. Oil and blood covered the decks and the people, and the smell was often intolerable. J. Ross Browne called the trying-out process "the most stirring part of the whaling business, and certainly the most disagreeable." He described the nighttime scene aboard the "*Styx*":

> Dense clouds of lurid smoke are curling up to the tops, shrouding the rigging from the view. The oil is hissing in the trypots. Half a dozen of the crew are sitting on the windlass, their rough, weather-beaten faces shining in the red glare of the fires, all clothed in greasy duck, and forming about as savage a looking group as ever was sketched by the pencil of Salvator Rosa. The cooper and one of the mates are raking up the fires with long bars of wood or iron. The decks, bulwarks, railing, try-works, and windlass are covered with oil and slime of black-skin, glistering with the red glare of the try-works. Slowly and doggedly the vessel is pitching her way through the rough seas, looking as if enveloped in flames.

Whalemen stood on "stages" extended from the sides of the ship to work on the carcasses. Here they are preparing to cut loose the tooth-studded lower jaw with long-handled cutting spades and flensing knives. (New Bedford Whaling Museum)

After peeling off huge strips of blubber from the carcass, the "blanket pieces" were lowered into the hold to be chopped into pieces small enough to fit into the cauldrons for boiling down. (Gordon Grand, *Greasy Luck*)

At the end of this description, he wrote, "Of the unpleasant effects of the smoke I scarcely know how any idea can be formed, unless the curious inquirer choose to hold his nose over the smoking wick of a sperm oil lamp, and fancy the disagreeable experiment magnified a hundred thousand fold. Such is the romance of life in the whale fishery."

Trying out. (New Bedford Whaling Museum)

One of the least romantic aspects of the whale fishery was the prospect of fire. Oil-soaked wooden ships upon whose decks fires are being encouraged do not lend themselves to a feeling of security, Care was taken to avoid conflagrations—water was pumped over the decks to keep the planks wet and cool—but occasionally the sails or rigging were ignited by flying sparks, and sometimes the ships burned to the waterline.

When the oil had cooled, it was ladled into the casks that had been assembled by the cooper. Each barrel held 302 gallons, and the figures for the fishery were almost always recorded in barrels. Starbuck's 1878 *History of the American Whale Fishery,* which contains the records of every American whaling ship from every American whaling port, "from its earliest inception to 1876" (insofar as these records were known), lists the result of every whaling voyage in sperm oil (barrels), whale oil (barrels), and whalebone (pounds).

A large female sperm whale might yield 35 barrels of oil, while the largest bulls gave up 75 to 90. As with the sometimes questionable lengths of large bulls, where there were reports of 90-footers (Clifford Ashley [1926] writes, "If these whalemen's records are accurate, it would appear that the hundred-foot Sperm Whale is not an impossibility"), the yield of these giants was the subject of occasional exaggeration. Because the reports were invariably made by men whose reputation would be enhanced by overstating the yield of individual

Trypot fires sent billows of oily smoke into the air; the smell carried for miles. Shown here is the bark *Jacob A. Howland* of New Bedford. (New Bedford Whaling Museum)

whales, many of the whales in the 100- to 150-barrel range must be questioned.*

The amount of oil that could be taken and stored was enormous, but it did not necessarily reflect the success of a voyage. The profits of a voyage could only be calculated when the ship reached port and sold the oil and bone at the prevailing prices. A 31-gallon cask was about five feet high and four feet in diameter at its bulging middle. On its maiden whaling voyage, which lasted from October 1841 to September 1843, the *Lagoda* brought home 600 barrels of sperm oil, 2,700 barrels

*If only the whalers' stories remained, we would have no way of verifying the size of the largest whales. There is something that they leave behind, however, and Ashley (1926) proposes a novel argument for the existence of gigantic bull sperm whales: he examines a particularly large pair of teeth, over 11 inches long, and suggests that "in the days before the Sperm Whale herds were depleted, there must have been exceptional whales, either larger or older than are found today." Mitchell (1983) finds this argument "well taken, but not conclusive," but a look at these teeth, which are on display in the New Bedford Whaling Museum, certainly gives one cause to wonder.

of whale oil, and 17,000 pounds of baleen. (Sperm oil was the stuff that was ladled out of the whale's case, and was of a finer quality than whale oil, which was rendered out of the blubber. Although whalers were not averse to taking an occasional right whale or humpback, most of the whales hunted by the Yankees were sperm whales.) The *Lagoda* was 108 feet long, with a beam of 27 feet. Hunting concluded when there was no more room for the storage of oil, but the whalers sometimes put into port, offloaded some of their greasy cargo, and set out again for the whaling grounds. Some of these sweaty, iron-bound vats were probably stored in the blubber room, but most were stored in the hold.

It was the mysterious head matter of the sperm whale that made it the primary object of this globe-girdling enterprise. The spermaceti was the ne plus ultra of this business, the pot of liquid gold that attracted the whalers to the Azores and the Galápagos, to Zanzibar and the Japan Grounds, to Kamchatka and the Sea of Okhotsk. The stuff is as poorly understood today as it was when some early beachcomber presumed that this vast reservoir in the whale's nose was its seminal fluid. Whatever its purpose to the whale (and it certainly is not its seminal fluid), the amber liquid that hardened into white wax as it was exposed to air was worth risking life and limb—and sometimes boat and ship—to the whaler. Kept free from contamination by other oils, sperm oil was worth from three to five times as much as whale oil. In *Nimrod of the Sea*, Davis (1874) records a whale that yielded 27 barrels of spermaceti from the case, and Clifford Ashley's (1926) research indicates that the largest bulls gave up something on the order of 30 barrels. At 31.5 gallons per barrel, that works out to 945 gallons of the mysterious liquid wax in the nose of a single whale.

To extract the spermaceti from the head, a much more direct method was used than the multistep process of turning blubber into oil. Because the spermaceti already *was* oil, the whalers only had to remove it from the whale and cask it. A hole was cut in the outer fabric of the whale, and a man lowered a bucket into it on a long pole, then turned it over to another man on deck, who would empty the bucket into a waiting tub—or as Melville put it, "Tashtego downward guides the bucket into the Tun, till it entirely disappears; then giving the word to the seamen at the whip, up comes the bucket again, all bubbling like a dairy-maid's pail of new milk."

When the oil had all been casked and the casks stowed, the decks were scrubbed down with lye, which had been leached from the cinders and ashes of the tryworks. The oily, smoky clothes of the whalemen were also scrubbed down, but the pernicious odor of smoked blubber could never really be removed, and until they could exchange their work clothes for new garments, the whalemen usually stank like disused tryworks.

X

The War on Whales

THROUGHOUT the long and bloody history of man's war on whales—where one side was heavily armed and the other didn't know there was a war going on—there has never been a battle that compares to the one staged by Jules Verne in *Twenty Thousand Leagues Under the Sea*. Verne's work was fiction, of course, but he was not one to let facts get in the way of the story, so his chapter called "Sperm Whales and Baleen Whales" ("Cachalots et Baleines" in the original) contains probably the craziest rationale for whale killing in all of literature.*

After rounding Cape Horn, the *Nautilus* heads "in the direction of the Antarctic" and encounters a school of baleen whales, identified as "only rorquals with dorsal fins" by Ned Land, the Canadian harpooner. Captain Nemo refuses to allow Ned Land to harpoon a couple ("leave the unfortunate whales in peace. . . . They already have enough problems with their natural enemies, the sperm whales, the swordfish, and the sawfish"), but when the inoffensive rorquals are attacked by a school of "cruel and destructive sperm whales," Captain Nemo changes his mind, and decides to kill these "natural enemies," which he describes thus:

> Mouth and teeth! There was no better way to depict the macro-cephalous being, which sometimes exceeds 25 metres in length.

*Verne's novel was published in 1870, some 20 years after *Moby-Dick*, but from his description of whales, it is obvious that Verne relied exclusively on his imagination for scientific details. Remember, in the novel, before they learn that it is Captain Nemo's submarine *Nautilus* that is sinking all those ships, Professor Arronax suggests that it might be a "sea-unicorn [narwhal] of colossal dimensions . . . armed with a ram like ironclad frigates."

The enormous head of this cetacean occupies about a third of its body. Better armed than the baleen whale, whose upper jaw is only equipped with bony plates, the sperm whale has 25 huge teeth: 20 centimetres long, cylindrical but with conical ends, and weighing two pounds each. . . . The sperm whale is a disgraceful animal, more tadpole than fish. . . . It is poorly constructed, being defective, so to speak, in the whole left-hand part of its framework, and hardly seeing except through its right eye.

Of course Nemo would want to kill these disgraceful, lopsided animals! "We'll give them no quarter," he says. "These ferocious whales are nothing but mouth and teeth." Verne describes the "Homeric massacre":

What a battle! The *Nautilus* was nothing but a formidable harpoon, brandished by the hand of its captain. It hurled itself against the fleshy mass, passing through from one part to the other, leaving behind it two quivering halves of the animal. It could not feel the formidable blows from their tails upon its sides, nor the shock which it produced itself, much more. One cachalot killed, it ran at the next, tacked on the spot that it might not miss its prey, going forward and backward, answering to its helm, plunging when the cetacean dived into the deep waters, coming up with it when it returned to the surface, striking it front or sideways, cutting or tearing in all directions, and at any pace, piercing it with its terrible spur. What carnage! What a noise on the surface of the waves! What sharp hissing, and what snorting peculiar to these enraged animals! . . . From the window we could see their enormous mouths studded with tusks, and their formidable eyes. Ned Land could not contain himself, he threatened and swore at them. We could feel them clinging to our vessel like dogs worrying a wild boar in a copse. But the *Nautilus*, working its screw, carried them here and there, or to the upper levels of the ocean, without caring for their enormous weight, nor the powerful strain on the vessel. At length, the mass of cachalots broke up, the waves became quiet, and I felt that we were rising to the surface. The panel opened, and we hurried on to the platform. The sea was covered with mutilated bodies.

When he is told that this was nothing but a massacre, Captain Nemo replies, "It was a massacre of harmful animals; the *Nautilus* is not a

In one of his most famous novels, Jules Verne portrayed sperm whales as vicious and "all mouth and teeth." (Illustration by Alphonse de Neuville from *Twenty Thousand Leagues Under the Sea*, Paris, 1870)

butcher knife." The narrator of the story, "the honorable Pierre Aronnax, Professor of the Paris Museum," manages to get virtually everything wrong about sperm whales, which is not unexpected, as Verne does not accept much of contemporary whaling lore, which he regards as "giving a lie to nature by admitting the existence of krakens, sea serpents, Moby Dicks, or other reports of delirious sailors."*

*In his 1998 translation of *Twenty Thousand Leagues Under the Sea*, William Butcher identifies Jules Verne as "the world's most translated writer." Since its original publication as a French magazine serial in 1869, *Vingt mille lieus sous les mers* has appeared in countless versions and languages. Verne's novel has been translated, interpreted, corrected, and even modified to fit what the translator knows to be an

Old whalemen knew the grounds where cachalots were to be found in the greatest numbers, "where the masters of ships usually resort for the purpose of fishing" (Beale 1835). The best-known grounds were as follows: the west coast of South America (Callao, Coast of Chile, and Galápagos grounds); the east coast of the same continent (Platte Ground and Brazil Banks); the entire North Atlantic (Charleston, Southern, and Western grounds); the west coast of Africa (Carroll and Tristan grounds); the east coast of Africa (Delagoa Bay, Zanzibar Ground, Mahe Banks); the Arabian Sea (Coast of Arabia Ground); the western North Pacific (Coast of Japan, Japan, and Sulu Sea grounds); the west coast of Australia (Coast of New Holland Ground); the Tasman Sea area between eastern Australia and New Zealand (Middle Ground); and the New Zealand area (Vasquez and French Rock grounds). All along the equator in the Pacific, from the Celebes east to Ecuador, was a particularly productive range known as On the Line.*

Ahab interrogated passing captains to learn if they had seen the white whale. One passing ship has lost five men to Moby Dick; Captain Gardiner of the *Rachel* has lost his son; and Captain Boomer of the *Samuel Enderby* has lost his hand, replaced by an ivory substitute, as was Ahab's leg. The final, fateful cruise of the *Pequod* took it into the western Pacific, "penetrating further and further into the heart of the Japanese cruising ground." Although we are not told exactly where the final encounter with Moby Dick took place, we know that Ahab consulted various charts where sightings of sperm whales had been recorded, and by chapter 109, the *Pequod* "was drawing nigh to Formosa and the Bashee Isles [the Philippines], between which lies one of the tropical outlets from the China waters into the Pacific."

Although the distribution of this species is not clearly understood, it

---

error, or to update Verne's story to make it more comprehensible to modern readers. In the original, this sentence appears as follows: "Un démenti à la nature, en admettant l'existence des Krakens, des serpents de mer, des 'Moby Dick' et autres élucubrations de marins en délire."

*Sperm whales can be—and are—found almost anywhere in the world's polar, temperate, and tropical waters. A list of where they have been seen would fill this book, but in a 2006 study of the whales' deep-diving foraging behavior, Watwood et al. tagged 37 individual sperm whales offshore in the western North Atlantic from Cape Cod to Cape Hatteras, the Gulf of Mexico south of New Orleans, and the Ligurian Sea (Mediterranean) west of Corsica and Sardinia.

is assumed to be affected by two factors: food and reproductive needs. The food of the sperm whale consists mostly of deep-water cephalopods, so the whale is likely to be found in those areas where conditions for squid are propitious. These conditions include deep water, usually not less than 3,300 feet (1,000 meters) (Watkins 1977), and locations of cold-water upwellings, such as can be found off the northwestern coast of South America, off the coast of Japan, and in the vicinity of various island groups, including the Azores, Canaries, Cape Verdes, Madeira, Galápagos, Seychelles, Comoros, and New Zealand.

Townsend (1935) plotted the world distribution of sperm whale catches as recorded in the logbooks of New England whalers, but as Gilmore (1959) pointed out, he did not have access to the records of the West Coast whalers from San Francisco and Seattle and "thus completely missed the extensive British Columbia and Kodiak grounds, both largely inhabited by summer males." Nevertheless, Townsend's charts represent a prodigious undertaking. When they are combined with Gilmore's additions of more recent statistics and records of the Pacific grounds that he claimed were lacking in Townsend, and with Holm and Jonsgård's 1959 review of sperm whale catches in the Antarctic, we can begin to see a picture of the cosmopolitan distribution of the sperm whale.

Townsend's (1935) records account for 36,908 sperm whales taken from 1761 to 1920, but the North Pacific Grounds, now considered to be the most productive of all (Nishiwaki 1967), are barren on his charts, as are the Antarctic waters. In the Southern Ocean, sperm whales, almost all of which are males, were not hunted as extensively as the whalebone whales, but when the stocks of blue, fin, sei, and humpback whales began to decline, the pelagic whalers turned toward the sperm. From 1910 to 1946, the annual world catch did not exceed 7,500 sperm whales per year, but postwar catches have increased to nearly 30,000 per year (Best 1974). The stocks of sperm whales are believed to be discontinuous—there are separate populations that do not integrate— but there is very little information on this subject. Sperm whales do not migrate with the predictable regularity of the baleen whales, and with the exception of the polar movements of the bulls, the populations may remain within the same general areas, occasionally moving en masse for unknown reasons. We do not know for certain whether the mature males that exist in the high latitudes ever return to the breeding herds (Gaskin 1972).

Sperm whales are smaller than they used to be, but not enough to account for the discrepancies between the old accounts and the more recent records. Numerous whales have been listed at "over 100 barrels." For example, Starbuck mentioned a 79-foot whale that yielded 107 barrels, and others that were good for 136 barrels and even 156 barrels. "In 1862 the *Ocmulgee*, of Edgartown, reported to have taken a 130-barrel sperm whale, with a jaw measuring 28 feet [8 meters] in length." Jaw measurements may be used to estimate size: according to the calculations of Fujino (1956), the lower jaw of an adult male sperm whale is approximately 20 percent of its total length. A whale with a 28-foot lower jaw would therefore be about 140 feet long.*

Nineteenth-century sperm whaling was never the epitome of efficiency because it took so long to complete a voyage. Trips two or three years in length were not uncommon, and by the time the fishery began its decline, even longer voyages were necessary to fill the holds. Townsend (1935) wrote, "At its best period, the great fleet probably captured less than 10,000 whales [of all species] per year," and his records showed a total of 36,908 sperm whales for the period 1761 to 1920 from New England logbooks. Obviously his figures are incomplete. The figures showed only whales captured by the ships whose logs Townsend examined. Not all logbooks survived, and not all whaleships were out of New England. It would not do, however, to take Yankee whaling too lightly; Scammon (1874) estimated that between 1835 and 1872, "there were no less than 292,714 whales captured or destroyed by the American whaler's lance," an annual average of 7,703. Of this annual catch, Scammon figured 4,253 ("or thereabouts") to be sperm whales; the rest were rights, bowheads, grays, and humpbacks.

The killing of sperm whales began to drop off during the first third of the 20th century, as the hunters concentrated on the vast herds of whalebone whales in the Antarctic. Sperm whales were the specialty of some southern whaling stations, such as Durban, South Africa, where

---

*Jonsgård (1960) has shown that the average sperm whale is smaller today: from 1937 to 1959, the average length decreased from 53.2 feet to 47.2 feet. This may be a function of "gunner selection," by which the larger whales were taken in the earlier seasons, leaving the smaller ones for subsequent years. However, this is unlikely, given the variability in whaling locations and the movement of the whales themselves. In any event, this information does not justify reports of 80- or 90-foot bull whales.

10,136 sperms were taken between 1904 and 1939, or those of Chile and Peru, where, in the same period, 8,039 sperms were killed. The total number of sperm whales killed at southern whaling stations during that 35-year period was 27,433. As an indication of the intensity of the southern baleen whale fishery, during the same period, 261,945 blue whales and 216,688 fin whales were killed (Mackintosh 1942). After World War II, the figure for sperm whales began to climb at an ever-increasing rate as the stocks of baleen whales—obviously unable to withstand this mammoth slaughter—declined, and the sperm whale again became the most heavily hunted of all the species. Although the peak years for blue and fin whale catches were 1931 and 1938, respectively, it was not until 1951 that the total kill of sperm whales reached 10,000 per year, and in 1963, the total climbed to over 20,000 per annum. It held at that level until 1971 (McHugh 1974).

In the sperm whale chapter of a 1984 report on "The Status of Endangered Whales," Gosho, Rice, and Breiwick summarized the catch statistics for every year from 1800 to 1982. In the numbers, we can see the history of the fishery: it peaks in 1846, with an estimated 2,428 whales killed; begins to decline in the ensuing decades as the number of whaleships falls (1,827 whales in 1871; 1,082 in 1883); and then begins to climb appreciably as new whaling grounds are discovered (particularly in the Antarctic), and mechanized whaling, with its catcher boats, grenade harpoons, and factory ships, replaces square-rigged whaleships, rowboats, and hand-thrown harpoons. The entry of the Soviet and Japanese fleets after World War II is obvious. A sampling from their survey is listed in Table 10.1.

Because of the discrepancy in size and in habitat, male and female sperm whales have been considered separately in the International Whaling Commission (IWC) quotas. In 1937, the International Conference on Whaling (the precursor of the IWC) reached an agreement that was supposed to protect the stocks of sperm whales by restricting the catch to animals over 35 feet (10.6 meters) in length. Because most females are smaller than this, such a restriction was ostensibly more important to the breeding stock. It was assumed that the capture of bulls, "if not excessive, would in no way damage the stock" (Matthews 1938). Later, the figures were adjusted so that whales 38 feet in length could be taken from pelagic factory ships in the Antarctic and the North Pacific (the only areas in which these ships operate), and the

TABLE 10.1. Sperm Whales Killed

Year	No. of Sperm Whales Killed
1910	155
1920	873
1930	1,311
1941	5,641
1948	9,850
1951	18,281
1959	21,298
1964	29,255
1972	18,895
1978	11,065
1979	8,536
1981	1,456
1982	526

Catch data for 1910–1937 from Clarke (1954), and for 1938–1982 from Committee for Whaling Statistics (1959–1983).

limit remained 35 feet for shore-based sperm whaling—those activities in which catcher boats set out from port to hunt whales and then return to the whaling station with the whales for processing.

Probably because the general public regards the sperm whale as the typical whale—unlike baleen whales, its mouth is located under its nose, where one might expect it to be—there has always been an awareness of the plight of the sperm whale. The recent history of the fishery has been much in the news and marked by confusion, emotion, and massive public outcries to save the whale. Approximately a century after the New England fishery declined (about 1865), hunting for sperm whales resumed on a large scale. (Throughout the history of Antarctic whaling, sperm whales were taken in comparatively low numbers until the stocks of baleen whales fell to levels that made their hunting uneconomical.) Because there were known sperm whale populations within range of Durban, South Africa, and Albany, Western Australia, these two fisheries, which practiced shore-based whaling, became the most profitable—and the most notorious.

THE SOUTH AFRICANS

Off the east coast of southern Africa were the sperm whale habitats known as the Delagoa Bay Grounds, the Zanzibar Grounds, and the Mahé Banks, in addition to the productive waters of the Mozambique Channel, which separates the island of Madagascar from the African continent. The first American whaleship to cross the equator was the Nantucket ship *Amazon*, which reported the discovery of the Brazil Banks in the South Atlantic in 1775. Even when they were whaling in the Indian Ocean, the American whalers overwintered at the Cape; by 1785, several Nantucket whalers were working the Delagoa Bay Grounds and returning to Cape Town for anchorage.*

Foreign whalers had hunted sperm whales in the waters of Delagoa Bay and Madagascar; they even had their own Moby Dick, a bull sperm whale called Madagascar Jack that menaced whaleships off southeast Africa. Many species of whales would be killed in South Africa's productive waters, but by the end, it would be the great cachalot whose life and death would define South African whaling. In 1867, the year diamonds were discovered along the Vaal and Orange rivers, Svend Foyn of Norway invented the harpoon cannon, a device that dramatically changed the method by which whales were killed. Hungry for profits, and finding themselves short of whales in their own waters, the Norsemen fanned out throughout the world's oceans as their Viking forebears had, only in these raids, their victims were whales. As the 19th century came to an end, the Norwegians were establishing footholds throughout the whaling countries of the world. In their *History of Modern Whaling*, Tønnessen and Johnsen (1982) wrote, "In the oceans on both sides of South Africa were to be found the richest whaling grounds outside the Antarctic and the North Pacific. . . . In the period 1908 to 1916 a total of some 33,200 whales were caught, processing some 962,000 barrels of oil."

In Africa between 1900 and 1914, the Norwegians established no fewer than 14 whaling stations, including one at Durban, which was under the supervision of Johan Bryde, a Sandefjord businessman and

---

*The east coast of southern Africa is one of the most dangerous coasts in the world. The Agulhas Current, moving southward along the coast, often meets a swell moving northeastward, causing what a 1773 mariner's guide referred to as "monstrous seas," which can break the backs of today's supertankers, never mind the comparatively tiny whaleships of the 18th and 19th centuries.

the man for whom Bryde's whale was named. Bryde formed a partnership with Jacob Egeland, the Norwegian consul in Durban, and in 1905, the first whale was taken in African waters using modern methods. Operating as the South African Whaling Company, Bryde then obtained permission to build another land station at Donkergat, Saldanha Bay (on the west coast of southern Africa), and from 1910 to 1913, Norwegian whalers scoured African waters from Angola to Mozambique, killing the thousands of humpbacks that were migrating northward.

In the early days of whaling, Durbanites took a particular interest in the new industry appearing off their shores. Of course there was the smell to contend with—described by Ommanney (1971) as "a particularly penetrating, clinging, fat sort of odour, which makes one anxious to vomit without delay, and hangs around one's nasal membranes for days"—but other aspects of the whale fishery also attracted the attention of the locals. Whale meat was eaten in some of Durban's best restaurants and clubs, where it met with mixed reactions. "Porpoises"— probably bottlenose dolphins—were considered a threat to fishing, and one of the duties of the fisheries inspector was to shoot them with a rifle, which led to a rumor that there was soon to be a company that would bottle the porpoise oil. (There was also a fishery for bottlenose dolphins off Cape Hatteras, North Carolina, and another at Cape May, New Jersey, in the late 19th century, specifically for the "jaw oil," which was used as a lubricant for watches and clocks, but no such enterprise seems to have been undertaken in South Africa.)

By 1914, because of the war in Europe and difficulties with British authorities (Bryde had sold whale oil and meal to the Germans, despite a British prohibition on trading with the enemy), the Norwegians were forced to lease the Saldanha Bay station to a South African company, the Southern Whaling and Sealing Co., and with that, along with the failure of the station at Plettenberg Bay, Norwegian operations in South Africa ceased. The Durban station closed down until 1922, and when it was reopened, there were two companies operating there, the Union and Premier whaling companies. (The Union Whaling Company was named in 1908 in honor of the forthcoming Union of South Africa.) The two companies used a common slipway to haul the whales out of the water.

After the first season, the citizens of Durban complained so vociferously about the smell that the factory was forced to move out of the

harbor and around the bluff to the ocean side. Because of the heavy
seas that sometimes attended this location, however, the whales could
not be safely dropped there, so the practice began of dropping the carcasses inside the harbor, then trucking them around the bluff to the
processing station by railroad, a system that was unique in the history
of whaling. In his *Lost Leviathan* (1971), F. D. Ommanney described
the process:

> The unique feature of the whaling stations at Durban was the man-
> ner in which the whales arrived at their last rites. They came by train.
> A single track railroad of wobbly appearance ran from the harbour
> mouth round the head-land beyond the lighthouse and then along a
> mile and a half of loose, driven sand to the whaling station, both of
> which had been built along the track like a railway station platform.
> The catchers dropped their whales at the harbour entrance over-
> night and a harbour tug hauled them up a slipway where they were in
> due course winched up on to specially built low-slung bogey trucks.
> They were then trundled off to their obsequies by tank engines with
> a cow-catcher and a huge square headlamp on the front. The har-
> bour authorities were very strict because of the complaints about the
> smell, so that no carcass was allowed to lie at the slipway for more

than a few hours. If it had not been trundled away within the stipulated time, the harbour tug towed it remorselessly out to sea. Owing to the wobbliness of the track, and the fact that it was often buried by wind-blown sand, the train with its monstrous cargo often got stuck or the trucks came off the rails. It then had to wait, with the engine hissing indignantly, until a breakdown crane was brought.

Because sperm oil was such an enormously versatile product, it was well worth the trouble. According to a study by Surmon and Ovendon (1962), it could be used "for cosmetics, pharmaceuticals, biodegradable detergents, candles, additives for heavy-duty lubricating oils, greases, precision-instrument lubricating oils, printing inks, wetting agents, fatty alcohols, fatty acids, and as a general chemical intermediate." Cold filtered and sulfurized, it was a superb lubricant, and it was considered the finest oil available for the lubrication of delicate machinery. When sulfated (combined with sulfuric acid and then neutralized with ammonia), it was used for tanning leather. Meat and bonemeal products were supplied for protein additives to animal feeds, whale meat extract served as a food-flavoring ingredient, and frozen whale meat was used for human and animal consumption. (Much of it ended up in pet food.) The teeth were exported by the ton to the Orient for ivory carving.

In 1968, the Union Whaling Company had to cut back again. The fleet of catchers was reduced from 12 to six, and the total staff, once as high as a thousand, was halved. Public pressure was being brought to bear on whaling nations, and even the South Africans, usually impervious to public opinion, felt the pressure. (In 1971, the United Nations Stockholm Conference on the Environment recommended that there be a 10-year moratorium on whaling, and although this recommendation was passed unanimously, it was not enacted by the IWC until 1982.) Fuel oil was one of the major expenses in running a whaling fleet, and the Arab oil embargo of 1974 made it increasingly uneconomical to bunker large ships for extended voyages. By 1975 the Union Whaling Company, once the largest shore-based whaling station in the world, shut down. The catcher boats were sold, some for scrap, and

A harpoon cannon, painted white, stands in the garden of a hotel at Plettenberg Bay, a historical memento of the whaling station that once flourished there. (Richard Ellis)

one, the *R. H. Hughes*, was towed out to sea and scuttled because it was in danger of sinking on its moorings.

When the whaling station closed, historian Cornelis de Jong proposed preserving "a set of harpoons, flensing knives, hooks and other tools, the furnace of the blacksmith who straightened harpoons bent by struggling whales, and above all, one of the six fast, storm-battered, dapper whaleboats." His suggestions were ignored, and although two of the whale catchers, the *Pieter Molenaar* and the *C. G. Hovelmeier*, were actually purchased by the van der Stel Foundation, the Durban city council did not recognize their importance, and the moment was lost. Although there was some talk about turning the whaling station into a museum or a maritime center, it sat untended at the foot of the bluff until the South African army claimed it for a military base and practice range.*

THE AUSTRALIANS

When it became obvious that the species was economically extinct, the IWC forbade all humpback whaling south of the equator, but again, it was a case of too little, too late. The only whales left for Australian whalers were the sperm whales, and the last whaling station in the country was the one at Albany, Western Australia. In 1947, the Albany Whaling Company was formed by a group of local businessmen, who used old air force rescue boats as chasers. As the Norwegians had predicted, this venture failed, as did the next attempt to start up a whaling operation at Cheynes Beach in 1949. The Norwegians returned in 1952, and with their advice and equipment, whaling resumed at Albany. They used proper catcher boats (the *Cheynes II, III,* and *IV*), and until the 1963

---

*In 1971, Peter Gimbel made the film *Blue Water White Death*, about his quest to obtain the first-ever underwater footage of the great white shark. The film begins in Durban, where Gimbel and his crew hoped to encounter the great white shark feeding on the carcasses of recently killed sperm whales 100 miles offshore. They would not meet white sharks until more than a year later, when they got to South Australia, but off Durban, they left the cages to photograph each other swimming among oceanic whitetip sharks that were feeding on a sperm whale carcass. The South African episodes incorporate footage of Durban's Union Whaling Company hunting whales, killing them, transporting them by flatbed railcar to the processing station, and butchering them. *Blue Water White Death* can now be seen not only as an extraordinary diving adventure, but also as a grisly photographic record of the South African sperm whale fishery.

In permanent drydock, the Australian catcher boat *Cheyne IV* serves as an introduction to the Whale World Museum, on the site of the old whaling station at Albany, Western Australia. (Richard Ellis)

ban on humpback whaling, they took an average of 86 whales a year. In 1955, however, they had begun to take sperm whales farther offshore, and as the humpback catch declined and eventually disappeared altogether, sperm whale catches rose accordingly. According to company records, a total of 14,695 sperm whales were taken during the period 1955 to 1978. During this same period, sperm whaling was coming under tighter scrutiny by the IWC, and various restrictions on size and sex were imposed. Despite the size of the sperm whale catches compared to the humpback numbers, the same era saw enormous catches in the North Pacific by Russian and Japanese whalers. For example, in the decade 1961–1971, the total number of sperm whales killed outside the Antarctic (which figure included the Western Australian catch) was 211,650.

The handwriting was on the wall by this time, and the days of Australian whaling were numbered. Curiously, the calligraphers were the Australians themselves. In 1977, in response to what he called "a natural community disquiet about any activity that threatens the extinction of any animal species . . . particularly when it is directed against a species as special and intelligent as the whale," prime minister Malcolm Fraser ordered an inquiry to examine "whether Australian whaling should continue or cease [and] the consequences for international whaling of Australia's decision." Under the chairmanship of Sir Sidney Frost, this board of inquiry held public hearings at Albany, Perth, Sydney, and Melbourne, and heard testimony from everyone from the Cheynes Beach Whaling Company and the Western Australian government to Project Jonah and Friends of the Earth. There was to have been a hearing at Albany to discuss the results of the inquiry, but as Sir Sidney wrote in the preface to the 1978 report, "this hearing was deprived of much of its substance by the announcement on the first day of the proceedings that the whaling company would cease operations in the near future." The Cheynes Beach Whaling Company took its last sperm whale on November 20, 1978, and quietly closed down. Notwithstanding the closure of Cheynes Beach, the board of inquiry concluded "that Australian whaling should end, and that, internationally, Australia should pursue a policy of opposition to whaling." And pursue it they did. From a past that included some of the bloodiest whaling traditions in history, the Australians lit out after the whaling nations with the shameless zeal that only reformed sinners can know. Along with the New Zealanders—whose whaling history was, if anything, even more sanguinary—the Aussies are now among the leaders in the battle to eliminate commercial whaling.

The whaling station at Albany fell into disrepair, its tanks deteriorated, and its wooden buildings began to crumble and rot. The great chains and winches that had clanked noisily as they hauled dead whales up the cement slipway silently rusted. Whalebones discolored in the surf at Frenchman Bay, and proud Albany, a center of Australian whaling for nearly two centuries, also began to crumble. Deprived of their noblest industry, the citizens of Albany searched for other ways to bolster their flagging economy. Whaling had been the ideal industry for this remote port; the humpbacks passed close to shore on their northward migrations, and not far to the south were the families of cachalots, following the instinctive urges that drove them from the poles to the equator. With whaling gone, Albany turned to wheat and apples. A tuna fishery sprang up when the unemployed seafarers discovered that the great schools of southern bluefin tuna passed close to the shores of Albany on their migratory routes to the Great Australian Bight. But even this was to be denied to Albany; the government ruled that the tuna were immature when they were in Western Australian waters, and that they could only be caught by South Australian fishermen.

Was this town to vanish like the whales and whalers on whose blubber and blood it had been built? Not yet. Enter John Bell and the Jaycees Foundation of Australia. Bell had been a spotter pilot for the Cheynes Beach Whaling Company, and when everyone else had departed, he stayed on, trying to maintain the station by himself. It was an impossible job for one man, and he eventually got help from the Jaycees. They bought the whaling station, and with government and community support, they renovated it, added new buildings and artifacts (including the drydocked *Cheynes III*, the last of the whaleships), and turned it into Whale World: The World's Largest Whaling Museum.*

From 1970 to 1972, a shore fishery was conducted at Dildo Bay, Nova Scotia, where a total of 105 sperm whales were killed. They were all males, captured in water up to 1,000 fathoms deep, off the continental shelf (Mitchell 1975). In the Azores, the open-boat whalers continued their traditional fishery until about 1990. In the latter decades of the 20th century, the major part of the sperm whale fishery was conducted by the Japanese and the Soviets, using gigantic factory ships and flotillas of catcher boats in the North Pacific and the Southern Ocean. For the seasons 1972–1973 and 1973–1974, the IWC global quotas were 23,000 sperm whales per year.

THE JAPANESE

With the Japanese surrender on August 14, 1945, General Douglas MacArthur, the supreme Allied commander, decreed that the Japanese ought to be encouraged to commence whaling again in order to provide much-needed meat for the vanquished and starving people. (This arrangement was not as altruistic as it appears on its face; while the Japanese were to get the meat, the Americans were to get the whale oil.) All the Japanese had available were two tankers of 11,000 gross tons apiece, and these were quickly converted to factory ships, *Hashidate Maru* and *Nisshin Maru*. Originally, the Japanese were supposed to have spent only the 1946–1947 season in the Antarctic, but somehow, this single season stretched into two, then three, four, five, and eventually 40 more.

*For my 1980 *Book of Whales* and 1982 *Dolphins and Porpoises*, I had painted portraits of all the whales and dolphins of the world. (I also painted all the seals and sea lions, but I never wrote that book.) The Smithsonian Institution's Traveling Exhibition Service (SITES) circulated the 106 paintings throughout the United States for two years, but in 1986, the museum at Albany bought the entire collection and made it a part of the Whale World permanent collection.

From 1946 to 1951, the Japanese Antarctic fleet took 3,119 blue whales, 5,292 fins, 76 humpbacks, and 584 sperm whales. In 1951, after six postwar years of unregulated whaling, Japan became a member of the IWC.

In that year, the Japanese increased their fleet by the addition of the revamped 19,209-ton *Tonan Maru*. This was followed by the addition of another *Nisshin Maru*, built from scratch, and the only factory ship that the Japanese constructed after the war; all the others were acquired from other nations. The *Olympic Challenger* was purchased from Aristotle Onassis's outlaw whaling company in 1957, and the South African *Abraham Larsen* (formerly the *Empire Victory*) became another *Nisshin Maru* in that same year.

Seiji Ohsumi (1980), Japan's foremost sperm whale biologist, wrote, "The North Pacific has been recognized as one of the world's major whaling grounds for sperm whales since the 19th century." In addition to their Antarctic activities, therefore, Japan's whalers worked their own front yard, the Aleutian arc of the North Pacific. According to Nishiwaki's 1967 study, most of the North Pacific, from the Bering Sea to the equator, contains sperm whales, but the large bulls "migrate to the northward, clockwise for feeding." Although we know very little about the migratory habits of the sperm whale, it stands to reason that whalers, whether 19th-century Yankees or 20th-century Japanese, would seek out the largest ones they could find, and male sperm whales are about a third again as large as females. The first Japanese factory ship to work the North Pacific was the *Kaiko Maru* in 1946, followed by the *Baikal Maru*, which was sent out to sea with a flotilla of catcher boats in 1952 after working around the Bonin Islands. A second fleet was added in 1954, and a third in 1962. From 1962 to 1975, the Japanese operated three fleets in the North Pacific in competition only with the Soviets.

In his 1965 *Stocks of Whales*, N. A. Mackintosh wrote, "It is hard to see any way at present of making even a wild estimate on the magnitude of the world population of sperm whales." By this time, however, it was necessary for the whalers to proclaim that they had some understanding of the population dynamics of the animals they were killing, so they began to count them. Whale counting is even more complicated than predicting migration patterns, especially for animals that spend most of their lives swimming below the surface of the ocean out of sight of those who would count them. In 1966, Nishiwaki estimated that there

Seiji Ohsumi and Richard Ellis, Tokyo, 1981. I had come to Japan to join a whaling voyage out of Taiji. (T. A. Ellis)

were some 150,000 male and female sperm whales migrating through the North Pacific. Within three or four years, as the controversy became more heated, the numbers went up proportionately. By 1971, Ohsumi was using complicated mathematical formulas that factored in pregnancy rates, age at sexual maturity, and other poorly understood elements, and estimated that there were 167,000 males and 124,000 females in the North Pacific.

The Japanese literature is (perhaps deliberately) vague on the uses made of sperm whales. Tønnessen and Johnsen (1982) wrote, "Even though sperm whale meat is used to some extent for human consumption in Japan, it is the least sought after. For this reason most of the sperm whale is reduced to oil," which is used for a variety of purposes. Yet ever since they began the massive hunt of sperm whales, the Japanese have insisted that they were taking the whales for human consumption. Sperm whale meat, like dolphin meat, is unappetizingly purplish black in color because of its high myoglobin content, and it is said to have a most unpleasant taste. The eating habits of the Japanese are different from those of many Westerners, however, and although they would probably have preferred to eat the meat of the baleen whales, they did indeed dine on the meat of the sperm whale.*

By 1988, Taiyo Fisheries and Nippon Suisan Kaisha, the two major producers of stewed whale meat, had stopped processing it. According to a Japanese news release of April 6, 1988, "Instead, the companies have tried to placate their customers by offering substitutes—barbecued mutton and stewed horse meat—prepared in the same sweet-spicy, soy-based sauce that made canned whale meat such a delicacy for Japanese taste buds for most of a century." In the same release, Furnio Imanaga, president of the whaling company Nippon Suisan, is quoted as saying, "In the heyday of whale-meat consumption in Japan, almost 20 percent of our total canned-food sales came from stewed

*When I asked Seiji Ohsumi what was done with the meat, he wrote, "Although sperm whale meat is not as delicious as the baleen whale meat, it is eaten by many Japanese people." He listed the methods of its preparation: steak, cooked, salted dry, bacon, and *matsuura-zuke* (the thinly sliced nasal cartilage, which is soaked in sake). Someone else liked the meat of the sperm whale, or so we are told by Thomas Beale in his 1835 book. He quotes Cuvier: "The Greenlanders are remarkably fond of its flesh, which they consider a delicate viand, when it is dried in smoke."

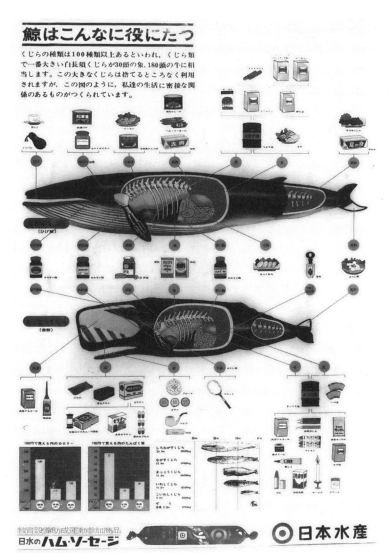

whale. During the years of postwar food shortages whale meat was a valued source of animal protein. To most Japanese who lived through those years, meat was synonymous with whale meat. Talk to any Japanese over 35. They'll tell you they will always remember the taste of whale meat stew with nostalgia." Perhaps it was the sperm whale meat that was stewed.

Whatever became of the meat, it was the sperm whale's oil that was important economically. In 1967, as the balance was shifting from the Antarctic (where they were simply running out of whales) to the North Pacific (which was an area that hadn't been regulated before), so too was the proportion of oil to other whale products. In 1962, sperm whales represented 57 percent of Japan's total catch; by 1975, they made up 93 percent of the total. Even as the numbers of available whales dwindled, Japan (along with the Soviet Union) continued whaling in the Antarctic, as well as in the North Pacific. With Norwegian whalers hors de combat (or seeking employment in South Africa or Australia), the field was wide open for the Japanese and the Soviets. For the next 10 years (1968–1978), they killed hundreds of thousands of fin and sei whales in the Antarctic (humpbacks were fully protected all around the world by that time, as were blue whales), as well as an ever-escalating number of sperm whales in the North Pacific.

The total number of sperm whales killed by the Japanese from 1951 (the year Japan joined the IWC) to 1976 was 124,458. During the same period, Soviet whalers in the North Pacific killed 102,314, for a grand total of 226,772. In Ohsumi's 1980 paper, from which the above figures were taken, he gives 60,842 as the number of sperm whales killed in the North Pacific for the period 1800–1909. The average per year for 1800–1909 is 558, while the average for 1951–1976 is 9,880. A more eloquent testimony to the efficacy of modern whaling—and its catastrophic effect on whales—would be hard to find. (Among the whaling statistics for 1957 was a record of a 35-foot-long male sperm whale that was completely white. Although he did not get to see it before it was flensed, and therefore could not observe its eye color, Seiji Ohsumi [1958] concluded that it was indeed an albino. "If it had not been killed in young generation," he wrote, "it would have reigned over the sea like the ancestor Moby Dick.")

It was during the period of intensified Japanese whaling that the antiwhaling movement was escalating throughout the rest of the world, and as part of the protest, a demonstration was arranged to coincide with the visit of Emperor Hirohito to Washington, D.C., on October 2, 1975. As the first Japanese emperor to ever set foot on the soil of the continental United States alighted from his limousine at the south portico of the White House to be greeted by President Ford, an airplane buzzed overhead, trailing a banner that read, "Emperor Hirohito, Please Save the Whales." It would appear that the emperor did not have much influence with the whalers, however, and despite the protests— or perhaps because of them—the Japanese campaigned for increased quotas.

Since the mid-1970s, world opinion has been solidly opposed to Japanese whaling. In response to the petitions, editorials, demonstrations, protests, magazine articles, and international condemnation, the Japanese dug in. They chose to interpret the antiwhaling attitudes as racially motivated—another instance of Europeans attempting to impose their customs on vulnerable Asians. At first, the whaling controversy seemed like a poker game, with each side holding cards that it dared the other side to play. The United States, the purported leader of the anti-Japanese movement, even went so far as to pass legislation (the Pelly Amendment and the Packwood-Magnuson Amendment) specifically designed to use against the Japanese. The Japanese threatened to pull out of the IWC and go whaling under their own rules. It was not the first time such threats would be expressed, nor would it be the last.

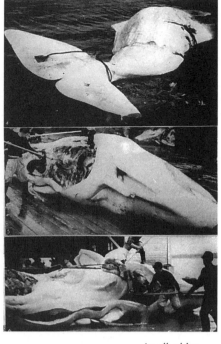

An all-white sperm whale captured by Japanese whalers in the North Pacific in April 1957. (Seiji Ohsumi)

Nineteen seventy-six was another interesting year for the Japanese whaling industry. Although the quotas for most whale species were curtailed at the IWC meeting held in Canberra, Japan surprised the world by using a tactic that would enable it to continue whaling without concern for IWC restrictions: Japan issued itself a scientific permit. The quota for Southern Hemisphere Bryde's whales for the 1976–1977 season was zero. Despite this, the Japanese whalers managed to kill 225 of them, and except for a loud public outcry, there was nothing that anybody could do about it. The IWC has no enforcement capabilities and depends on the willingness of the member nations to comply with its resolutions, but in this instance, the Japanese discovered a loophole large enough to drive a factory ship through.

Written into article 8 of the *International Convention for the Regulation of Whaling* is this seemingly innocuous paragraph:

Notwithstanding anything contained in this Convention, any Contracting Government may grant to any of its nationals a special permit authorising that national to kill, take, and treat whales for purposes of scientific research subject to such other conditions as the Contracting Government thinks fit, and the killing, taking, and treating of whales in accordance with the provisions of this article shall be exempt from the operation of this Convention.

I went a'whaling aboard the Japanese catcher boat *Toshi Maru* 18. Note the crow's nest and the harpoon cannon mounted on the sharply raked bow. They killed a Bryde's whale. (Richard Ellis)

Obviously, this language was written into the convention to allow contracting governments to perform scientific experiments without falling under restrictions imposed on commercial whalers. In other words, if a country's scientists believed that they could find a cure for cancer by using whale oil, they shouldn't have to qualify as whalers to collect material for their experiments. (Before the passage of the Marine Mammal Protection Act in 1972, which, among other things, shut down the last American whaling station, United States whaling companies had regularly awarded themselves scientific permits for the taking of gray whales and sperm whales off the California coast, and they did indeed conduct research on the whales. For a study published in 1971, American cetologists Rice and Wolman examined 316 gray whales that were collected by catcher boats off the coast of central California.) It was nowhere written that the government had to tell anyone what they had in mind if and when they issued themselves a scientific permit, however, and when the Japanese awarded their whalers the right to harvest 240 Bryde's whales in the Southern Ocean, they were bound by no restrictions to identify the nature of the science that was the ostensible purpose of the hunt. In fact, their scientists wrote reports about the Bryde's whales, and almost 2,000 tons of whale meat went right into the freezers of the whaling company and eventually into Japanese stomachs. By the following year, the IWC believed it had closed this loophole by recommending that future applications be subject to scrutiny by the Scientific Committee, but as we shall see, it was not closed, only lightly papered over.

At the 57th annual meeting of the IWC, held in Ulsan, South Korea, from June 20 to June 24, 2005, despite much backroom politicking (and threats to withdraw financial aid to countries that did not support its revised whaling plan), Japan's plan was soundly defeated. Arguing that Japan's "scientific" whaling provided little science and a lot of whale meat, opponents voted down the Japanese proposal and upheld the continuation of the 19-year-old moratorium. Not surprisingly, Japan threatened to quit the IWC altogether, automatically rendering any of its regulations moot and allowing the Japanese to kill as many whales as they wanted and to do whatever they wanted with the meat. It appears that Japan will continue its "scientific" whaling program, and despite universal condemnation of their repudiation of the moratorium, they intend to kill protected humpbacks and fin whales anyway.

Estimates of whale populations depend largely on who is doing the estimating. Pro-whaling countries usually produced higher estimates to allow them to continue killing whales, whereas those who would reduce or eliminate whale killing for profit produce lower estimates, to show that the hunted whales were already in trouble. In the past, population estimates were based on the number of whales killed, the number of whales sighted in a given area, and other seat-of-the-pants methods, none of which really produced satisfactory results. Whale counting is a difficult business at best, but to set the annual quotas, the IWC had to have some idea of the size of the stocks being fished. Sperm whales, hunted since the early 18th century by Yankee whalers and later by almost every other whaling nation, are poorly understood, especially with regard to population size.

It was easier to tally the number of sperm whales killed, especially in modern times. Although Yankee whaling captains may have kept meticulous logbooks, many of these have been lost, so the totals for the 18th and 19th centuries are estimates at best. For his 1935 study, *The Distribution of Certain Whales as Shown by Logbook Records of American Whaleships,* Charles Townsend was able to find records of

36,909 sperm whales killed between 1753 and 1914. For this 161-year period, that would average 229 whales per year for the entire whaling fleet, which at its peak in around 1850 numbered over 800 ships. Townsend found some 1,500 logbooks; obviously, more than a few have gone missing.

Townsend's 1935 monograph is accompanied by four charts that show the plattings of the whale catches listed in the book. There is one chart for humpbacks and bowheads, one for right whales, and two for sperm whales, the first for October to March and the second for April to September. For the most part, the records coincide with the old whalers' grounds—those areas where they could expect to find sperm whales with some regularity. However, neither of the two sperm whale charts shows any whale catches at all in the Pacific north of 40°N latitude. Sperm whales are thought to have first been taken by modern whaling in the North Pacific at the earliest in 1905, and by 1976, a grand total of 268,972 whales had been killed there by the whaling fleets of Japan and the USSR (Ohsumi 1980). By the start of the 20th century, hand-thrown harpoons had been replaced by cannons firing exploding grenades, and factory ships could process the carcasses at sea, factors that account for the great increase in numbers. In his 1974 history of the IWC, J. L. McHugh reported that 532,392 sperm whales had been killed by whalers from 1920 to 1971, a large proportion of them in the North Pacific.

About a century after the New England fishery declined in about 1865, hunting for sperm whales resumed on a large scale, primarily from the shore-based fisheries in Western Australia and South Africa. In the 1960s and 1970s, however, the major part of the sperm whale fishery was conducted by the Japanese and the Soviets using gigantic factory ships and flotillas of catcher boats in the North Pacific and the Southern Ocean. During 1972–1973 and 1973–1974, the IWC global quotas were 23,000 sperm whales per year.

To establish these quotas, the whaling nations needed an idea of the number of sperm whales and their reproduction rates so that the Scientific Committee of the IWC could determine how many could be killed without damaging the population's ability to regenerate. It had always been assumed that females were critical to the breeding populations, and therefore, the numbers of females included in the yearly quotas were always lower than the numbers of males. In 1977, E. D. Mitchell questioned this concept and suggested that within the complex social structure of the sperm whale, the removal of the single bull (the "harem master") servicing a group of females "could reduce the pregnancy rate in this school drastically." This was indeed a remarkable observation, for the early fishery emphasized the taking of the largest bulls (they naturally yielded more oil), and for years, the decline of the fishery was attributed to economics, the Civil War, the discovery of petroleum, and other factors. Until Mitchell's suggestion, it seems no one believed that the whales might not have been able to reproduce enough to maintain the species. "It is possible," Mitchell continued, "that the techniques of the early American fishery, coupled with the complex social behavior of the sperm whale, might have resulted in reduction of the population over decades far out of proportion to reduction judged from the landed catch or oil yield alone."

If determining the relative importance of males and females to the population seems difficult, estimating the population itself turns out to be almost impossible. In 1965, N. A. Mackintosh, an authority on whale populations, wrote, "It is hard to see any way at present of making even a wild estimate on the magnitude of world populations of sperm whales." Despite his warning, experts in population dynamics labored hard, and by using the marking of whales, the sighting of whales, and the application of various mathematical formulas, they tried to come up with some workable numbers. Estimates varied widely depending on who was doing the estimating and what methods were used. (There is also the variable of whether the estimator is in the whaling business; Japanese estimates of sperm whale populations always appear higher than those of, say, British or Americans.) Best (1975) reviewed the various estimates of world sperm whale stocks and reported as follows:

> Division II. (East Atlantic). The least squares method gave estimates of 18,000 exploitable males (from Donkergat data) or 27,000 exploitable males (pelagic data) for the 1963 stock, and a rounded average of 22,000 was used to get an initial stock of 34,000 males. From this the original mature female stock was calculated to be 44,000 and the 1972 stock 42,000. An independent estimate of the mature female population size for the period 1957 to 1964 by Best (1970) gave values of 15,550 or 31,940, using fishing mortality rates and catches.

From this, it would appear that Mackintosh was correct, and even the most sophisticated methods were not producing a particularly strong foundation for setting quotas. Since 1968, the Scientific Committee of the IWC has been unable to devise an accurate method for estimating sperm whale stocks, and in addition to the annual meetings of the IWC, a number of special meetings were convened, at which the primary topic was the problem of sperm whale numbers. At one of these special meetings at La Jolla, California, in December 1976, despite the presentation of 29 papers on sperm whale population biology, the scientists were forced to admit that they had very little real knowledge of the sheer numbers or the reproduction rate of the sperm whale. In short, the Scientific Committee was setting quotas for these whales with virtually no idea of how many there were or how to figure it out. In a report on the La Jolla meeting, population biologist Tim Smith (1976) made the following comments: "We must assume that the harvesting of sperm whales is having an impact, even though we cannot determine the magnitude of that impact with certainty. . . . The one thing we do know is that harvesting of large whales can cause rapid and extensive reductions in abundance." Not a great revelation, to be sure, but for the first time, the scientific community publicly conceded the possible adverse effects of harvesting whales without knowing their population figures.

In spite of their admitted ignorance, the Scientific Committee—under pressure from the sperm-whaling nations—recommended raising the 1977–1978 quota to 13,037, as compared with 12,676 for the previous season. At another special meeting convened at Cronulla, Australia, in June 1977, the IWC Scientific Committee recommended a quota of 763 sperm whales for the North Pacific (the area where the Japanese and the Soviets concentrated their efforts, and where, only four years previously, the quota had been 10,000 animals), but in a move that shocked the world and the whaling community as well, the IWC general meeting set the quota at 6,444 sperm whales—nearly 10 times the number recommended by their own scientists. Reaction to this maneuver was a worldwide outcry, as well as directly and indirectly applied pressure on the IWC. At the 1978 meeting in London, no quota at all was set for sperm whales; confusion and indecision seemed to be the order of the day. It took still another special meeting in Tokyo in December 1978 to establish the North Pacific quota of 3,800 animals, a reduction of almost 40 percent from the previous year's calamitous 6,444. It was now apparent that the IWC could no longer continue to function as a gentleman's club for the whaling nations of the world, and atten-

International Whaling Commission, June 1984, Buenos Aires. The author (head bowed, dark jacket) is at the center of the photograph. (International Whaling Commission)

tion had to be given to the pressures being administered from without, especially from the numerous conservation and environmental groups that had mounted such massive "save the whale" campaigns.

The estimates for the total number of sperm whales still vary widely. In an article in *National Geographic,* Victor Scheffer (1976b) provided the following estimates for sperm whales: 212,000 males and 429,000 females, or 69 percent of the estimated pre-exploitation population of 922,000. As a further example of the difficulties inherent in estimating populations, M. R. Clarke (1977) offered the figure of 1.25 million sperm whales, a number considerably higher than Scheffer's estimate. In *Sea Guide to Whales of the World*, published in 1981, Lyall Watson wrote, "The best estimates suggest that there may be 350,000 in the southern hemisphere and another 175,000 in the north. These figures are based on approximations which may be hopelessly optimistic." In what is probably the most accurate assessment for its time, Dale Rice (1989) wrote, "Prior to extensive exploitation by modern-style whaling, the world population of the sperm whale was probably close to three million. It has since been reduced by about 31% to less than two million."

Townsend's (1935) figure of 36,909 sperm whales killed during the 161 years of Yankee whaling was seriously flawed. The estimates were probably too low by at least an order of magnitude. McHugh's (1974) half million sperm whales killed between 1920 and 1971 is an accurate compilation of what the whalers reported—but they were lying. After Soviet whaling had ceased—and in fact, after the Soviet Union itself had ceased—Russian cetologist Alexey Yablokov (1994) revealed in an article in *Nature* that the Soviet whalers had greatly underreported the number they killed. "It was also known in the 1960s," he wrote, "that a Soviet factory ship illegally operated for a couple of weeks in the Okhotsk Sea, and caught several hundred right whales. It was also well known in the Soviet Union that blue whales continued to be killed after they were protected by the IWC." Citing Yablokov, Phil Clapham (1996) noted that "former Soviet biologists revealed that the USSR had conducted a massive campaign of illegal whaling beginning shortly after World War II. Soviet factory fleets had killed virtually all the whales they encountered, irrespective of size, age, or protected status." In the Southern Hemisphere, while reporting a total catch of 2,710 humpbacks, the Soviets had actually taken more than 48,000. They exported all the meat from the illegally killed whales to Japan.

Further evidence of underreporting in Japanese coastal whaling operations shows that the problem was not limited to pelagic whaling operations and occurred despite the presence of onboard inspectors and observers. Isao Kondo, a former executive of Nihon Hogei (Japan Whaling Company) and director of whaling stations at Ayukawa, Taiji, and Wakkanai, published *Rise and Fall of the Japanese Coast Whaling* in 2001, in which he detailed the methods by which Japanese whalers misreported the whales they killed:

> Cheating on the taking of sperm whales began in 1950. And the degree became more terrible after 1955. According to a record of some company, the catch number made public was 326 while the true number of captured whales was 464 (the company concealed 138 whales), which was still more conscientious compared with concealment. Then, the cover-up of captured whales escalated. It was said that according to the data of one whaling company, the captured number made public was 30% of the actual number—in other words, they captured three times the number made public. We do not have the numbers of "mis-reported" whales from Japan, but the numbers for the Soviet sperm whales—74,834 reported; 89,493 actually killed—are probably closer to reality. The deviation of the actual numbers killed from the reported numbers renders estimating the actual number of whales remaining increasingly difficult.

Even if they killed 10 times the number that Townsend found in the logbooks, the Yankee whalers would not have made that much of a dent in the world population of sperm whales. As far as we knew,

Townsend's estimate of 700,000 remaining would reproduce and multiply as New England whaling declined. (The *Wanderer*, the last of the square-rigged whalers, set sail from New Bedford on August 27, 1924, and was wrecked the next day on Cuttyhunk Island.) In the 1960s, the IWC set merciless quotas for the North Pacific, where the Soviet and Japanese catcher boats and factory ships operated. But as long as the estimated world population of sperm whales was believed to number a million or more, even the most severe critics of sperm whaling were not really worried about the disappearance of *Physeter macrocephalus*.

THE SOVIETS

When it was recognized that the North Pacific was as important an area for sperm whales as the Antarctic was for rorquals, the whalers adjusted accordingly. Until 1954, the Soviets operated only the flotilla of the factory ship *Aleut* in the vicinity of the Kamchatka Peninsula and the Commander Islands, but because other whaling nations also had access to these grounds, the whales were quickly fished out, and the whaling changed from an inshore operation to a purely pelagic one. The hunting area was extended throughout the North Pacific, all along the arc of the Aleutians to the Gulf of Alaska. As the quotas were increased, the Soviets added men and ships to their fleets. In 1956, they announced their five-year plan (1956–1960) for whaling, which included the building of five factory ships. By the late 1960s, they were operating the *Aleut*, the *Sovietskaya Ukrania* (built in 1959), the *Yuri Dolgorukiy* (1960), the *Sovietska Rossiya* (1961), the *Vladivostok* (1962), and the *Dal'nii Vostok* (1963), all gigantic factory ships with diesel catcher boats to do the hunting. During the years 1956–1964, the Soviets built 67 new catcher boats, averaging 843 gross tons, their 3,600-horsepower engines capable of powering them at more than 19 knots. It was obvious that the Soviet Union was planning to become a major factor in the business of pelagic whaling. For this design, they benefited enormously from the support of the state; other whaling countries, such as Britain, the Netherlands, and Norway, were trying to compete as privately run companies. To this day, it is not known to what extent the Japanese whaling industry was subsidized by the government. There seemed to be no way of stopping the Soviet juggernaut; it appeared that this late-comer to pelagic whaling was going to dominate it so completely that there would be no whales left for anyone else. By 1968, only the Soviet Union and Japan would be sending whaling ships to the Antarctic.

At the 1959 IWC meeting, the Antarctic whaling nations were unable to arrive at national quotas for baleen whales. The Netherlands and Norway quit the commission, leaving it up to the individual governments to plan their own quotas. The Soviet Union refrained from participating in the discussions, but when the quotas had been worked out, it had ended up with 20 percent of the total. The same thing happened the following year, and the USSR was on the way to carving out a sizable portion of the Antarctic pie for itself. It was becoming clear that the dominant whaling nations were Japan and the USSR. Britain retired from whaling in 1963, the Netherlands in 1964. Only Norway was left to compete with the behemoths of the whaling industry, and the Norwegians could not contend with the socialized industries of Japan and the USSR. By 1967, Norway had ended its whaling, and the two North Pacific nations were left to argue over the number of whales they ought to be allowed to kill in the Antarctic.

By the time the field had been cleared, however, the whales of the Antarctic had been so reduced by the excesses of the competing nations that the whalers realized that they would have to look elsewhere. They decided to put their men and matériel to use in the North Pacific. (One of the problems with a reduction in quotas has always been the employment of the workers. Because they are different types of societies, Japan and the USSR have different approaches to this problem, but the solution is the same: don't lay off the whalemen.) When scientists realized that there weren't enough blue whales and humpbacks in the North Pacific to sustain an industry, they turned to the sperm whale. As of 1970, there were no quotas for this species, but by the following year, the whaling commission had set a limit of 10,481 for the Japanese and the Soviets, the only nations still in the pelagic whaling business.

Other whalers continued to kill other whales. The Norwegians still hunted minke whales; the Canadians were killing fin whales out of Newfoundland stations; there were various Japanese-financed operations in South America; and the Australians were taking sperm whales off their west coast. Sperm whaling out of Durban was winding down, and "research" whaling was being conducted by the Americans in California on such a small scale as to make comparison with the whaling flotillas absurd. All these operations combined could not compare with the numbers of sperm whales being killed by the pelagic whalers.

In 1966, the Soviets brought up the *Slava*, the last Soviet factory ship to operate in the Antarctic, and in that year, the Soviet fleets took

At sea, whale carcasses were delivered for processing to Soviet factory ships like the 700-foot-long *Slava*. (P. Golubovsky)

three times as many sperm whales as the Japanese (the actual count was 9,436 to 3,000), even though a large proportion of the whaling was taking place in Japanese waters. The Soviet cetologists B. A. Zenkovich and V. A. Arsen'ev (1955) have calculated that 86,000 sperm whales were taken in the 14 years from 1950, the year the Soviets began sperm whaling. Soviet cetologists conducted intensive research on the whales they killed—obviously in the interest of continuing this highly profitable industry—and were more than a little surprised to discover that the whales were in trouble; they were much smaller and much younger than they had been in past seasons. A. A. Berzin (1972) said: "By the end of 1963 it had become clear that there was an urgent need of protective measures for sperm whales that come to subarctic latitudes for the summer. Absence of stringent international restrictions may prove fatal for the North Pacific sperm whales." After this realization, the Soviets and the Japanese voted to increase the quotas for North Pacific sperm whales.

Mechanized sperm whaling was not very different from the old open-boat methods; it was just more efficient. The whales were spotted from a lookout, and then the catcher boat approached them, under

Peeling the blubber from the head of a sperm whale aboard a Soviet factory ship in the North Pacific. (Victor Scheffer)

minimum or no power so as not to frighten them. The gunners tried to shoot the whale on the left side, to get at the heart. When the whale was dead, it was brought alongside the catcher boat, and its body cavity was inflated with compressed air to keep it afloat. (Nineteenth-century whalers did not have the ability to pump dead sperm whales full of air, and evidently they didn't have to. Sperm whales are not supposed to sink, but the Soviets inflated them anyway, claiming that only the adults, with their thicker blubber layer, remained afloat after they were killed.)

The carcasses were delivered to the factory ships and then hauled up the slipway, where the blubber was stripped off and the spermaceti ladled into kettles. The bones, meat, and viscera were boiled together, and the teeth were removed from the lower jawbone. In the 19th century, only the blubber, the spermaceti, and the teeth were kept, but the modern Soviet whalers used every part of the whale except the smell. The skin was removed and tanned aboard the factory ship, and if it was not marred by parasite or harpoon holes, it was cut into sheets that measured 70–120 by 50–80 centimeters (27–46 by 19–31 inches). A

52-foot whale yielded a hide that was 78 meters (255 feet) square and weighed 4,030 kilograms (2,266 pounds)—more than a ton. Whale leather was used to make shoe soles and heavy work gloves. Soviet cetologist A. A. Berzin (1972) summarized the use of the whale oil and other products:

> The fatty substances obtained in sperm whale processing are unsuitable for human consumption because of the high content of unsaponifiable substances; they are being used as technical oils. . . . The fatty acids are used in the soap industry and the unsaponifiable substances in manufacturing detergents; the high molecular aliphatic alcohols in the leather and rubber industries; in the manufacture of cosmetics; in the degreasing [of] wool; and flotation of ores; the stock after fat extraction is used for preparation of gelatin. . . . Spermaceti oil is used as lubricants for fine mechanisms. . . . Solid spermaceti is used as carriers in manufacturing many medical and cosmetic products, mainly face creams and ointments, and for the production of lithographic ink. The therapeutic properties of spermaceti have been known for a long time; for instance, it is very good for the treatment of burns. . . . The meat of sperm whales is inedible because it contains adipocere, but it is rich in proteins and is therefore valuable for the production of feed meal. . . . Boiled sperm whale flesh can be used for feeding fur-bearing animals and in the preparation of dry protein. The liver of whales, particularly of the sperm, is the most valuable raw material for the vitamin industry. . . . The liver of one sperm whale contains as much carotene as 50 tons of carrots.

A whale-liver extract was also used in the manufacture of a preparation called Campolon MG, which the Soviets use to treat anemia. The pancreas yielded a substance used to make insulin, and the pituitary produced an adrenocorticotrophic substance for the treatment of arthritis and gout. A surgical sponge made from the collagen of the whale's flukes was used as a temporary replacement for donor skin in the treatment of burns. The tendons were used in the manufacture of glue. Ambergris is usually encountered in smallish lumps in the whale's lower intestines, but sometimes great boulders of the stuff are revealed when the whale is eviscerated. The largest piece found by the Soviet whaling fleet weighed 595 pounds, and was extracted from a 50-foot male on the *Sovietskaya Rossiya* in 1967.

As described by the Soviet cetologists, sperm whales sounded like the panacea for all of mankind's ills. They provided food, medicine, leather, perfume, oil, and numerous other products necessary for the good life in Petropavlovsk. Unfortunately, the whale supply was running low, and neither the USSR nor any other whaling nation was willing to stop the killing. Perhaps they would have gotten the picture if the whales also laid golden eggs. Public opinion, rarely a factor in the Soviet scheme of things, would eventually put the Soviets out of the whaling business, but like every other whaling country, they would not go quietly, and before they muzzled their harpoon cannons, thousands more sperm whales would die.

In the mid-1960s, as the stocks of baleen whales fell to levels where the whalers found it uneconomical to search for them (in the whaler's parlance, the "catch per unit effort" was unsatisfactory), they noticed that there were still lots of whales around; they were just not the proper ones. The right whales were gone altogether, and blues, finners, seis, and humpbacks had been reduced to such low levels that they were classified as protected or endangered. The economics and politics of whaling had reduced the killer fleets to those of Japan and the Soviet Union, two countries that were fully prepared to change their targets from baleen whales to sperm whales. Both nations needed the oil and fertilizer, and although the Soviets didn't eat the meat, their minks and sables did, and besides, if there was any left over, they could always sell it to the Japanese. For the Japanese, these previously untapped resources represented a vast new area to be plundered.

After ignoring sperm whales for so long, the IWC finally recognized their existence—or their way out of existence. The only whale that most people had ever heard of or imagined had finally achieved the ultimate recognition. The species could not have had a worse endorsement; the IWC's imprimatur was the kiss of death. In the annual conferences, whenever the subject of *Physeter macrocephalus* was raised, it was to determine how many of them might be killed.

Like a trickle that eventually becomes a raging torrent, the sperm whale made its appearance in the annual reports of the IWC in marginal references. In a prepared statement to the IWC in 1965, amid its concerns about the "depletion of baleen whale stocks," the USSR delegation "agreed . . . that it is necessary to reduce the catches of female sperm whales." (Male and female sperm whales would be treated almost as separate species because of the variations in their range—the females remain separated from the males for most of the year and inhabit different areas—and because of the size differential. So that neither gender would be unduly taxed and thus become unavailable for breeding, the sexes were assigned separate quotas.)

Because it was becoming evident that sperm whales were going to replace baleen whales as the major objects of the fishery, a special session of the IWC was called for 1966. Among its conclusions: "a special group should be set up to undertake over-all stock assessments as soon as possible. To this end, complete effort statistics should be made available by each country to the International Whaling Commission." If sperm whales had been able to read, they would have departed for the planet Pluto. The mechanized whale navy was getting ready to train their warheads on the sperm whales, as soon as they conducted the research that they believed was necessary to justify the impending slaughter.

First they divided the world into convenient sectors, so as to make it easier to monitor catches. It didn't matter that the whales did not respect these divisions and freely wandered from one sector to another; nor did it matter that there was a strong possibility that the whales might even migrate from one hemisphere to the other. The Scientific Committee apportioned the Southern Hemisphere into nine divisions and cut the North Pacific in half, under the (probably erroneous) assumption that there were two separate stocks of sperm whales in that area. It also decided that North Atlantic sperm whales constituted a single population, so for the benefit of the Norwegians, the Icelanders, and the Canadians, that entire ocean was deemed a single management area. By 1983, scientists estimated that there were some 111,400 male sperm whales and 162,600 females in the eastern North Pacific, and 61,000 males and 137,000 females in the western sector.

For the year 1968, North Pacific sperm whalers took 12,740 males and 3,617 females. The next year, the figures were 11,239 and 3,605. In the IWC report for 1970, the Scientific Committee "agreed that it is

Bumper sticker designed by the author in 1978. (Richard Ellis)

desirable to slow down the decrease of male stock in view of apparent excessive catches," and they recommended that "a further reduction in the catch of male sperm whales is desirable." The IWC therefore set the first quotas on sperm whales in 1970 for the 1971 season. It allocated 5,760 sperm whales to Japan, 7,716 to the USSR, and a surprising 75 to the United States, which was in the process of closing down the last of its whaling operations in California.* For the next three seasons, the commission set quotas of 6,000 males and 4,000 females for the North Pacific, but not once did the Soviet Union or Japan meet these quotas. At the same time, the two nations were taking some 13,000 sperm whales in the Antarctic.

With savage intensity, the sperm whalers raised their catch figures, and the IWC, helpless against the powerful whaling nations, could only watch as the statistics soared to astronomical levels. From 1964 to 1974, 267,194 sperm whales were killed in the northern and southern oceans, an average of 24,270 per year. With the passage of the United Nations resolution of 1972 on the cessation of whaling, the eyes of the world were on the IWC. Would this body conform to world opinion? Would it reduce the quotas of whales? When a moratorium was introduced to the IWC by the United States delegation at the 1972 meeting, the commission rejected it, and it did so again in 1973. Meanwhile, the United States passed its own Marine Mammal Protection Act in 1972, which marked the end of American whaling. The slaughter continued. In 1974, 21,217 sperm whales died.

Better systems for estimating whale populations—and populations of other marine creatures as well—are critical to understanding and preserving marine biodiversity. In the introduction to their 2004 paper entitled "Modelling the Past and Future of Whales and Whaling," Scott Baker and Phillip Clapham wrote, "Historical reconstruction of the

*In 1970, when Kenneth Norris wanted to examine a sperm whale's head to try to identify the source of its sounds, he visited the whaling station at Richmond in San Francisco Bay. In *The Porpoise Watcher* (1974), he describes the whaling station: "Down at the end of one such track was a gaunt building whose sign dimly proclaimed, 'Del Monte Fishing Company.' A single dirty bulb depended from a metal yardarm over the dirt street. From inside, through the grease-covered windows, came a hum of activity. A dump truck, comfortable in its rusting old age, stood stumped under a chute that pierced the building wall. From the chute came a mass of formless whale parts and oily fragments of blubber, sluiced into the recesses of the truck by a man standing in hip boots deep in the gurry."

population dynamics of whales before, during and after exploitation is crucial to marine ecological restoration and for the consideration of future commercial whaling. . . . At present, demographic and genetic estimates of pre-exploitation abundance differ by an order of magnitude, and, consequently, suggest vastly different baselines for judging recovery." They continued: "Estimating the former abundance of whale populations and 'stocks' and reconstructing the historical trajectory of their decline are essential to make an accurate assessment of the true impact of whaling on the marine ecosystem, and to establish a baseline for judging the recovery of whale stocks." Modeling—the subject of their study—is the major tool for untangling and then predicting population dynamics. Modeling depends on the assignment of mathematical values to such factors as hunting pressure, birth and death rates, and environmental fluctuations to estimate the population size during a given period. When the IWC was setting quotas on various species, models were used by members of the Scientific Committee to determine the maximum sustainable yield for whales. High-speed computers now enable modelers to insert various values into the model, permitting them to anticipate results of best-case or worst-case scenarios, and any variations in between.

The year between the 1978 and 1979 meetings was marked by even greater pressure on the whaling industry, and by 1979, the stage was set for substantial changes. Not even the most hopeful of the observers, however, was prepared for the results. All factory-ship whaling was banned, effectively putting the Soviets out of the whaling business* and relegating the Japanese to the Antarctic, where they could only hunt the minke whale. The entire Indian Ocean was declared a whale sanctuary for 10 years. The totals for worldwide sperm whale quotas, from 1974–1975 to 1979–1980, for male and female sperm whales, including North Pacific, North Atlantic, and Southern Season oceans, are listed in Table 10.2.

In 1980, the IWC published a compilation of studies of the North Pacific populations called *Sperm Whales: Special Issue.* The book, which was the report of a special meeting held in Australia in November 1977,

*Those who hoped that the Soviets would immediately retire their whaling fleets were in for a disappointment. Only one month after the July 1979 IWC meeting, the USSR proposed—by postal vote—that they be allowed to take 1,508 male sperm whales. The proposal was soundly defeated, but the Soviets took 201 sperm whales anyway, claiming that they had misunderstood the IWC directive.

TABLE 10.2. Worldwide Sperm Whale Quotas for
Male and Female Sperm Whales, Including North Pacific,
North Atlantic, and Southern Oceans

Year	Quota
1974–1975	23,000
1975–1976	19,040
1976–1977	12,676
1977–1978	13,037
1978–1979	9,921
1979–1980	2,203

contained 42 papers by cetologists and population biologists from Australia, Japan, Great Britain, and the United States, and although Soviet scientists attended the meeting, they did not contribute to the report. With titles like "Catches of Sperm Whales by Modern Whaling in the North Pacific" (Ohsumi), "Size Distribution of Male Sperm Whales in Pelagic Catches" (Allen), "Biases in a Time Budget Model for Modern Whaling" (Rørvik), and even "Two Concerns About the Sperm Whale Model" (Holt), you might assume that the assembled body would have come up with an idea of how many sperm whales might be swimming around the North Pacific, if only to be able to determine the maximum sustainable yield, but they didn't. Despite the abundant equations, no population estimate appears in this report, and by the next IWC meeting (in 1978), the quotas for sperm whales in all oceans were set at 3,796 males and 898 females.

On July 23, 1982, at its 32nd annual meeting, the IWC passed a resolution calling for a moratorium on commercial whaling, putting an end—for 10 years, anyway—to the killing of whales for profit.* The

*When the moratorium vote passed, those delegates who believed that we were present at the historic moment when commercial whaling had ended broke into raucous cheers and applause. I think some people actually threw their papers into the air. The secretary of the meeting, Ray Gambell of the United Kingdom, banged his gavel and called for order, and when it was not forthcoming, he announced that the noisy delegates would have to leave the room. We did, convening in the hotel's pub for a celebratory pint or two.

paragraph of the resolution, passed by a vote of 27 for, 7 against, with 5 abstentions, reads as follows:

> Notwithstanding the other provisions of paragraph 10 [which provide for otherwise allowable commercial whaling], catch limits for the killing for commercial purposes of whales from all stocks for the 1986 coastal season and for the 1985–86 pelagic seasons and thereafter shall be zero. This provision will be kept under review, based upon the best scientific advice, and by 1990 at the latest the Commission will undertake a comprehensive assessment of the effects of this decision on whale stocks and consider modification of this provision and the establishment of other catch limits.

July 23, 1982, the day the moratorium was passed. Shown celebrating are the author; Tom Garrett, deputy U.S. commissioner; and Sir Peter Scott, president of the World Wildlife Fund and member of the British delegation. (Richard Ellis)

Of course, it wasn't really that simple, and whale killing didn't exactly end. The bylaws of the IWC allow a country that takes exception to a resolution to file a protest, rendering that resolution nonbinding on that country, so as soon as the moratorium was passed, Norway took exception to it and continued whaling. Japan, Peru, and the Soviet Union also objected to the moratorium and could therefore have continued whaling, but Peru and the USSR intended to quit whaling anyway, and the Japanese withdrew their objection because they found the "scientific" whaling loophole.

Many people believe that whaling is cruel and unnecessary, but there is no legal objection to Norway's continued whaling, as it takes place squarely within IWC regulations. There are those, however, who object to Japan's "research" whaling, claiming that the science—mostly consisting of finding out how old the whale was when it died—is spurious and that the main reason for the Japanese whale hunt is to get whale meat into the markets. In a 2000 article in *Science,* Baker et al. analyzed 700 "whale products" purchased in Japanese markets and, by means of molecular genetic methods, identified the meat of baleen whales, sperm whales, beaked whales, killer whales, dolphins, porpoises, domestic sheep, and horses. They concluded:

> Scientific hunting of an abundant population can also act as a cover for continued exploitation of a protected or endangered population

of the same species. Using population-level molecular markers, we estimate that up to 43% of market products from the North Pacific minke whales do not originate from the reported scientific hunt in pelagic waters but, instead, from the illegal or unregulated exploitation of a protected population in the Sea of Japan. At this rate of exploitation, the genetically unique Sea of Japan population is predicted to decline toward extinction over the next few decades.

Undeterred, the Japanese in 2000 awarded themselves a scientific research permit for minke whales, Bryde's whales, and sperm whales, and then went out and killed them. In early 2002, the Japanese decided to double their self-assigned quota for "research" whaling. In addition to the 440 minkes that Japanese whalers kill every year in the Antarctic, on February 22, 2002, they notified the IWC that they were going to take 50 minkes and 50 sei whales in the North Pacific.

The IWC has repeatedly passed resolutions critical of the research whaling program, but these are not binding. Critics point to the fact that meat from the "research" program is sold in restaurants and supermarkets as evidence that the program is commercial whaling under a different name. The Japanese government is lobbying for an end to the commercial whaling moratorium, claiming that whale population numbers are increasing rapidly and therefore threatening the recovery of fish stocks, a claim widely dismissed by most independent scientists. Japan proposed to expand its "research whaling" and take 900 minke whales in the Antarctic and 1,600 humpbacks and fin whales in the Southern Ocean. Despite the diplomatic protests of numerous countries (including Australia, in whose Antarctic waters the Japanese whaleships will be operating), the Japanese are taking a hard line, even going so far as to drop the "research" component and claim that they have to kill whales because their people have always eaten whale meat, and no country has the right to tell them what to eat.

XI

"Can Leviathan Long Endure So Wide a Chase?"

WE'RE NOT SUPPOSED to need a reason to save a species from extinction; that they made it this far should be more than enough justification for their continued existence. But now it appears that sperm whales are making a previously unsuspected contribution to the health of the planet. At the 2009 Biennial Conference on the Biology of Marine Mammals, held in Quebec City in October 2009, Trish Lavery of Flinders University in Adelaide, Australia, said that all those whales exhaling all that CO_2 into the atmosphere probably contributed to the buildup of greenhouse gases in the atmosphere, but the nutrients—particularly iron—that they bring up from their deep feeding excursions encourage plankton growth, with the newly nourished organisms drawing in CO_2, thus preventing it from escaping into the atmosphere. It is therefore possible that sperm whales are among the few mammals with a net zero carbon footprint: they are able to balance the amount of carbon released with an equivalent amount sequestered. Carbon neutrality, which seems to come naturally to *Physeter macrocephalus*, is a goal intensively sought by *Homo sapiens* in the effort to reduce greenhouse gases in the atmosphere, which are contributing to the warming of the planet (Milius 2009).

In 1712, Christopher Hussey may have been blown off the Nantucket shore in a storm and may have killed the first sperm whale in New England waters. Even if the story is apocryphal, there is no doubt that the Nantucketers shortly thereafter began to roam the world in search of the mighty square-headed cachalots and developed an industry that would change the way the Western world was lit and lubricated. When the British shipped their first load of convicts to Botany Bay in 1788, they could not have known that the captains would find the waters of Australasia thick with whales. Quick to capitalize, the whalers rounded the Horn in 1789 and discovered the rich whaling grounds of the eastern Pacific. The Nantucket whaler *Maro* encountered more than the riches of Cipango in 1820 when the heavy concentrations of sperm whales were found on the Japan Grounds. Of these distant whaling grounds, Melville wrote,

> That great America on the other side of the sphere, Australia, was given to the enlightened world by the whaleman. After its first blunder-born discovery by a Dutchman, all other ships long shunned those shores as pestiferously barbarous; but the whale-ship touched there. The whale-ship is the true mother of that now mighty colony. Moreover, in the infancy of the first Australian settlement, the emigrants were several times saved from starvation by the benevolent biscuit of the whale-ship luckily dropping an anchor in their waters. The uncounted isles of all Polynesia confess the same truth, and do commercial homage to the whale-ship, that cleared the way for the missionary and the merchant, and in many cases carried the primitive missionaries to their first destinations. If that double-bolted land, Japan, is ever to become hospitable, it is the whale-ship alone to whom the credit will be due; for already she is on the threshold.

Just when it seemed that the world's whales were destined to provide their oil to lubricate the industrial revolution, Colonel Edwin Drake drilled the Western world's first oil well at Titusville, Pennsylvania, in 1859. Did the discovery of petroleum save the whales? Hardly. In fact, it provided the impetus for the whalers to mechanize and modernize their industry, and, armed with the exploding grenades of Svend Foyn, they took out after the whales with a vengeance that was fueled by equal portions of greed, bloodlust, and technology. The great rorquals, long considered too fast and too powerful for the whalers in their open rowing boats, were now in firing range. They were harpooned, shot, exploded, poisoned, and electrocuted in numbers that defy the imagination. Millions of tons of whales were reduced to their components for the lights, machines, wars, fashions, and tables of the world. Deep in the bone-chilling cold of the Antarctic, the great whales had remained unmolested since the morning of the world. In 50 years, the rapacious whalers found them and slaughtered them to near extinction. They shot them under the lowering skies of the Ross Sea and hauled them aboard factory ships with gaping maws that swallowed these 100-ton creatures and reduced them to oil and fertilizer in an hour.

When it appeared that the whalers would run out of whales if they kept up the carnage, they convened to figure out a way whereby they could preserve their industry before their source material disappeared. On May 30, 1949, representatives from 15 nations met in London, and the International Whaling Commission was born. For the next 40 years, this organization, which was supposed to preserve whales for the industry, sat and watched as the whales vanished and the industry deteriorated before their uncomprehending eyes. Compared to the millennium that it took the whalers to reduce the whale stocks to vestigial scattered populations, the end came remarkably quickly. One by one, the whaling nations quit their deadly, costly, anachronistic business. In 1972, the United Nations passed a unanimous resolution calling for a complete cessation of worldwide whaling. That year, the United States passed its own Marine Mammal Protection Act, which protected all whales, dolphins, and seals in American waters, and closed down the last of the American whaling stations. South Africa shut down Durban in 1975, and the Australians conducted an inquiry in 1978 that resulted in the elimination of Australian whaling. In 1982, only the Soviets and the Japanese were killing whales on the high seas.

On July 23, 1982, probably the most important date in the thousand-year history of whaling, the IWC voted for a moratorium on all commercial whaling. Killing whales for money was almost over. Many of the whaling countries protested, objected, and litigated; they invented myriad subterfuges and excuses to continue their unnecessary and wasteful business. Faced with declining profits, declining whales, and a manifold increase in global criticism, however, the Soviets quit in 1987 and the Japanese in 1988.

We will not know for years whether the end of whaling came too late. The "great" whales were all decimated to the point where they may never recover. Right whales, humpbacks, and bowheads have been reduced to sparse shadow populations throughout the world. Even with worldwide protection, the rorqual species are struggling to survive. (Only the Pacific gray whale seems to have recovered to its former plenitude.) When the IWC convened in Brighton in 1983, Peru had withdrawn its objection to the moratorium in exchange for a quota of 165 Bryde's whales that otherwise would not have been allocated. This set the stage for the 1984 meeting, which was ostensibly to be the last meeting of the IWC at which whaling quotas would be set. In 1985, the primary subjects would be the moratorium itself (already passed) and the relatively small numbers of the whales to be taken by various aboriginal peoples for their own consumption. The days in which the delegates from the whaling nations sat around a table and parceled out thousands of the world's whales to their killer ships are over.

It is now no longer clear what purpose, if any, the IWC is supposed to serve. During its tortured and terminally compromised existence, it didn't save the whales—not even for the industry. It didn't save the industry as a whole either. Assuming that the whaling nations do not take over the forum and reestablish commercial whaling for profit, the watchdog commission has outlived its usefulness. It began as an organization that encouraged the slaughter of the whales, passed through a stage where some of its member nations opposed others on basic whaling issues, and came out of the tunnel into an era of confused and conflicted conservation. At the 2009 meeting, held on the island of Madeira, Norway announced that it was suspending whaling in midseason because there was no market for the meat. The pro-whaling nations (Norway, Iceland, Japan, and the newly independent country of Greenland, which requested a quota of 50 humpbacks) were unable to resolve their differences with the antiwhaling countries, and the meeting ended in a stalemate, with the stated intention of negotiating for another year outside the confines of formal meetings. If a compromise cannot be reached between those countries that want to keep killing whales and those that believe that killing is unnecessary and wrong, the very existence of the IWC is in question. On this, at least, the pro- and antiwhalers agreed. "There's only one more year," said U.S. commissioner William Hogarth. "If not then it's over and we'll have to look at the IWC and see how it functions in the future." Joji Morishita, a Japanese delegate, said that if the commission does not approve limited commercial whaling, "the future of the IWC is seriously in doubt."

When I watched the passage of the moratorium on that July day in 1982, membership in the IWC numbered 37 countries: 25 voted for the moratorium and seven voted against, with five abstentions. As pro- and antiwhaling member nations enlisted more countries to join, attempting to pack the commission so that votes would go in their favor, the number has grown to 88. (Neither whaling history nor even a shoreline is a prerequisite for membership in the IWC, which explains the presence of such nations as Estonia, Israel, Mongolia, and Switzerland.) To date, neither side is close to achieving the two-thirds majority required to make binding decisions, but the whaling nations have enough votes to prevent their opponents from blocking their self-allocated quotas (Simmonds and Fisher 2010).

In March 2010, rumors began to circulate throughout the whaling (and antiwhaling) communities that the June 2010 IWC meeting in Agadir, Morocco, would see the moratorium on commercial whaling overturned and a return to limited whale hunting by Japan, Norway, and Iceland. The 1983 moratorium would remain on the books, but the ongoing confrontation between the pro- and antiwhaling nations would cease, as some nations, more interested in reconciliation than in whale conservation, would propose new quotas for the first time in 25 years. Naturally, conservation organizations that lobbied for the passage of the moratorium in 1983 were aghast at this reversal of what was considered the environmental movement's greatest achievement, but the United States and New Zealand, among the leaders in fighting for the moratorium, had evidently reversed their positions, and in what appeared to be an overwhelming interest in maintaining good relations with Japan, they were prepared to sacrifice the ban on commercial whaling—and the lives of hundreds of whales.*

It didn't happen. On June 24, 2010, the day before the IWC meeting was scheduled to close, the controversial attempt to scrap the 24-year-old international moratorium on commercial whaling collapsed, to the delight of antiwhaling campaigners and the frustration of Japan, Norway, and Iceland, the countries that continue to hunt whales in out-

---

*Japan torpedoed a 2010 campaign to list the bluefin tuna as an endangered species. At the 2010 Convention on International Trade in Endangered Species of Wild Fauna and Flora (CITES) meeting in Doha, Qatar, a proposal forwarded by Monaco, and strongly supported by the United States, to list the bluefin tuna as an endangered species was soundly defeated. The plan to grant the fish stronger protection drew little support, with developing countries joining Japan in opposing a measure that they were warned would affect fishing economies. The United Kingdom, the Netherlands, and possibly other European nations voted in favor of the Monaco proposal, which was against the European Union's official position. Campaigners complained that debate on the fate of the Atlantic bluefin fishery was cut short and an immediate vote pushed through by Libya. Seventy-two of 129 CITES members voted against the trade ban and 43 voted in favor, with 14 abstentions.

right defiance of world opinion. After two days of talks behind closed doors, delegates from the 88 member nations were unable to reach agreement on the three-year-old proposal to abandon the official whaling ban in exchange for smaller, agreed kills by the whaling states. The issue is now off the agenda for at least a year, until the next meeting of the IWC, but the result was greeted as a triumph by some environmental groups who feared that the deal would put the future of the great whales in jeopardy once again.

Yasue Funayama, the Japanese commissioner, said that her country had offered major concessions to reach a compromise and blamed antiwhaling countries that refused to accept the killing of a single animal. The deal that failed was originally proposed by the United States (and supported by President Obama), seeking agreement with Japan to secure whaling permissions for its Inuit native peoples in Alaska, without the Japanese making trouble because of American support for the moratorium—something that had happened in 2002. At this meeting, however, no quotas were actually established, and many of the antiwhaling countries thought that such a deal would be virtually impossible to police. Further, it would open up commercial whaling to potential new participants, such as South Korea. Without the deal, however, Japan, Norway, and Iceland may continue to kill whales for "research" or any other rationalizations they can come up with.

Some people believe that killing whales for profit is cruel, mercenary, and utterly unnecessary, but recently, a new and persuasive argument for the elimination of commercial whaling has been introduced into the dialogue: sentient mammals with large brains, a high level of intelligence, a complex communications capability, and acknowledged self-awareness probably ought not to be killed for food. Many years ago, at an IWC meeting, Roger Payne said to me, "It's as if intelligent aliens arrived from outer space, and because we couldn't understand their language, we cooked and ate them." In a June 27, 2010, *New York Times* article that appeared shortly after the IWC had decided to postpone a decision on rescinding the moratorium, Natalie Angier wrote,

> Many biologists who study whales and dolphins urge that negotiators redouble efforts to abolish commercial whaling and dolphin hunting entirely. As these scientists see it, the evidence is high and mounting that the cetacean order includes species second only to humans in

mental, social, and behavioral complexities, and maybe we shouldn't talk about what we're harvesting or harpooning, but whom.

Hal Whitehead, quoted in the same article, said, "At the very least you could put it [hunting whales] in line with hunting chimps. . . . When you compare relative brain size or levels of self-awareness, sociality or the importance of culture, cetaceans come out on most of these measures in the gap between chimps and humans. They fit the philosophical definition of personhood." Angier concludes:

> But mostly, whales and dolphins apply their minds to weighty matters like learning what to eat and how to forage—and where to find the tastiest fish, and what your personal whistle will sound like, and where your mother, sisters, aunts, grandmother and great aunts and your second cousin once removed can be found, and who your fast friends and half-friends may be, and which male ally deserted you when you needed a posse to help herd a fertile but recalcitrant female around. Whaling not only kills individual whales, scientists said. It can disrupt social networks an ocean wide and tens of thousands of Moby and Mabel Dicks strong.

Historically, men saw whales as products, a perpetual source of matériel for commerce, industry, and fashion. There has always been a deadly intermingling of the lives of whales and humans. Men who put to sea in pursuit of Leviathan were defined by this pursuit; they left their lives behind, became professional sea hunters, and regularly risked their lives during the dangerous business of whaling. Whale hunting often resulted in the death of the hunter. Many of these brave lads never returned to terra firma. Were not the great whales also defined by their hunters? To be sure, we hunted wolves to the edge of extinction, but that was because we saw them, however mistakenly, as a threat to human life and livelihood. But was ever anyone threatened by a whale? The number of whales that ever threatened humans (excluding a certain malevolent albino) can be counted on the fingers of one hand. Many men were injured during the business of whaling, but that was a result of the sort of accident that might occur when a 60-ton animal thrashed about after having been speared by men afloat in a fragile cockleshell whaleboat. At its heyday, the war on whales looked like nothing more than a concerted attempt to rid the seas of cetaceans.

Many men died, but for each man who was maimed or killed, a thousand or more whales would suffer and die.

From that moment, long lost in history, when a sperm whale beached itself and died on a European beach, this magnificent creature has enthralled and mystified us. The giant whale with the oil-filled nose may very well be the weirdest animal on earth, but that is only part of our fascination with the cachalot. It is the largest predatory animal that has ever lived, with the largest brain; the deepest diver and breath-holding champion; the creator of the loudest sounds ever made in nature; and the embodiment of more unsolved biological mysteries than any other known animal. The sperm whale played a major (albeit unwilling) role in early American economics; it provided the impetus for global marine exploration; it was the object of a massive slaughter of innocent creatures in pursuit of a dollar; and it resides at the acknowledged pinnacle of American literature. The sperm whale is the powerful quintessence of life on earth. In its pursuit, it became a concentrated symbol of the eternal struggle of man versus untamed nature, and although modern technology had awarded most of the "victories" to *Homo sapiens,* some natural forms have stood fast against encroachment and conquest.

What Melville called the "humped herds of buffalo" have gone with the prairie winds, their once-dominant, thundering hordes reduced to penned populations being raised for beef; the wolf, the enduring symbol of free-range predation, now skulks in the shadows as cattlemen and farmers, fearful of the loss of their hardscrabble dominance of the land, hunt them toward extinction once again, lest they reestablish the life force that has characterized their troubled relationship with man for millennia; in the air, the California condor and the whooping crane, the largest birds in North America, were mercilessly hunted until only a few of those soaring symbols of freedom remain, and they have to be coddled and pampered to protect their dwindling numbers, lest they achieve the dubious fate of the passenger pigeon, once the most numerous bird in America, now gone into the irreversible black hole of extinction. And how shall we speak of the majestic ivory-billed woodpecker, that raucous denizen of southern first-growth forests, now that its echoing machine-gun pounding has been silenced forever? With considerable help from man, the all-conqueror, prairie dogs no longer fear the black-footed ferret, and in fact, the prairie dogs themselves, considered vermin by farmers and ranchers, are in big trouble too.

We live on the land, and we have taken it by force from those lesser beings who had the temerity to try to coexist with us. But we cannot live in the ocean, although we gratefully use it as a highway. Hundreds of thousands of living beings can and do rely on the salty sea as the fundament of their lives; it is the largest and most densely populated biome on earth. Indeed, because some 90 percent of the world's oceans are more than two miles deep, the deep ocean is the earth's predominant environment. Neither depth nor distance has protected the denizens of the deep from the predations, poisons, and pollutions of man. At one time or another, almost every living thing in the oceans has been under attack. First, of course, we went after the fishes: 25,000 species, living in every watery habitat imaginable, from sun-spangled coral reefs to almost freezing black abyssal depths, from ice-choked polar waters to boiling hydrothermal vents. Fisheries are older than mankind, and we have harvested almost every kind of fish, from the supremely edible tunas to the dangerously inedible puffer fish known in Japan as *fugu*, whose toxic tissues can kill if not prepared properly. We have repaid the ocean's bountiful offerings by overfishing almost everything that lives there. For some species—bluefin tunas, for a painful example—we have been so careless in our greedy overexploitation of the world's tuna stocks that we have had to classify the bluefin as an endangered species.

Moby-Dick was written and published at the zenith of the worldwide sperm whale fishery; its home port of New Bedford was the richest city per capita in America. As Melville described it, "nowhere in all America, will you find more patrician-like houses, parks and gardens more opulent, than in New Bedford. Whence came they? . . . Go and gaze upon the iron emblematical harpoons round yonder lofty mansion, and your question will be answered. Yes, all those brave houses and flowery gardens came from the Atlantic, Pacific and Indian oceans. One and all, they were harpooned and dragged from the bottom of the sea." In 1850, as Melville was composing his great novel, the last thing on anybody's mind was a shortage of whales.

In the early days of the whale fishery, very few people recognized the delicate nature of the relationship between men and whales. Occasionally, however, a small voice was heard to wonder about this one-sided affair, in which one side gave everything while the other took. In 1804, Comte Bernard-Germain de Lacépède, a French anatomist and naturalist, published *Histoire Naturelle des Cétacés,* in which he wrote,

Man, attracted by the treasure that the victory over the whales might afford him, has troubled the peace of their immense solitary abodes, violated their refuges, sacrificed all those which the icy, unapproachable polar deserts could not screen from his blows; and the war he has made on them has been especially cruel because he has seen that it is large catches that make his commerce prosperous, his industry vital, his sailors numerous, his navigators daring, his pilots experienced, his navies strong and his power great.

Thus it is that these giants among giants have fallen beneath his arms; and because his genius is immortal and his science now imperishable, because he has been able to multiply without limit the imaginings of his mind, they will not cease to be the victims of his interest until they have ceased to exist. In vain do they flee before him; his art will transport him to the ends of the earth; they will find no sanctuary except in nothingness.

DURING A VACATION in the Virgin Islands, I became intrigued with an old queen conch shell that I found on the beach. It was encrusted with coral growths and grit, but I decided to clean it, so I sat for hours, chipping away at it with a penknife. Finally I had cleaned it up, and for no particular reason (except that I had expended so much time on it), I decided to pack it in my suitcase and take it home. My wife and I lived in a New York apartment and really didn't need a doorstop, and we didn't have a mantel (because we certainly didn't have a fireplace), so I just put it on my desk. I picked it up and moved it around. I stared at it. I liked the way it looked. When I like the way something looks, whether it's a porpoise, a panther, or a peregrine, I draw it, so I made some sketches of that big old shell. Then I decided to paint its portrait. I knew hardly anything about shells, but I was hooked on them. I learned that there were shell dealers in New York, so I visited them to buy particularly interesting-looking specimens. I bought shell books and subscribed to shell magazines. I would set up the shells on my desk, light them carefully, then paint pictures of them. What was I going to do with these shell portraits? I had no idea.

There was a shell dealer named Elsie Malone in Sanibel, Florida, and I was ordering exotic shells from her. For the most part, I was interested in the especially gaudy and heavily ornamented murexes, which were beautifully decorated with swooping spines and branches, not unlike the antlers of deer. In fact, some of them were known as deer-horn murexes, latinized to *Cervicornis*. I painted a lot of these, as well as assorted shells of the genus *Strombus*, which included my old friend *Strombus gigas,* the queen conch. Of course the original shell was bleached and faded, but recent queen conchs have a showy, almost obscene pink lip. Elsie wrote and told me that a friend of hers named Clarice Fox was opening an art gallery on Sanibel, and asked whether I would like to have Clarice look at my paintings. Sanibel Island is the shell-collecting capital of America, so a gallery there willing to exhibit my paintings was

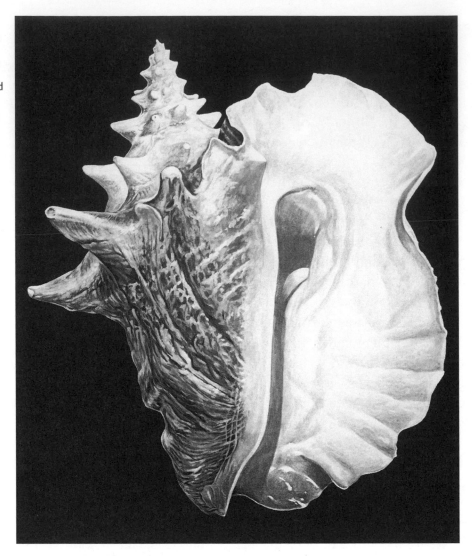

a natural. I sent a couple of slides to Clarice, and before I knew it, I was scheduled for a one-man show at the Schoolhouse Gallery. I had never done this sort of thing before, but I had the paintings framed, wrote up a catalog, and shipped myself and the paintings to Florida.

The show was a great success, and much to my surprise, every painting and drawing was sold. I had become a successful artist my first time out of the gate. Did that lead to a long and prosperous career as a shell painter? Not exactly. In fact, soon after my gallery success, I resigned from the shell-painting business. You see, shells are made by living gastropods (snails), and they are not some waste product that they discard. Except for shells that you find washed up on the beach, such as those at Sanibel, the animal has to be killed to get the shell. There is a huge industry devoted to harvesting shells, which are often obtained by chemically poisoning reefs or actually blowing them up to kill the animals. Did you really think that those baskets and baskets of similar shells in souvenir shops were collected by beachcombers? I realized that in some small way, my glorification of shells could only encourage the dynamiters, so I wrote an article in 1975 for *Audubon* magazine that I called "Why I Became an Ex–Shell Painter," in which I explained my decision to quit.

During my shell-painting period, I had become involved with a fledgling conservation organization known as RARE, for Rare Animal Relief Effort. It was the brainchild of David Hill, who wanted to raise money for the short-range rescue of species in danger because it was thought that the larger organizations, such as the World Wildlife Fund, while ultimately helpful, took too long to swing into action. At this time, there was a groundswell of concern for the whales that were being slaughtered all over the world, and nobody seemed to be doing anything about it.

While working with bottlenose dolphins at America's first oceanarium in St. Augustine, Florida, in the late 1940s, curator F. G. Wood noticed the variety of sounds they made and speculated as to whether they might be able to echolocate. Ken Norris, then at Marineland of the Pacific, observed that dolphins could easily locate objects blindfolded, but not if their lower jaws were covered, demonstrating that these amazing mammals relied almost entirely on sound. John Lilly began his research on dolphins in 1955, trying to understand what they were "saying," leading to the 1969 film *Day of the Dolphin,* in which the dolphins actually talk. Researchers Roger and Katy Payne first listened to the haunting songs of the humpback whale (and produced the record in 1967); these recordings crystallized "whale consciousness" and raised the painful question, "Who gave the whalers permission to kill these sensitive, intelligent, echolocating, leaping, smiling, singing creatures?" At about this time, Scott McVay wrote two popular articles, "The Last of the Great Whales" (published in *Scientific American*) and "Can Leviathan Endure So Wide a Chase?" (in *Natural History*), which brought the question before people who had not thought about a whale since they (reluctantly) read *Moby-Dick* in high school.

At the United Nations Conference on the Human Environment in Stockholm in 1972, there was a unanimous call for a 10-year moratorium on all whaling, which was studiously ignored by the International Whaling Commission. By this time, however, the movement had gained so much momentum that numerous conservation organizations were devoting more and more time and energy to the business of whale preservation. Curiously, there is no organization actually named "Save the Whale," although there are many devoted solely to whales, such as the American Cetacean Society. Other organizations, such as the Humane Society of the United States, the National Audubon Society, the World Wildlife Fund, the Sierra Club, and the National Wildlife Federation, devoted much time and energy to the problems of whales. There is something about cetaceans—a uniqueness that has elevated them far above other mammals. They may be superintelligent or they may not be; we only know that they have very large brains—but in any case, they are masters of an environment that we can only enter as jealous, clumsy aliens. Whales have captured the emotions of millions in a manner unprecedented in the history of humans and wild animals. It is not surprising that so many people were so deeply offended by the arrogance of the International Whaling Commission signatories in deciding how many of these creatures will be turned into pet food and fertilizer every year. I was asked to join the board of RARE specifically to address the issues of whale conservation.

Although most people think that the heyday of whaling was the time of the square-rigged whalers out of Nantucket and New Bedford in the mid-19th century, with somebody shouting "Thar she bloooows!" from the crosstrees and brave lads harpooning fierce whales from a whaleboat, it was the 1960s that saw the most massive slaughter of sperm whales in history. In the North Pacific, an area virtually unknown to Yankee whalers, Japanese and Soviet sperm whalers were taking some 25,000 whales every year—as contrasted with the New England fishery, where it has been estimated that no more than 10,000 whales were ever taken by the entire fleet. Our first whale-saving exercise consisted of a few RARE members (I remember me, David Hill, his wife, Marty, a lawyer named Ken Berlin, and the bird painter Guy Tudor, but I'm sure there were others) taking to the streets of New York with petitions that we asked people to sign, so that we could present the signatures to the Japanese and Soviet embassies to show them how much America disapproved of their whale-killing policies. We stood on street corners in front of Fifth Avenue department stores, shouting, "Save the whales! Boycott Japanese and Russian goods!" (It made little difference that there were hardly any Soviet goods to boycott; we wanted the makers of Nikons, Panasonics, and Hondas to know that the world disapproved of their whale killing.) We did manage to collect thousands of signatures, which we presented to the respective embassies. Although we may have raised the consciousness of many people on the streets, I'm not sure that the petitions got past the clerks we handed them to, who assured us that they would be put in the hands of the ambassadors. I thought that there must be a way to spread the message more effectively.

David, a longtime birder, was friendly with Les Line, the editor of *Audubon,* and suggested that we might submit an article to Line's magazine on the subject of the plight of the great whales. David would write it, and I would provide the illustrations. (At that time, my whale-painting experience consisted mostly of painting the little portraits for the *Encyclopedia Britannica,* one of which was printed upside down, but that's a story for another time.) There were few guidelines for a whale painter, and even fewer references. Commercial whaling had been practiced by various peoples for more than 1,000 years and had reached its zenith in the 20th century, but few photographers had recorded the enterprise, and when they did, it was usually to photograph dead or chopped-up whales. I wanted to paint them swimming in their natural habitat, to show that they were not products or harvestable resources, but living, breathing animals. How could you engender support for endangered animals if people didn't know what they looked like? This meant searching the literature for drawings of the various whales so I could get an idea of their size and shape, and then (figuratively) reassembling them, lifting them off the flensing deck, and putting them back in the water. Once again, I began in the American Museum of Natural History (AMNH) library, but I soon learned that there were people who were studying various whales at sea, and although they might not have underwater photographs, they surely knew more than I did about what the whales looked like.

In the chapter entitled "Of the Monstrous Pictures of Whales," Melville wrote:

> I shall ere long paint to you as one can without canvas, something like the true form of the whale as he actually appears to the eye of the whaleman when in his own absolute body the whale is moored alongside the whale-ship so that he can be fairly stepped upon there. It may be worth while, therefore, previously to advert to those curi-

ous imaginary portraits of him which even down to the present day confidently challenge the faith of the landsman. It is time to set the world right in this matter, by proving such pictures of the whale all wrong." He then catalogs ancient illustrations, recent books and engravings, and even drawings of the sperm whale's skeleton ("his skeleton gives very little idea of his shape"), and concludes that it is not possible to paint a whale.

Melville was unfamiliar with Jacques Cousteau, underwater cameras, snorkels, scuba equipment, and other devices that would enable humans to get a look at whales in a way completely unavailable to the old whalers, who could only see whales in the distance or moored alongside a ship as they were being butchered. I could avail myself of resources Melville never dreamed of.

There were 10 species of great whales that I wanted to paint: the rorquals (from the Norwegian word for "grooved whale"), consisting of the blue, fin, sei, Bryde's, and minke whales; the closely related bowhead and right whales; and the humpback, gray, and sperm whales. It seemed that every species of great whale had its champion, as with Melville and the sperm whale in *Moby-Dick*. (By the way, Rockwell Kent's illustrations for the 1930 Random House version of *Moby-Dick* were surprisingly good, and I consulted them often, but how could he possibly have portrayed a right whale so accurately?) Because of a (probably incorrect and politically motivated) Japanese identification of a "pygmy" species of blue whale in the mid-1960s, there were many articles about the larger and smaller versions, often well illustrated. Fin whales were fairly common off the very coast I lived on, and they were occasionally photographed by sailors. Sei and Bryde's whales, not native to New England waters, often washed ashore in other parts of the world, and amateur photographers sometimes took pictures of the stranded leviathans. (In 1972, Canadian author Farley Mowat had written *A Whale for the Killing*, the heartbreaking story of a fin whale trapped in a bay in Newfoundland that was shot repeatedly by local "sportsmen" until it died. In 1981, when they made a movie of Mowat's novel, because there was no underwater footage of fin whales, they substituted Hawaiian humpbacks, so you see a whale begin its dive in the cold, gray waters of Newfoundland, but by the time it submerges, it's in the aquamarine waters of Hawaii.) It would not be until 1977 that photographer Jim Hudnall would swim with the humpbacks of Maui, producing the first underwater pictures of whales in the wild, so when I was starting the paintings, there were no humpback photographs available either.

I had seen minke whales in Newfoundland, but only at a distance, and then only the dorsal fin and part of the arching back. Right whales were no longer common anywhere; they, along with humpbacks, were usually the first whales killed when the whalers arrived at a particular spot because they both breed in inshore, protected waters, perfect for whaling. Bowheads, the object of a concentrated Dutch and British fishery in the 17th and 18th centuries in the eastern Arctic, lived in areas so remote and inhospitable that they would not be photographed underwater until 1998, so any hope of getting an image of living bowhead was as likely as getting an underwater photograph of the Loch Ness monster. The California gray whale, as befits its name, migrates every year along the west coast of North America, so there are lots of pictures of their barnacled backs, but little to give an indication of what the whole animal looks like. Which leaves the obvious whale, the one every kid draws, a square-headed creature—like the one in Disney's *Pinocchio*—with a water fountain spouting out of its head. The sperm whale was the object of the fabled Yankee fishery, the eponymous protagonist of Melville's 1851 novel, and, like the great white shark, the occasional star of Hollywood movies.

I worked on the sketches and paintings for six months. I still wasn't sure about the whales, because I'd never actually seen one, except in the movies of *Moby-Dick* and the distant view of a minke's dorsal fin in Newfoundland. (This was long before those *National Geographic* and *Nature* television programs.) I knew it would be better to have them critiqued before publication rather than after, so I decided to submit the paintings to the scrutiny of a person who knew as much about whales as anyone in America. I packed the paintings in my car and drove to Woods Hole, Massachusetts, where Bill Schevill held court. Schevill, born in 1906 (he died in 1994), was a bioacoustician, and with his wife, Barbara Lawrence, had been among the first to record the sounds of whales. He had recorded sperm whales, and more importantly, he had actually seen them. Affiliated with Harvard at the time, he had a reputation for not suffering fools gladly, so it was with some trepidation that I brought my paintings to him. I was afraid that he'd take one look at them and tell me that I should have stuck to painting seashells. Schevill, who was born in Brooklyn, had acquired the mannerisms of a crusty old Yankee sea captain, right down to an Ahab-like beard, and to my enormous relief, when I showed him the 10 paintings, he approved. (He did point out that he had never seen a humpback "so gaily spangled," which was because I had gotten somewhat carried away with the reflections of light on the whales' backs.) With his approval, I felt confident in turning in the paintings to *Audubon,* and with David Hill's text, "Vanishing Giants" ran in the January 1975 issue. My sperm whale, not so gaily spangled, was used as the cover illustration, the first painting to appear on the cover of *Audubon* in 30 years.

The response to the *Audubon* article was amazing. All extra copies in the press run were sold out almost immediately, and in an editorial, the *New York Times* wrote that "this issue should help, through its text, to illuminate the role of these extraordinary creatures and, through its illustrations, to arouse concern for their fate." In 1975–1976, the 10 paintings were exhibited at the Newark Museum, South Street Seaport, the New Bedford Whaling Museum, Mystic Seaport, the Field Museum in Chicago, and the AMNH. (Normally, the AMNH has a policy of not exhibiting the work of living artists, but they made an exception in this case because of the relevance of the subject matter.) A man from Quadrangle Press called us to ask if we would like to turn "Vanishing Giants" into a book. It was very flattering, and David and I would have loved to have made a book out of the magazine article, but the 10,000 words that he had written and the 10 paintings that I had done were all we had. A 10,000-word book would have been about 50 pages long, even if it was bulked up with color plates. We countered with an offer to do a field guide to the marine mammals of the world (whales, dolphins, seals, sea lions, sea otters, and so on) and he agreed. David would write the text, and I would do the illustrations.

There were a few marine mammals—mostly seals and sea lions—that had been extensively photographed, but by and large, there wasn't much in the way of pictorial reference material on, say, Burmeister's porpoise or Shepherd's beaked whale. I headed back to my favorite library and spent almost a year researching the appearance of the various whales, dolphins, seals, and sea lions. There was no single source. I had to look up reference material for every one of the 100-odd species, hope that there was a drawing or a photograph to work from, make notes and sketches in the library, and then paint a portrait of the creature that I hoped would be accurate and definitive. During the process, I was constantly in touch with David, whose day job was airline pilot, and who lived in New Jersey. He assured me that his research was going

Part of the one-man show of the paintings for "Vanishing Giants," at the American Museum of Natural History, New York, 1976. (American Museum of Natural History)

well, and as the deadline approached, we prepared for the delivery of the manuscript and drawings. When the fateful day arrived, I packed up the paintings and headed for the meeting at Quadrangle.

Present were various editors, art directors, me and David, and our agent. I plopped the drawings on a table (they were all painted on 11 × 14–inch illustration boards), and unwrapped them so everybody could see what a spectacled porpoise or a Ross seal looked like. Then everybody turned to David to have a look at the manuscript. "Um," he said. "There is no manuscript." A stunned silence fell over the room. How could there be a book with just drawings and no manuscript? What could be done? In an utterly recklesss demonstration of overconfidence, I said that because I had spent the past year in the library, I was totally familiar with the literature, and that I would write the book. Everybody seemed greatly relieved, and the meeting adjourned with David agreeing to give me that portion of the advance that he had received to write the book. When they asked, I told them that my experience with the whale and dolphin literature would enable me to write the book in a year and a half. I packed up my drawings, and I planned another 18 months in the library of the AMNH.

I was wrong in my estimate by 18 months. It took the better part of three years, and when I finally emerged from the library, white as the belly of a fish from living in artificial light for so long, I had written 250,000 words on the cetaceans of the world. (The seals and sea lions were earmarked for another book, which never got written.) The manuscript and drawings were resubmitted to Quadrangle, and we waited. And waited. Seems like the folks at Quadrangle had expected to publish a field guide to the marine mammals of the world (well, the cetaceans, anyway), and they couldn't figure out how to do that with the mass of material I had submitted. I guess they envisioned a "field guide" as being something you could put in the pocket of your anorak, take it aboard your sailboat, and identify what kind of whale belonged to that dorsal fin, kind of like the Peterson bird guides. But with 250,000 words, a dozen large color paintings (many from the 1975 *Audubon* article), 76 illustrations, and innumerable drawings, the resulting book would have been the size of a small phone book. Another complication not foreseen when I agreed to write the book was that I thought I could just write the natural history of a given animal by covering size, food, distribution, gestation, and so on. But I soon realized that whaling and politics played an integral part in the natural history of many of the whale species, for some of them had been hunted to the brink of extinction, and whatever else it may be, incipient extinction is certainly a part of the "natural history" of an animal species.

Quadrangle finally gave up. They couldn't figure out how to make a book out of the material I had submitted, but that didn't mean I had to give the money back. I had, after all, given them what they asked for—it was just a little larger and more complex. So my agent and I now had a huge pile of material and no one to publish it. For their 10 percent, agents are supposed to be able to solve problems like this, and he said he would submit the whole kit and kaboodle to Ash Green, an editor at Knopf. Editors rarely see a finished manuscript and all the illustrations when they are considering a book project, but that's what happened this time, and Ash agreed to publish it. "Only one problem," he said. "It's too big. Can't we figure out a way to break this into two volumes?" "Of course," I said. "We can put all the 'great' whales and the beaked whales into one volume, and all the dolphins and porpoises in the other." I was now to be the author of a two-volume work, which struck me as very sophisticated and literary.

I set out to divide the work into the two sections, but it was a lot more complicated than I thought it would be, because there were a lot of places where cross-references were made from whales to dolphins that would work in a single volume but not in separate volumes. Also, at the time I was revising, much new material on whales and dolphins was published. For example, I had originally written that very little was known about the Chinese river dolphin, but as I was revising, Western scientists in conjunction with resident Chinese cetologists were publishing new material on the ecology of this heretofore poorly known animal.* This was also a period of great turmoil in the International Whaling Commission, to which I had been named a delegate in 1980, so I was calling in bulletins daily. (Eventually Ash told me that he was an editor at a publishing house, not at a newspaper, and I had to cease these hysterical "stop the presses" calls.) *The Book of Whales* was published in 1980, and *Dolphins and Porpoises* in 1982.

All those one-man museum shows were enormously gratifying, but even the world's foremost (and only) whale painter has to eat, and in 1976, I had a one-man show at a Manhattan gallery called Sportsman's Edge, where the paintings were for sale. I had been storing the shark paintings in a closet, and I brought them out for this show, along with the well-traveled whale paintings. Without planning it, I had become a painter who specialized in large marine critters. It was the height of the "save the whale" movement, and by this time, the novel and movie of *Jaws* were at the top of their respective best-seller lists, so it was a pretty good time to be painting whales and sharks. I sold most of the paintings in the show, and I even got a couple of commissions. One came from Richard Wehle in Buffalo, who owned a prosperous electric company, and who, because of his name, collected whale and whaling memorabilia. His company, Wehle Electric, had just opened a new office building, and he wanted to put a private museum in it, to display his scrimshaw, harpoons, logbooks, and so on. He asked me whether I would paint a mural for him.

I went to Buffalo to look at the space—the mural was to be 20 feet long by six feet high—and to discuss the details of painting it. Dick Wehle wanted me to paint it in place, but the thought of spending six weeks (or however long it would take) in Buffalo was too much for me.

---

*I incorporated that new material into *Dolphins and Porpoises* in 1982, but the story of the Chinese river dolphin has taken a terrible turn, and as of 2007—roughly three years before I wrote this note—*Lipotes vexillifer* has been declared extinct, the first cetacean species to be eliminated in recent times.

We decided that sperm whales would be the subject, and I returned to New York to try to figure out how and where to paint a 20-foot mural. (This would not be the last time I would face this problem; later I would have to figure out how and where to paint a 100-foot-long mural.) The living room in my apartment was just over 18 feet long, so with a minor rearrangement of the furniture (which, for some reason, my wife did not regard as minor), I was able to tack up a 20 × 6–foot canvas. (She didn't much like the tacks, either.) First I painted in the background. I then drew the whales on brown paper, cut them out with scissors, and, moving them around until I arrived at a composition I liked, taped them into place. Then I traced them in chalk on the canvas and began to fill in the whales. It took about six weeks, and my decision to do it at home was vindicated by what happened in Buffalo. That was the year of the monster blizzard, where Buffalo had snow piled as high as 18 feet, people lost their cars, and freight trains were sent to cart the snow to other places, because Buffalo had no room for all of it. (When Dick Wehle died in 1990, the contents of his museum were auctioned off, and the mural was bought by Mystic Seaport.)

New Bedford, originally an 18th-century Pilgrim settlement, quickly grew to a position of prominence in the whaling business. It supported such industries as shipbuilding and the processing and sale of whale oil, but it also had the requisite support industries, such as ship fitting, rope making, and barrel making (coopers), as well as the farmers and green-grocers that provisioned the hundreds of whaleships that sailed for and returned to New Bedford. In 1848, New Bedford resident Lewis Temple invented the toggle harpoon, a device that revolutionized the whaling industry. By 1850, New Bedford—population 22,000—was the richest city per capita in America. California's 1851 gold rush attracted many young men (and often older ones, too) who forsook the low pay, long hours (sometimes years), inedible food, strict discipline, and dangerous quarry that defined life aboard a whaleship. The economy depended almost exclusively on the oil of the sperm whale, but when petroleum was discovered in Pennsylvania in 1859, the demand for whale oil dropped. New Bedford was able to remain prosperous because of its textile industry, which by 1881 had grown large enough to sustain the city's economy. The creation of the New Bedford Textile School in 1895 ushered in an era of textile prosperity that began to decline in the Great Depression and ended in the 1940s. At its height, more than 30,000 people were employed by the 32 cotton-manufacturing companies that owned the textile factories of New Bedford (which were worth $100 million in total). By 1960, New Bedford's population had risen to over 100,000, and the only business left was fishing. Then New Bedford's commercial fishing community was devastated by government restrictions on the catch of halibut, redfish, and—until recovery plans can work—haddock and yellowtail flounder. Although New Bedford is still listed among the top fishing ports in the United States (Dutch Harbor, in the Aleutians on the Bering Sea, is first), the fishery is now in decline, and the city's economy, like the economy of almost every American city, is in a tailspin.

The Old Dartmouth Historical Society was founded in 1903 to create and foster an interest in the history of the territory included in Old Dartmouth, promote historical research, and collect documents and artifacts and provide for their proper custody. The society established the Whaling Museum in 1907 to tell the story of American whaling and to describe the role New Bedford played as the whaling capital of the world in the 19th century. In 1916, the largest ship model in the world was built in a special building at the museum, a half-sized model of the New Bedford whaleship *Lagoda*. The actual ship was 108 feet long, so the model is 54 feet, 9 inches on the waterline. (From the tip of the flying jib boom to the tip of the spanker boom, however, the model is 82.5 feet long.) Today, the museum is the largest museum in America devoted to the history of the American whaling industry and its greatest port.

By the time I moved to Little Compton, Rhode Island—a half hour from New Bedford—in 1974, I was already involved in the politics of whale conservation, and I had completed the whale paintings for *Audubon* magazine. I spent a lot of time in and around the Whaling Museum. It was a pleasant surprise—but not a little intimidating—when I was asked to paint a 100-foot mural for the Lagoda Room of the Museum. A hundred feet! A third of the length of a football field! The room takes its name from the model of the whaleship *Lagoda*, which put to sea out of New Bedford for the first time in 1841. The room in which the *Lagoda* sits is 155 feet long, 54 feet wide, and high enough for a fully rigged—albeit half-sized—whaleship. What better complement to a half-sized whaleship than half-sized whales? ("Half-sized" always sounds a little diminutive to me, but because the whales I was going to paint can reach a length of 60 feet, the largest whale in the painting—Moby Dick himself—was going to be 30 feet long).

The New Bedford Whaling Museum is an old building, with brick exterior walls faced with plaster within, and because of the dampness of the New England maritime atmosphere, there was no possibility of actually painting the mural on the interior wall. I then began to research the myriad alternatives. I could paint it on canvas. Or on boards. Or on some sort of canvas-covered panels. It could be hung like a giant picture, or applied directly to the wall. I even investigated the possibility of having a small painting of mine reproduced by a computer-generated painting process, but the museum rejected this high-tech solution. (Thirteen years later, when the process had been perfected, I used this method to produce another mural for New Bedford.) After months of research, consultation, and experimentation, I concluded that the old way was best—with some modern modifications, as will be seen. I decided to paint the mural on canvas. In the museum, the space available for the mural was 100 feet long by 13 feet high, but the 100-foot-long wall is broken in the middle by a doorway, so I only had to resolve the problem of painting two 50 × 13–foot murals.

First I made two 24-inch sketches of the different kinds of whales that had traditionally been hunted out of New Bedford. Half of the mural was going to show right whales, humpbacks, and gray whales,

Preliminary sketches for
the New Bedford murals.
Each one was 6 by 24 inches.
(Richard Ellis)

and the other was going to depict the primary object of the Yankee
whale fishery: the mighty sperm whale. This seemed a little too ambig-
uous, because New Bedford is best known for its sperm whale industry, so I revised the sketches to include only sperm whales. Then, in
what I regard as an almost divine inspiration, I added the most famous
whale in the world, the one whale that everyone knows, the White
Whale, Moby Dick. In the novel, Captain Ahab's ship *Pequod* actually
departs from New Bedford, so the transition from fact to fiction was
not particularly troublesome. The white whale was a product of Herman Melville's imagination, but albino sperm whales have been taken,
so I thought that I could add this most infamous of cetaceans without
compromising the authenticity of the painting. Moreover, the thought
of painting a giant portrait of Moby Dick seemed to be the most exciting prospect in my two-decade career as a painter of whales.

When the museum had approved my suggestions and sketches, I
returned to New York to work out the details. There were several problems. I didn't know what to paint the mural on (all I knew was that it
would be done on canvas and then installed in the museum); I didn't
know what to paint it with; and I didn't know where to paint it. Of
course, I could have done it in situ in New Bedford, but I had a sort
of business to run—not to mention a family to maintain—and however long this project was going to take (I originally estimated three
months, but I was short by a full month), I couldn't afford to be away
from New York for that amount of time. My own studio was not nearly
large enough for a project of this magnitude, so I had to find a proper
space. Luckily, a film cameraman with whom I had worked in the past
lived in a loft in the Tribeca area of lower Manhattan, and he agreed to

allow me to rent some space in his loft, which measured approximately 100 by 36 feet, and most importantly, it had 13-foot ceilings.

Now I had a place to paint *in*, but nothing to paint *on*. I located an organization in Brooklyn that manufactured stretchers, and they agreed to meet with me to discuss my problem. I met with the president of the company, and we decided (although it was much more his decision than mine; I was a complete novice at this business) to build eight 12 × 12–foot stretchers, bring them to the loft along with the canvas, and build the panels there. Of course, I knew virtually nothing about canvas in this size either, because I usually work on normal-sized canvases—perhaps 30 inches by 40 inches. I had to locate a source for the linen canvas in widths that would allow for the construction of a 12 × 12–foot panel—no easy task, even in New York.

Each of the eight stretchers had to be built in two halves because the stairway to the loft was much too small to permit a 12-foot-high object to be carried up. Even then, when they finally arrived with the 6 × 12–foot sections, they had to remove the end pieces because they couldn't turn the corner to get the sections through the door. For the next 10 days or so, I learned more than I thought there was to know about stretchers. I watched the crew assemble these beautifully crafted examples of the joiner's art, then stretch 144 square feet of canvas on each one. These stretchers (which, when uncanvased and horizontal, looked more like the components of a barn raising than anything to do with painting) were furnished with expansion bolts, so they could be minutely adjusted when the canvas was stapled in place, to ensure a tight, smooth surface. When the eight canvases were finally stretched, I was on my own. I was facing 1,152 square feet of raw Belgian linen, and I now had to transform this vast expanse of beige blankness into a mural.

First, of course, the canvases had to be primed. I used acrylic primer (which I bought in five-liter buckets), and I painted each canvas by hand with a five-inch brush, in order to work the primer well into the canvas. At this stage, I was working on the top half of the canvas from an industrial scaffold that I rented from a commercial ladder-and-scaffold company. When the component parts of this scaffold were delivered—a "sidewalk delivery," by the way, where they dropped all the pieces off in front of the loft, and I had to carry all the platforms, crosspieces, and 20-pound wheels up three flights of stairs—I then had to assemble it. Because I had never even seen one of these things before, it was like constructing a particularly complicated kid's toy on Christmas Day, without instructions, and with some of the pieces being 10-foot sections of iron pipe.

I could only work on two canvases at any given time, because this was, after all, my friend's house, and he had to live there while I clattered around with my stretchers, staple guns, and scaffolding. I therefore developed a sort of assembly-line technique, where I could work on one canvas (I always worked from left to right), then remove the completed left-hand canvas, slide the right-hand one into its place, and begin to work on a new right-hand one. (A stretched 12-foot-square canvas is too large for one person to move, so I always had to enlist somebody's help when the time came to change canvases.) After the primer had dried, I had to sand it down to remove all

The whale painter at work. I am about to start painting in the teeth. (Richard Ellis)

the rough threads, lumps, and bumps that linen canvas is heir to. Of all the myriad aspects of this project, the sanding was probably the most physically demanding. It didn't help that I had to do a lot of this work in an un-air-conditioned space while the outside temperatures in New York were in the high 90s, either. (On the other hand, I can recommend sanding and scaffold-climbing as a surefire weight-loss program. In the four months that it took me to paint this mural, I lost 30 pounds.) The smoothed canvases (and the surrounding floors) were vacuumed to remove the pervasive primer dust, and then I began the actual painting.

The first order of business was to paint the background color. It is a sort of tropical turquoise, and I put it on with a roller. Compared to the hand painting of the thick primer into the canvas or the sanding process, this was a piece of cake. Now that I had beautiful, flat, blue canvases, I had to figure out how to get the whales drawn onto them so I could paint them. I vaguely remembered something about drawing a grid and then enlarging the sketch, but while I could easily draw a grid on my 6 × 24–inch sketches, I had no idea how to draw a grid on a 12-foot-square canvas. So I said to hell with it, and, holding

Patti Forkan of the Humane Society of the United States during the filming of a documentary about the New Bedford murals. The twisted harpoons have not been added to Moby Dick. (Richard Ellis)

the sketch in front of me, I simply drew the whales (in sections) onto the canvas with ordinary blackboard chalk. When the whale—or half of it—was sketched in to my satisfaction, I began to paint it.

Have I mentioned that I was working in latex house paints? I ordinarily work in watercolors or acrylics, but neither of these seemed appropriate. I selected several shades of blue from the paint store (I used Pratt & Lambert "Vapex" flat wall paints throughout), and in what the army used to call OJT—"on the job training"— I learned how to paint with this unusual medium. At first, I thought I would be able to blend the colors on the canvas the way one does with oils or acrylics, but because latex house paints dry even faster than acrylics—not to mention the problem of working wet on a painting that is 12 feet long—I had to come up with another solution. I developed a unique "palette" composed of plastic Dannon yogurt cups (the 32-ounce size for the larger amounts of paint and the 16-ounce size for the smaller amounts), with the tops painted the color of the paint within. I mixed many variations of the blues (and then the whites and beiges for Moby Dick), and then I painted in a sort of updated pointillist manner, shading the colors from dark to light, and adding the scars and scratches of squid suckers on the faces of the adult whales. The babies, who are too young to engage in battle with their lunches, are unscarred.

It took me four months of 12-hour days to complete the murals. (I quit for the day only when my legs gave out from standing or scaling the scaffold). When the painting was completed, each of the eight canvases was removed from its stretcher and rolled carefully on a 12-foot, 16-inch-diameter heavy-gauge cardboard tube. (Each of the tubes cost $137.50.) Unfortunately, the 12-foot-long tubes wouldn't round the turn to get them down the stairs, so we were stuck with eight giant tubes that we couldn't get out of the loft. We lowered the tubes out the window and got them down three stories by using a specially constructed scaffold. We assembled the exterior scaffold without benefit of a building permit, and the police closed us down when they saw we were conducting an illegal and dangerous activity. I told them we would take it down

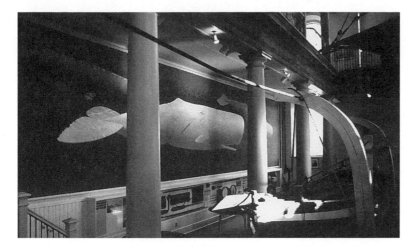

Moby Dick in place at the New Bedford Whaling Museum. At a length of 100 feet, the mural was too large to photograph in its entirety, and besides, there was a whaleship in the way. (Richard Ellis)

immediately, and as soon as they left, we continued lowering the rolled canvases, loaded them in the truck, and escaped. The whole business (including the artist and his yogurt cup collection) was then trucked up to New Bedford, and the entire process was reversed. A total of 1,521 square feet of canvas was restretched on the reassembled stretchers, which were then tuned and tightened by the use of the expansion bolts in the back of each one. Special support rails had been constructed on the walls of the Lagoda Room so that the canvases could be hung.

Almost two and a half years after the project was initiated, the Moby Dick murals were installed. At the installation, aside from seeing this leviathan of a painting hung on the wall, another, totally unexpected, benefit occurred: I was able to see the entire painting for the first time.

References

Aguilar de Soto, N., M. P. Johnson, P. T. Madsen, F. Diaz, I. Dominguez, A. Brito, and P. Tyack. 2008. Cheetahs of the sea: deep foraging sprints in short-finned pilot whales off Tenerife (Canary Islands). *Journal of Animal Ecology* 77:936–947.

André, M., and C. Kamminga. 2000. Rhythmic dimension in the echolocation click trains of sperm whales: a possible function of identification and communication. *Journal of the Marine Biological Association of the U.K.* 80:163–169.

Andrews, R. C. 1911. Shore-whaling: a world industry. *National Geographic* 22(5):411–442.

———. 1916. *Whale Hunting with Gun and Camera.* D. Apppleton.

———. 1921. A remarkable case of external hind limbs in a humpback whale. *American Museum Novitates* 9:1–6.

———. 1933a. Floating gold: the romance of ambergris. Part 1. *Natural History* 33(2):117–130.

———. 1933b. Floating gold: the romance of ambergris. Part 2. *Natural History* 33(3):303–310.

Angier, N. 2010. Save a whale, save a soul, goes the cry. *New York Times,* June 27.

Anonymous. 1851. Thrilling account of the destruction of a whale ship by a sperm whale—sinking the ship—loss of the boats and miraculous escape of the crew. *New York Times,* November 5.

Aristotle. N.d. *Historia Animalium.* Translated by D'Arcy Wentworth Thompson. Reprint, Clarendon Press, 1910.

Ash, C. E. 1962. *Whaler's Eye.* Macmillan.

Ashford, J. R., P. S. Rubilar, and A. R. Martin. 1996. Interactions between cetaceans and longline fishery operations around South Georgia. *Marine Mammal Science* 12(3):452–457.

Ashley, C. W. 1926. *The Yankee Whaler.* Halcyon House.

Backus, R. H., and W. E. Schevill. 1966. *Physeter* clicks. Pp. 510–528 in K. S. Norris, ed., *Whales, Dolphins and Porpoises.* University of California Press.

Baird, A. 1999. *White as the Waves: A Novel of Moby Dick.* Tuckamore.

Baird, R. W. 2000. The killer whale: foraging specializations and group hunting. Pp. 127–153 in J. Mann, R. C. Connor, P. L. Tyack, and H. Whitehead, eds., *Cetacean Societies: Field Studies of Dolphins and Whales.* University of Chicago Press.

Bajpai, S., and P. D. Gingerich. 1998. A new Eocene archaeocete (Mammalia, Cetacea) from India and the time of origin of whales. *Proceedings of the National Academy of Sciences of the United States of America* 95:15464–15468.

Bajpai, S., and J. M. G. Thewissen. 1998. Middle Eocene cetaceans from the Harudi and Subathu Formations of India. Pp. 213–234 in J. M. G. Thewissen, ed., *The Emergence of Whales.* Plenum.

Bajpai, S., J. M. G. Thewissen, and A. Sahni. 1996. *Indocetus* (Cetacea, Mammalia) endocasts from Kachchh (India). *Journal of Vertebrate Paleontology* 16(3):582–584.

Baker, C. S., and P. J. Clapham. 2004. Modelling the past and future of whales and whaling. *Trends in Ecology and Evolution* 19(7):365–371.

Baker, C. S., G. M. Lento, F. Cipriano, M. L. Dalebout, and S. R. Palumbi. 2000. Scientific whaling: source of illegal products for market? *Science* 290:1695–1696.

Barnes, L. G. 1976. Outline of eastern North Pacific fossil cetacean assemblages. *Systematic Zoology* 25(4):321–343.

———. 1984. Search for the first whale: retracing the ancestry of cetaceans. *Oceans* 17(2):20–23.

———. 1985. Review: G. A. Mchedlizde, General features of the paleobiological evolution of Cetacea, 1984 [English translation]. *Marine Mammal Science* 1(1):90–93.

Barnes, L. G., and E. D. Mitchell. 1978. Cetacea. Pp. 582–602 in V. J. Maglio and H. B. S. Cooke, eds., *Evolution of African Mammals.* Harvard University Press.

Barnes, L. G., D. P. Domning, and C. E. Ray. 1985. Status of studies on fossil marine mammals. *Marine Mammal Science* 1(1):15–53.

Beale, T. 1835. *A Few Observations on the Natural History of the Sperm Whale.* Effingham Wilson, London.

Bel'kovich, V. M., and A. V. Yablokov. 1963. The whale—an ultrasonic projector. *Yuchnyi Teknik* 3:76–77.

Bennett, F. D. 1840. *Narrative of a Whaling Voyage Round the Globe from the Years 1833–1836: Comprising Sketches of Polynesia, California, the Indian Archipelago, Etc.* Richard Bentley.

Berkaw, M. K. 1989. Melville's borrowings. *Log of Mystic Seaport* Summer:35–44.

Bernardini, C. 2009. Heavy Melville: Mastodon's *Leviathan* and the popular image of Moby-Dick. *Leviathan* 11(2):27–44.

Berta, A. 1994. What is a whale? *Science* 263:180–181.

Berta, A., and T. A. Deméré. 2009. Mysticetes, evolution. Pp. 749–753 in W. F. Perrin, B. Würsig, and J. G. M. Thewissen, eds., *Encyclopedia of Marine Mammals.* Academic Press.

Berta, A., and J. L. Sumich. 1999. *Marine Mammals: Evolutionary Biology.* Academic Press.

Berzin, A. A. 1972. *The Sperm Whale.* [Izdatgel'stvo "Pischevaya Promyshlennost" Moskva, 1971.] Translated from the Russian by Israel Program for Scientific Translation, Jerusalem.

Berzin, A. A., and G. M. Veinger. 1976. Investigations of the population morphology of sperm whales, *Physeter macrocephalus* L. 1758, of the Pacific Ocean. Pp. 259–268 in *Mammals in the Seas,* vol. 3: *General Papers and Large Cetaceans.* FAO Fisheries Series no. 5. FAO, Rome.

Best, P. B. 1967. The sperm whale (*Physeter catodon*) off the west coast of South Africa. I. Ovarian changes and their significance. *Division of Sea Fisheries Investigational Report* 61:1–27.

———. 1968a. A comparison of the external characters of sperm whales off South Africa. *Norsk Hvalfangst-tidende* 57(6):146–164.

———. 1968b. The sperm whale (*Physeter catodon*) off the west coast of South Africa. II. Reproduction in the female. *Division of Sea Fisheries Investigational Report* 66:1–32.

———. 1969a. The sperm whale (*Physeter catodon*) off the west coast of South Africa. III. Reproduction in the male. *Division of Sea Fisheries Investigational Report* 72:1–20.

———. 1969b. The sperm whale (*Physeter catodon*) off the west coast of South Africa. IV. Distribution and movements. *Division of Sea Fisheries Investigational Report* 78:1–12.

———. 1970. The sperm whale (*Physeter catodon*) off the west coast of South Africa. V. Age, growth and mortality. *Division of Sea Fisheries Investigational Report* 79:1–27.

———. 1974. The biology of the sperm whale as it relates to stock management. Pp. 257–293 in W. E. Schevill, ed., *The Whale Problem.* Harvard University Press.

———. 1975. Review of world sperm whale stocks. *FAO Marine Mammals Symposium.* ACMRR/MM/EC/8.

———. 1979. Social organization of sperm whales (*Physeter macrocephalus*). Pp. 227–289 in H. E. Winn and B. L. Olla, eds., *Behavior of Marine Animals,* vol. 3, *Cetaceans.* Plenum.

Beston, H. 1928. *The Outermost House.* Holt, Rinehart and Winston.

Bianucci G., and W. Landini. 2006. Killer sperm whale: a new basal physeteroid (Mammalia, Cetacea) from the Late Miocene of Italy. *Zoological Journal of the Linnean Society* 148(1):103–131.

Bianucci, G., W. Landini, and A. Varola. 2004. First discovery of the Miocene Northern Atlantic sperm whale *Orycterocetus* in the Mediterranean. *Geobios* 37:569–573.

Bohannon, J. 2004. A toxic odyssey. *Science* 304:1584–1586.

Boschma, H. 1938. On the teeth and some other particulars of the sperm whale (*Physeter macrocephalus* L.). *Temminckia* 3:261–262.

Boyer, W. D. 1946. Letter to the editor. *Natural History* 55:96.

Boyle, P., and P. Rodhouse. 2005. *Cephalopods: Ecology and Fisheries.* Blackwell.

Browne. J. R. 1846. *Etchings of a Whaling Cruise, with Notes of a Sojourn on the Island of Zanzibar. To Which Is Appended a Brief History of the Whale Fishery, Its Past and Present Condition.* Harper & Brothers. Reprint, Harvard University Press, 1968.

Brownlee, S. 2003. Blast from the vast. *Discover* 24(12):51–57.

Bryant, P. 1979. The Baja sperm whale mass-stranding. *Whalewatcher* 13(2):10.

Buchanan, J. Y. 1896. The sperm whale and its food. *Nature* 1367(53):223–225.

Budker, P. 1959. *Whales and Whaling.* Macmillan.

———. 1971. *The Life of Sharks.* Columbia University Press.

Buijs, D., and D. van Heel. 1979. Bodyplan of a male sperm whale (*Physeter macrocephalus*) stranded near Breskens, Netherlands. *Aquatic Mammals* 7(1):27–32.

Bullen, F. T. 1898. *The Cruise of the "Cachalot": Round the World after Sperm Whales.* Appleton.

———. 1902. *Deep-Sea Plunderings.* Appleton.

———. 1904. *Denizens of the Deep.* Revell.

Bush, S. L., B. H. Robison, and R. L. Caldwell. 2009. Behaving in the dark: locomotor, chromatic, postural and bioluminescent behaviors of the deep-sea squid *Octopoteuthis deletron* Young, 1972. *Biological Bulletin* 216:7–22.

Caillet-Bois, T. 1948. Las pseudorcas de Mar del Plata. *Revista Geograficas Americana* 28(172):5–10.

Caldwell, M. C., and D. K. Caldwell. 1966. Epimeletic (care-giving) behavior in cetacea. Pp. 755–789 in K. S. Norris, ed., *Whales, Dolphins, and Porpoises.* University of California Press.

Caldwell, D. K., M. C. Caldwell, and D. W. Rice. 1966. Behavior of the sperm whale, *Physeter catodon* L. Pp. 678–717 in K. S. Norris, ed., *Whales, Dolphins, and Porpoises.* University of California Press.

Carrier, D. R., S. M. Deban, and J. Otterstrom. 2002. The face that sank the *Essex:* potential function of the spermaceti organ in aggression. *Journal of Experimental Biology* 205:1755–1763.

Carroll, R. L. 1998. *Vertebrate Paleontology and Evolution.* Freeman.

Cheever, H. T. 1850. *The Whale and His Captors; or, The Whalemans' Adventures, and the Whale Biography, as Gathered on the Homeward Cruise of the Commodore Preble.* Reprint, Ye Galleon Press, 1991.

Christal, J., and H. Whitehead. 1997. Aggregations of mature male sperm whales on the Galápagos Islands breeding grounds. *Marine Mammal Science* 13(1):59–69.

Christal, J., H. Whitehead, and E. Lettevall. 1998. Sperm whale social units: variation and change. *Canadian Journal of Zoology* 76:1431–1440.

Church, A. C. 1938. *Whale Ships and Whaling*. Bonanza Books.

Clapham, P. 1996. Too much is never enough: can the whaling industry be trusted? *Whalewatcher* 30(1):4–7.

Clarke, M. R. 1962. Stomach contents of a sperm whale caught off Madeira in 1959. *Norsk Havalfangst-tidende* 173–191.

——. 1966a. A review of the systematics and ecology of oceanic squids. *Advances in Marine Biology* 4:91–300.

——. 1970. Function of the spermaceti organ of the sperm whale. *Nature* 228:873–874.

——. 1976. Observations on sperm whale diving. *Journal of the Marine Biological Association of the U.K.* 56:809–810.

——. 1977. Beaks, nets, and numbers. *Symposium Zoological Society of London* 38:89–126.

——. 1978a. Structure and proportions of the spermaceti organ in the sperm whale. *Journal of the Marine Biological Association of the U.K.* 58:1–17.

——. 1978b. Physical properies of spermaceti oil in the sperm whale. *Journal of the Marine Biological Association of the U.K.* 58:19–26.

——. 1978c. Buouyancy control as a function of the spermaceti organ in the sperm whale. *Journal of the Marine Biological Association of the U.K.* 58:27–71.

——. 1979. The head of the sperm whale. *Scientific American* 240(1):128–141.

——. 1980. Cephalopoda in the diet of sperm whales of the Southern Hemisphere and their bearing on sperm whale biology. *Discovery Reports* 37:1–324.

——. 1983. Cephalopod biomass—estimation from predation. *Memoirs of the National Museum of Victoria* 44:95–107.

——. 1986. *A Handbook for the Identification of Cephalopod Beaks*. Clarendon.

——. 1997. Beaks, nets and numbers. *Symposium Zoological Society of London* 38:89–126.

Clarke, R. 1953. A great haul of ambergris. *Norsk Hvalfangst-tidende* 43(8):450–453.

——. 1955. A giant squid swallowed by a sperm whale. *Norsk Hvalfangst-tidende* 44(10):589–593.

——. 1956. Sperm whales of the Azores. *Discovery Reports* 28:237–298.

——. 1966b. The stalked barnacle *Conchoderma* ectoparasitic on whales. *Norsk Hvalfangst-tidende* 55(8):154–168.

Clarke, R., and O. Paliza. 2000. The food of sperm whales in the Southeast Pacific. *Marine Mammal Science* 17:427–429.

Clarke, R., A. Aguayo, and O. Paliza. 1968. Sperm whales of the southeast Pacific. Part 1: Introduction. *Hvalrådets Skrifter* 51:1–80.

——. 1988. Sperm whales of the southeast Pacific. Part 4: Fatness, food, and feeding. *Investigations on Cetacea* 21:53–195.

——. 1994. Sperm whales of the southeast Pacific. Part 6: Growth and breeding in the male. *Investigations on Cetacea* 25:93–224.

Cockrum, E. L. 1956. Sperm whales stranded on the beaches of the Gulf of California. *Journal of Mammalogy* 37(2):288.

Connor, R. C., J. Mann, P. L. Tyack, and H. Whitehead. 1998. Social evolution in toothed whales. *Trends in Ecology and Evolution* 13(6):228–232.

Cope, E. D. 1884. The Creodonta. *American Naturalist* March:255–267; April:344–353; May:478–485.

Cousteau, J.-Y., and P. Diolé. 1972. *The Whale: Mighty Monarch of the Sea*. Doubleday.

———. 1973. *Octopus and Squid: The Soft Intelligence.* Doubleday.

Cranford, T. 1999. The sperm whale's nose: sexual selection on a grand scale? *Marine Mammal Science* 15(4):1133–1157.

Dalebout, M. L., J. G. Mead, C. S. Baker, A. N. Baker, and A. L. van Helden. 2002. A new species of beaked whale *Mesoplodon perrini* sp. n. (Cetacea: Ziphiidae) discovered through phylogenetic analyses of mitochondrial DNA sequences. *Marine Mammal Science* 18(3):577–608.

Dalebout, M. L., G. J. B. Ross, C. S. Baker, R. C. Anderson, P. B. Best, V. G. Cockcroft, H. L. Hinsz, V. Peddemors, and R. L. Pitman. 2003. Appearance, distribution, and genetic distinctiveness of Longman's beaked whale, *Indopacetus pacificus. Marine Mammal Science* 19(3):421–461.

Dannenfeldt, K. H. 1982. Ambergris: the search for its origin. *Isis* 73(3):382–397.

Daugherty, A. 1972. *Marine Mammals of California.* California Department of Fish and Game.

Davis, D. 1976. Story of a baby sperm whale. *Aquasphere* 10(2):18–23.

Davis, E. Y. 1946. Man in whale. *Natural History* 55(6):241.

Davis, W. M. 1874. *Nimrod of the Sea, or, The American Whaleman.* Christopher Publishing House.

De Jong, C. 1976. A whaleboat the first museum ship in South Africa? *Restorica (Bulletin of the Simon van der Stel Foundation, Pretoria)* 17(33):77–78.

De Smet, W. M. A. 1981. Evidence of whaling in the North Sea and English Channel during the Middle Ages. Pp. 3:301–309 in *Mammals in the Seas.* FAO Fisheries Series No. 5. FAO, Rome.

Dewhurst, W. H. 1835. *The Natural History of the Order Cetacea and the Oceanic Inhabitants of the Arctic Regions.* London.

Donoghue, M., R. R. Reeves, and G. S. Stone, eds. 2003. *Report of the Work-shop on Interaction between Cetaceans and Longline Fisheries.* New England Aquarium Press.

Dudley, P. 1725. An essay upon the natural history of whales, with particular account of the ambergris found in the sperma ceti whale. *Philosophical Transactions of the Royal Society of London* 33(387):256–259.

Dudzinski, K. M., J. A. Thomas, and J. D. Gregg. 2009. Communication in marine mammals. Pp. 260–269 in W. F. Perrin, B. Würsig, and J. G. M. Thewissen, eds., *Encyclopedia of Marine Mammals.* Academic Press.

Dufault, S., and H. Whitehead. 1995a. The geographic stock structure of female and immature sperm whales in the South Pacific. *Report of the International Whaling Commission* 45:401–405.

———. 1995b. An encounter with recently wounded sperm whales (*Physeter macrocephalus*). *Marine Mammal Science* 11(4):560–563.

———. 1998. Regional and group-level differences in fluke markings and notches of sperm whales. *Journal of Mammalogy* 79(2):514–520.

Dufresne, F. 1946. *Alaska's Animals and Fishes.* A. S. Barnes and Company.

Du Pasquier, T. 1982. *Les Baleiniers Français au XIXᵉ Siecle, 1814–1868.* Terre et Mer.

Ellis, R. 1980. *The Book of Whales.* Knopf.

———. 1981a. A visitor from inner space. *Animal Kingdom* 84(4):5–11.

———. 1981b. Observations on a captive sperm whale, *Physeter macrocephalus,* at Fire Island, New York [abstract]. Fourth Biennial Conference on the Biology of Marine Mammals, San Francisco.

———. 1982. *Dolphins and Porpoises.* Knopf.

———. 1991. *Men and Whales.* Knopf.

———. 1993. *Physty: The True Story of a Young Whale's Rescue.* Running Press.

———. 1998. *The Search for the Giant Squid.* Lyons.

———. 2001. *Aquagenesis: The Origin and Evolution of Life in the Sea*. Viking.

Ely, B.-E. 1849. *"There She Blows": A Narrative of a Whaling Voyage in the Indian and South Atlantic Oceans*. James K. Simon. 1971 edition, edited by Curtis Dahl, Wesleyan University Press.

Emlong, D. R. 1966. A new archaic cetacean from the Oligocene of northwest Oregon. *University of Oregon Natural History Bulletin* 3:1–51.

Fadiman, A. 1979. Will we kill the last whale? *Life*, July, 18–26.

Ferber, D. 2005. Sperm whales bear testimony to worldwide pollution. *Science* 309:1166.

Fichtelius, K.-E., and S. Sjölander. 1972. *Smarter Than Man? Intelligence in Whales, Dolphins, and Humans*. Ballantine Books.

Fiscus, C. H., and D. W. Rice. 1974. Giant squids, *Architeuthis* sp., from stomachs of sperm whales captured off California. *California Fish and Game* 60(2):91–93.

Fiscus, C. H., D. W. Rice, and A. A. Wolman. 1989. Cephalopods from the stomachs of sperm whales taken off California. *NOAA Technical Report NMFS* 83:1–12.

Fish, F. E. 1998. Biomechanical perspective on the origin of cetacean flukes. Pp. 303–324 in J. G. M. Thewissen, ed., *The Emergence of Whales*. Plenum.

Fitzgerald, E. M. G. 2006. A bizarre new toothed mysticete (Cetacea) from Australia and the early evolution of baleen whales. *Proceedings of the Royal Society B* 273:2955–2963.

Fleisher, K. J., and J. F. Case. 1995. Cephalopod predation facilitated by dinoflagellate luminescence. *Biological Bulletin* 189(3):263–271.

Flower, W. H. 1883. On whales, present and past and their probable origin. *Proceedings of the Zoological Society of London* 1883:466–513.

Ford, J. K. B., G. Ellis, and K. Balcomb. 1994. *Killer Whales*. University of British Columbia/ University of Washington Press.

Fordyce, R. E. 1977. The development of the Circum-Antarctic Current and the evolution of the Mysticeti (Mammalia: Cetacea). *Palaeogeography, Palaeoclimatology, Palaeoecology* 21:256–271.

———. 1980. Whale evolution and Oligocene southern ocean environments. *Palaeogeography, Palaeoclimatology, Palaeoecology* 31:319–336.

———. 1982. A review of Australian fossil cetacea. *Memoirs of the National Museum of Victoria* 43:43–58.

———. 2009a. Cetacean evolution. Pp. 201–207 in W. F. Perrin, B. Würsig, and J. G. M. Thewissen, eds., *Encyclopedia of Marine Mammals*. Academic Press.

———. 2009b. Cetacean fossil record. Pp. 207–214 in W. F. Perrin, B. Würsig, and J. G. M. Thewissen, eds., *Encyclopedia of Marine Mammals*. Academic Press.

Fordyce, R. E., and L. G. Barnes. 1994. The evolutionary history of whales and dolphins. *Annual Review of Earth and Planetary Sciences* 22:419–455.

Frank, S. M. 1989. "Unequal cross-lights": Melville's pictures and the aesthetics of a sometime whaleman. *Log of Mystic Seaport* Summer:45–55.

Fristrup, K. M., and G. R. Harbison. 2002. How do sperm whales catch squids? *Marine Mammal Science* 18(1):42–54.

Frost, S., chairman. 1978. *Whales and Whaling*. Australian Government Printing Service.

Fujino, K. 1956. On the body proportions of the sperm whale (*Physeter catadon*). *Scientific Reports of the Whales Research Institute* 11:47–83.

Gambell, R. 1966. Foetal growth and the breeding season of sperm whales. *Norsk Hvalfangst-tidende* 55(6):113–118.

———. 1968. Aerial observations of sperm whale behavior. *Norsk Hvalfangst-tidende* 57(6):126–138.

———. 1970. Weight of a sperm whale, whole and in parts. *South African Journal of Science* 66:225–227.

———. 1972. Sperm whales off Durban. *Discovery Reports* 35:199–358.

Gambell, R., C. Lockyer, and G. J. B. Ross. 1973. Observations on the birth of a sperm whale calf. *South African Journal of Science* 69:147–148.

Gaskin, D. E. 1964. Recent observations in New Zealand waters on some aspects of behaviour of the sperm whale (*Physeter macrocephalus*). *Tuatara* 12:106–114.

———. 1967. Luminescence in a squid *Moroteuthis* sp. (probably *ingens* Smith) and a possible feeding mechanism in the sperm whale *Physeter catodon* L. *Tuatara* 15:86–88.

———. 1982. *The Ecology of Whales and Dolphins*. William Heinemann.

Gaskin, D. E., and M. W. Cawthorne. 1967. Diet and feeding habits of the sperm whale *Physeter catodon* L. in the Cook Strait region of New Zealand. *New Zealand Journal of Marine and Freshwater Research* 1(2):159–179.

Gatesy, J. 1998. Molecular evidence for the phylogenetic affinities of Cetacea. Pp. 63–112 in J. G. M. Thewissen, ed., *The Emergence of Whales*. Plenum.

Gero, S., and H. Whitehead. 2007. Suckling behavior in sperm whale calves: observations and hypotheses. *Marine Mammal Science* 23(2):398–413.

Gero, S., D. Engelhaupt, L. Rendell, and H. Whitehead. 2009. Who cares? Between-group variation in alloparental caregiving in sperm whales. *Behavioral Ecology* 20(4):838–843.

Gesner, C. 1551–1558. *Historia Animalium*. Zurich.

Gibbes, R. W. 1845. Description of the teeth of a new fossil animal found in the Green Sand of South Carolina. *Proceedings of the Academy of Natural Sciences of Philadelphia* 2:254–256.

———. 1847. On the fossil genus *Basilosaurus*, Harlan (Zeuglodon, Owen), with a notice of specimens from the Eocene Green Sand of South Carolina. *Journal of the Academy of Natural Sciences of Philadelphia* 1:2–15.

Gihr, M., and G. Pilleri. 1979. Interspecific body length–body weight ratio and body weight–brain weight ratio in cetacea. *Investigations on Cetacea* 10:245–253.

Gillis, J. 2010. Giant plumes of oil forming under the Gulf. *New York Times*, May 15.

Gilmore, R. 1959. On mass strandings of sperm whales. *Pacific Naturalist* 1(10):9–16.

Gingerich, P. D. 1991. Partial skeleton of a new archaeocete from the earliest middle Eocene Habib Rahi Limestone, Pakistan. *Journal of Vertebrate Paleontology* 11:31A.

———. 1997. The origin and evolution of whales. *LSA magazine*, University of Michigan 20(2):4–10.

———. 1998. Paleobiological perspectives on Mesonychia, Archaeoceti, and the origin of whales. Pp. 423–450 in J. G. M. Thewissen, ed., *The Emergence of Whales*. Plenum.

Gingerich, P. D., and D. E. Russell. 1981. *Pakicetus inachus*, a new archaeocete (Mammalia, Cetacea) from the early–middle Eocene Kuldana Formation of Kohat (Pakistan). *Contributions to the Museum of Paleontology of the University of Michigan* 25(11):235–246.

———. 1991. Dentition of Early Eocene *Pakicetus* (Mammalia, Cetacea). *Contributions to the Museum of Paleontology of the University of Michigan* 28(1):1–20.

———. 1995. Unusual mammalian limb bones (Cetacea? Archaeoceti?) from the early-to-middle Eocene Subathu Formation of Kashmir (Pakistan). *Contributions to the Museum of Paleontology of the University of Michigan* 29:109–117.

Gingerich, P. D., and M. D. Uhen. 1996. *Analectus simonsi*, a new dorudontine archaeocete (Mammalia, Cetacea) from the early late Eocene of Wadi Hitan, Egypt. *Contributions to the Museum of Paleontology of the University of Michigan* 29:359–401.

———. 1997. The evolution of whales. *LSAmagazine,* University of Michigan 20(2):4–10.

Gingerich, P. D., N. A. Wells, D. E. Russell, and S. M. I. Shah. 1983. Origin of whales in epicontinental remnant seas: new evidence from the Early Eocene of Pakistan. *Science* 220:403–406.

Gingerich, P. D., B. H. Smith, and E. L. Simons. 1990. Hind limbs of Eocene *Basilosaurus:* evidence of feet in whales. *Science* 246:154–157.

Gingerich, P. D., S. M. Raza, M. Arif, M. Anwar, and X. Zhou. 1993. Partial skeletons of *Indocetus ramani* (Mammalia, Cetacea) from the lower middle Eocene Domanda Shale in the Sulaiman Range of Punjab, Pakistan. *Contributions to the Museum of Paleontology of the University of Michigan* 16:393–416.

Gingerich, P. D., D. P. Domning, C. E. Blane, and M. D. Uhen. 1994. Cranial morphology of *Protosiren fraasi* (Mammalia, Sirenia) from the Middle Eocene of Egypt: a new study using computed tomography. *Contributions to the Museum of Paleontology of the University of Michigan* 29(2):41–67.

Gingerich, P. D., M. Arif, and W. C. Clyde. 1995a. New archaeocetes (Mammalia, Cetacea) from the middle Eocene Domanda Formation of the Sulaiman Range, Punjab (Pakistan). *Contributions to the Museum of Paleontology of the University of Michigan* 29:291–330.

Gingerich, P. D., M. A. Bhatti, H. A. Raza, and S. M. Raza. 1995b. *Protosiren* and *Babiacetus* (Mammalia: Sirenia and Cetacea) from the Middle Eocene Drazinda Formation, Sulaiman Range, Punjab (Pakistan). *Contributions to the Museum of Paleontology of the University of Michigan* 29(12):331–357.

Gingerich, P. D., M. Arif, M. A. Bhatti, and W. C. Clyde. 1998. Middle Eocene stratigraphy and marine mammals (Mammalia: Cetacea and Sirenia) of the Sulaiman Range, Pakistan. *Bulletin of the Carnegie Museum of Natural History* 34:239–259.

Gingerich, M. ul-Haq, W. von Koeningswald, W. J. Sanders, B. H. Smith, and I. S. Zalmout. 2009. New protocetid whale from the Middle Eocene of Pakistan: birth on land, precocial development, and sexual dimorphism. *Public Library of Science* 4(2):e4366.

Gordon, J. C. D. 2009. Evaluation of a method for determining the length of sperm whales (*Physeter catadon*) [*sic*] from their vocalizations. *Journal of Zoology* 224(2):301–314.

Gosho, M., D. W. Rice, and J. M. Breiwick. 1984. The sperm whale, *Physeter macrocephalus.* Special report of *Marine Fisheries Review,* "The Status of Endangered Whales," ed. J. M. Breiwick and H. W. Braham, 46(4):54–64.

Gould, S. J. 1994. Hooking Leviathan by its past. *Natural History* 103(5):8–16.

Gowans, S. W., and L. Rendell. 1999. Head-butting in northern bottlenose whales (*Hyperoodon ampullatus*): a possible function for big heads. *Marine Mammal Science* 15(4):1342–1350.

Graves, E., ed. 1977. *Dangerous Sea Creatures.* Time-Life Films.

Green, R. G. 1974. Teuthids of the Late Cretaceous Niobrara Formation of Kansas and some ecological implications. *Compass* 51(3):53–60.

———. 1977. *Niobrarateuthis walkeri,* a new species of teuthid from the Upper Cretaceous Niobrara Formation of Kansas. *Journal of Paleontology* 51(5):992–995.

Griggs, K. 2003. Super squid surfaces in Antarctic. *BBC News.* http://news.bbc.co. uk/1/hi/sci/tech/2910849.stm.

Grimm, D. 2010. Is a dolphin a person? *Science* 327:1070–1071.

Haley, N. C. 1948. *Whale Hunt: The Narrative of a Voyage by Nelson Cole Haley, Harpooner in the Ship "Charles W. Morgan," 1849-1853*. Ives Washburn. Reprint, Mystic Seaport, 1990.

Hanna, G. D. 1924. Sperm whales at St. George Island, Bering Sea. *Journal of Mammalogy* 5(1):64.

Harmer, S. F. 1928. The history of whaling. *Proceedings of the Linnean Society of London* 140:51-95.

Hart, J. C. 1834. *Miriam Coffin; or, The Whale-fishermen*. G., C. & H. Carvill, New York. Reprint, Mill Hill Press, Nantucket, 1995.

Harvey, M. 1874. Gigantic cuttlefishes in Newfoundland. *Annual Magazine of Natural History* 13:67-70.

Hass, H. 1959. *We Come from the Sea*. Doubleday.

Hastie, G. D., R. J. Swift, J. C. D. Gordon, G. Slesser, and W. R. Turrell. 2003. Sperm whale distribution and seasonal diversity in the Faroe Shetland Channel. *Journal of Cetacean Research and Management* 5(3):247-252.

Hawes, C. B. 1924. *Whaling*. Doubleday, Page.

Heezen, B. C. 1957. Whales entangled in deep-sea cables. *Deep-Sea Research* 4:105-115.

Heezen, B. C., and C. D. Hollister. 1971. *The Face of the Deep*. Oxford University Press.

Herman, L. M., and W. N. Tavolga. 1980. The communication systems of cetaceans. Pp. 149-209 in L. M. Herman, ed., *Cetacean Behavior: Mechanism and Functions*. Wiley.

Heyning, J. E. 1997. Sperm whale phylogeny revisited: analysis of the morphological evidence. *Marine Mammal Science* 13(4):596-513.

———. 1999. Whale origins—conquering the seas [book review]. *Science* 283:943.

Hill, D. O. 1975. Vanishing giants. *Audubon* 77(1):56-107.

Hirota, K., and L. G. Barnes. 1994. A new species of Middle Miocene sperm whale of the genus *Scaldicetus* (Cetacea; Physeteridae) from Shiga-mura, Japan. *Island Arc* 3:453-472.

Hochberg, F. G. 1974. Southern California records of the giant squid, *Moroteuthis robusta*. *Tabulata* 7:83-85.

———. 1986. Of beaks and whales. *Bulletin of the Santa Barbara Museum of Natural History* 95:1-2.

Hochberg, F. G., and W. G. Fields. 1980. Cephalopoda: the squids and octopuses. Pp. 429-444 in R. H. Morris, D. P. Abbott, and E. C. Haderlie, eds., *Intertidal Invertebrates of California*. Stanford University Press.

Hockett, C. P. 1978. In search of Jove's brow. *American Speech* 53(4):243-313.

Hof, P. R., and E. Van der Gucht. 2006. The structure of the cerebral cortex of the humpback whale, *Megaptera novaeangliae* (Cetacea, Mysticeti, Balaenopteridae). *Anatomical Record* (290):1-31.

Hof, P. R., R. Chanis, and L. Marino. 2005. Cortical complexity in cetacean brains. *Anatomical Record* 287A:1142-1152.

Hohman, E. P. 1928. *The American Whaleman*. Reprint, Augustus M. Kelley, 1972.

Holm, J. L., and A. Jonsgård. 1959. Occurrence of the sperm whale in the Antarctic and the possible influence of the moon. *Norsk Hvalfangst-tidende* 48(4):161-182.

Hope, P. L., and H. Whitehead. 1991. Sperm whales off the Galápagos Islands from 1830-50 and comparisons with modern studies. *Report of the International Whaling Commission* 41:273-283.

Hopkins, W. J. 1922. *She Blows! And Sparm at That!* Houghton Mifflin.

Housby, T. 1971. *The Hand of God: Whaling in the Azores*. Abelard Schuman.

Husson, A. M., and L. B. Holthius. 1974. *Physeter macrocephalus* Linnaeus 1758, the valid name for the sperm whale. *Zoologische Mededelingen* 48(19):206-217.

International Whaling Commission. 1980. *Sperm Whales: Special Issue.* Cambridge.

Jacques, T. G., and R. H. Lambertsen. 1997. *The North Sea Sperm Whales, One Year After.* Institut Royal des Sciences Naturelles de Belgique, Brussels.

Jaquet, N., and H. Whitehead. 1996. Scale-dependent correlation of sperm whale distribution with environmental features and productivity in the South Pacific. *Marine Ecology Progress Series* 135:1–9.

Jaquet, N., H. Whitehead, and M. Lewis. 1996. Coherence between 19th century sperm whale distributions and satellite-derived pigments in the tropical Pacific. *Marine Ecology Progress Series* 145:1–10.

Jaquet, N., S. Dawson, and E. Slooten. 2000. Seasonal distribution and diving behaviour of male sperm whales off Kaikoura: foraging implications. *Canadian Journal of Zoology* 78:407–419.

Jaquet, N., S. Dawson, and L. Douglas. 2001. Vocal behavior of male sperm whales: why do they click? *Journal of the Acoustical Society of America* 109(5):2254–2259.

Jardine, W. 1839. *The Naturalist's Library.* Vol. 26: *Mammalia, Whales, Etc.* W. H. Lizars, Edinburgh.

Jenkins, J. T. 1921. *A History of the Whale Fisheries.* Reprint, Kennikat Press, 1971.

Jenkins, T. H. 1902. *Bark "Kathleen" Sunk by a Whale. To Which Is Added an Account of Two Like Occurrences, the Loss of Ships "Ann Alexander" and "Essex."* Hutchinson.

Jerison, H. J. 1978. Brain and intelligence in whales. Pp. 2:161–197 in S. Frost, ed., *Whales and Whaling.* Australian Government Publishing Service.

———. 1980. The cetacean brain. IWC Conference on Cetacean Intelligence and Behavior and the Ethics of Killing Cetaceans. International Whaling Commission.

Jonsgård, A. 1960. On the stocks of sperm whales (*Physeter macrocephalus*) in the Antarctic. *Norsk Hvalfangst-tidende* 49(7):289–299.

Josselyn, J. 1672. *New England Rarities Discovered.* London. Reprint, 1972.

Kahn, B., H. Whitehead, and M. Dillon. 1993. Indications of density-dependent effects from comparisons of sperm whale populations. *Marine Ecology Progress Series* 93:1–7.

Kasuya, T., and S. Ohsumi. 1966. A secondary sexual characteristic of the sperm whale. *Scientific Reports of the Whales Research Institute* 20:89–94.

Kawakami, T. 1980. A review of sperm whale food. *Scientific Reports of the Whales Research Institute* 32:199–218.

Kellogg, R. 1922. Two fossil physeteroid whales from California. *University of California Bulletin of Geological Sciences* 13(4):23–132.

———. 1928. The history of whales—their adaptation to life in the water. *Quarterly Review of Biology* 3(1):29–76, 174–208.

———. 1936. *A Review of the Archaeoceti.* Carnegie Institution of Washington.

———. 1940. Whales, giants of the sea. *National Geographic* 77(1):35–90.

———. 1944. Fossil cetaceans from the Florida Tertiary. *Bulletin of the Museum of Comparative Zoology* 44(9):432–471.

Kemp, B., and H. A. A. Oelschaläger. 2000. Evolutionary strategies of odontocete brain development. *Historical Biology* 14(1–2):41–45.

Klima, M. 1995. Cetacean phylogeny and systematics based on the morphogenesis of the nasal skull. *Aquatic Mammals* 21(2):79–89.

Koefoed, E. 1950. Bleksprutter. Pp. 420–427 in B. Føyn, G. Ruud, H. Røise, and H. Christensen, eds., *Norges Dyreliv IV* (Animal Life of Norway, Vol. 4.) Cappelens.

Köhler, R., and R. E. Fordyce. 1997. An archaeocete whale (Cetacea: Archaeoceti) from the Eocene Waihao Greensand, New Zealand. *Journal of Vertebrate Paleontology* 17(3): 574–583.

Kojima, T. 1951. On the brain of the sperm whale. *Scientific Reports of the Whales Research Institute* 6:49–72.

Kondo, I. 2001. *Rise and Fall of the Japanese Coast Whaling.* Sanyosha.

Kooyman, G. L. 1985. Physiology without restraint in diving mammals. *Marine Mammal Science* 1(2):166–178.

Kozak, V. A. 1978. Receptor zone of the video-acoustic system of the sperm whale (*Physeter catodon* L. 1758). *Fisiologishnyy Zhurnal Academy Nauk Ukrayns'koy RSR* 120(3):1–6. National Technical Information Service.

Kubodera, T., and K. Mori. 2005. First-ever observations of a live giant squid in the wild. *Proceedings of the Royal Society B* 272:2583–2586.

Kugler, R. C. 1976. The historical records of American sperm whaling: what they tell us and what they don't. *FAO Scientific Consultation on Marine Mammals.* ACMRR/MM/SC/105.

———. 1980. The whale oil trade, 1750–1775. *Publications of the Colonial Society of Massachusetts* 52:153–173.

Kumar, K., and A. Sahni. 1986. *Remingtonocetus harudiensis,* new combination, a middle Eocene archaeocete (Mammalia, Cetacea) from western Kutch, India. *Journal of Vertebrate Paleontology* 6(4):326–349.

Lacépède, B.-G. 1804. *Histoire Naturelle des Cétacés.* Chez Plassan.

Laitman, J. T. 2007. Thar she blows . . . and dives, and feeds, and talks, and hears, and thinks: the anatomical adaptations of aquatic mammals. *Anatomical Record* 290:504–506.

Lambert, O., G. Bianucci, and C. de Muizon. 2008. A new stem-sperm whale (Cetacea, Odontoceti, Physeteroidea) from the Latest Miocene of Peru. *Comptes Rendus Palevol* 7(6):361–369.

Lambert, O., G. Bianucci, K. Post, C. de Muizon, R. Salas-Gismondi, M. Urbina, and J. Reumer. 2010. The giant bite of a new raptorial sperm whale from the Miocene epoch of Peru. *Nature* 466:105–108.

Lavery T., J. G. Mitchell, and L. Seuront. 2009. Whales: a net sink or source of carbon in the atmosphere? Presented at the 18th Marine Mammal Biennial Conference, October 12–16, Quebec City, Canada.

Lawrence, D. H. 1923. *Studies in Classical American Literature.* Reprint, Cambridge University Press, 2002.

Laws, R. M. 1961. Laminated structure of bones in some marine mammals. *Norsk Hvalfangst-tidende* 50(12):499–507.

Leidy, J. 1867. Orycterocetus. *Proceedings of the Academy of Natural Sciences of Philadelphia* December:144–145.

Letteval, E., C. Richter, N. Jaquet, E. Slooten, S. Dawson, H. Whitehead, J. Christal, and P. M. Howard. 2002. Social structure and residency in aggregations of male sperm whales. *Canadian Journal of Zoology* 80(7):1189–1196.

Ley, W. 1987. *Exotic Zoology.* Bonanza.

Lilly, J. C. 1961. *Man and Dolphin.* Doubleday.

———. 1967. *The Mind of the Dolphin.* Doubleday.

Lindberg, D. R., and N. R. Peynson. 2007. Things that go bump in the night: evolutionary interactions between cephalopods and cetaceans in the Tertiary. *Lethaia* 40(4):335–343.

Liu, Y., J. A. Cotton, B. Shen, X. Han, S. J. Rossiter, and S. Zhang. 2010. Convergent sequence evolution between echolocating bats and dolphins. *Current Biology* 20(2):R53–R54.

Lockyer, C. 1976. Estimates of growth and energy budget for the sperm whale, *Physeter catodon. FAO Scientific Consultation on Marine Mammals.* ACMRR/SC/38.

Lowenstein, J. M. 1983. Very like a whale. *Oceans* 16(5):65.

Luo, Z. 1998. Homology and transformation of cetacean ectotympanic structures. Pp. 269–302 in J. G. M. Thewissen, ed., *The Emergence of Whales.* Plenum.

———. 2000. In search of the whales' sisters. *Nature* 404:235–238.

Luo, Z., and K. Marsh. 1996. Petrosal (periotic) and inner ear of a Pliocene kogiid whale (Kogiinae, Odontoceti): implications on relationships and hearing evolution of toothed whales. *Journal of Vertebrate Paleontology* 16(2):328–348.

Mackintosh, N. A. 1942. The southern stocks of whalebone whales. *Discovery Reports* 22:197–300.

———. 1965. *The Stocks of Whales.* Fishing News (Books) Ltd.

MacLeod, C. D., J. S. Reidenberg, M. Weller, M. B. Santos, J. Herman, J. Goold, and G. J. Pierce. 2007. Breaking symmetry: the marine environment, prey size, and the evolution of asymmetry in cetacean skulls. *Anatomical Record* 290:539–545.

Madsen, P. T., R. Payne, N. U. Kristiansen, M. Wahlberg, I. Kerr, and B. Møhl. 2002. Sperm whale sound production studied with ultrasound time/depth recording tags. *Journal of Experimental Biology* 205:1899–1906.

Madsen, P. T., M. Wilson, M. Johnson, R. T. Hanlon, A. Bocconcelli, N. Aguilar de Soto, and P. L. Tyack. 2007. Clicking for calamari: toothed whales can echolocate *Loligo pealeii. Aquatic Biology* 1:141–150.

Malloy, M. 1989. Whalemen's perceptions of the "high and mighty business of whaling." *Log of Mystic Seaport* (Summer 1989):56–67.

Mann, J., R. C. Connor, P. L. Tyack, and H. Whitehead, eds. 2000. *Cetacean Societies: Field Studies of Dolphins and Whales.* University of Chicago Press.

Marino, L. 2007. Cetacean brains: how aquatic are they? *Anatomical Record* 290:694–700.

———. 2009. Brain size evolution. Pp. 149–152 in W. F. Perrin, B. Würsig, and J. G. M. Thewissen, eds., *Encyclopedia of Marine Mammals.* Academic Press.

Marino, L., and P. Hof. 2005. Nature's experiments in brain diversity. *Anatomical Record* 287A:997–1000.

Marino, L., D. W. McShea, and M. D. Uhen. 2004. Origin and evolution of large brains in toothed whales. *Anatomical Record* 281:1247–1255.

Marino, L., R. C. Connor, R. E. Fordyce, L. M. Herman, P. R. Hof, L. Lefebvre, D. Lusseau, B. McCowan, E. A. Nimchinsky, A. A. Pack, L. Rendell, J. S. Riedenberg, D. Reiss, M. D. Uhen, E. Van der Gucht, and H. Whitehead. 2007. Cetaceans have complex brains for complex cognition. *PLoS Biology* 5(5):966–972.

Masaki, Y. 1969. A malformed sperm whale with two nostrils. *Journal of the Mammal Society of Japan* 4(4–6).

Mate, B. 1985. A mass stranding of sperm whales: all was not lost! *Whalewatcher* Summer:18–20.

Mathias, D., A. Thode, J. Straley, and K. Folkert. 2009. Relationship between sperm whale (*Physeter macrocephalus*) click structure and size derived from videocamera images of a depredating whale (sperm whale prey acquisition). *Journal of the Acoustical Society of America* 125(5):3444–3453.

Matsen, B. 2009. *Jacques Cousteau: The Sea King.* Pantheon.

Matsumoto, H. 1936. Some fossil cetaceans of Japan. *Scientific Reports* Tohoku Imperial University 10:18–27.

Matthews, L. H. 1938. The sperm whale. *Discovery Reports* 27:93–168.

McAuliffe, K. 1998. When whales had feet. *Sea Frontiers* 40(1):20–33.

McCann, C. 1974. Body scarring on cetaceans—odontocetes. *Scientific Reports of the Whales Research Institute* 26:145–155.

McCarthy, N. 2010. Baby sperm whale washes ashore in Cannon Beach. *Daily Astorian*, January 4.

McCormick, J. G., E. G. Weaver, G. Palin, and S. H. Ridgway. 1970. Sound conduction in the dolphin ear. *Journal of the Acoustical Society of America* 48:1418–1428.

McCormick, J. G., E. G. Wever, S. H. Ridgway, and J. Palin. 1979. Sound reception in the porpoise as it is related to echolocation. Pp. 449–467 in R. G. Busnel and J. G. Fish, eds., *Animal Sonar Systems*. Plenum.

Mchedlidze, G. A. 1984. *General Features of the Paleobiological Evolution of Cetacea*. Amerind.

———. 2009. Sperm whales, evolution. Pp. 1097–1098 in W. F. Perrin, B. Würsig, and J. G. M. Thewissen, eds., *Encyclopedia of Marine Mammals*. Academic Press.

McHugh, J. L. 1974. Role and history of the IWC. Pp. 305–335 in W. E. Schevill, ed., *The Whale Problem*. Harvard University Press.

McIntyre, J., ed. 1974. *Mind in the Waters*. Scribners/Sierra Club.

McVay, S. 1966. The last of the great whales. *Scientific American* 215(2):13–21.

———. 1971. Can Leviathan long endure so wide a chase? *Natural History* 80(1):36–40, 68–72.

Mead, J. G. 1975. A fossil beaked whale (Cetacea: Ziphiidae) from the Miocene of Kenya. *Journal of Paleontology* 49(4):745–751.

———. 2009. Beaked whales: overview. Pp. 94–97 in W. F. Perrin, B. Würsig, and J. G. M. Thewissen, eds., *Encyclopedia of Marine Mammals*. Academic Press.

Mellinger, D. K., K. M. Stafford, and C. G. Fox. 2004. Seasonal occurrence of sperm whales (*Physeter macrocephalus*) sounds in the Gulf of Alaska. *Marine Mammal Science* 20(1):48–62.

Melville, H. 1847. Book review of *Etchings of a Whaling Cruise*. *Literary Journal*, March 6.

———. 1851. *Moby-Dick*. Reprint, Norton Critical Editon, H. Hayford and H. Parker, eds., W. W. Norton, 1967.

———. 1930. *Moby-Dick*. Random House.

Mesnick, S. L., B. L. Taylor, R. G. LeDuc, S. E. Traviño, G. M. O'Corry-Crowe, and A. E. Dizon. 1999. Culture and genetic evolution in whales. *Science* 284:2055a.

Middleton-Kaplan, R. 2009. Play it again, Herman: Melville at the movies. *Leviathan* 11(2): 55–71.

Milinkovitch, M. C., G. Ortí, and A. Meyer. 1993. Revised phylogeny of whales suggested by mitochondrial ribosomal DNA sequences. *Nature* 361:346–348.

Milinkovitch, M. C., M. Bérubé, and P. J. Palsbøll. 1998. Cetaceans are highly derived artiodactyls. Pp. 113–132 in J. G. M. Thewissen, ed., *The Emergence of Whales*. Plenum.

Milius, S. 2009. Sperm whales as a carbon sink. *Science News* 16(9):43.

Miller, H. W., and M. V. Walker. 1968. *Enchoteuthis melanae* and *Kansasteuthis lindneri*, new genera and species of teuthids, and a sepiid from the Niobrara Formation of Kansas. *Transactions of the Kansas Academy of Science* 71(2):176–183.

Mitchell, E. D. 1975. Preliminary report on Nova Scotia fishery for sperm whales (*Physeter catodon*). *Report of the International Whaling Commission* 25:226–235.

———. 1977. Sperm whale maximum length limit: proposed protection of "harem masters." *Report of the International Whaling Commission* 27:224–227.

———. 1983. Potential of logbook data for studying social structure in the sperm whale, *Physeter macrocephalus*, with an example—the ship *Mariner* to the Pacific, 1836–1840. *Report of the International Whaling Commission* Special Issue 5:63–80.

———. 1989. A new cetacean from the Late Eocene La Meseta Formation, Seymour Island, Antarctic Peninsula. *Canadian Journal of Fisheries and Aquatic Sciences* 46:2219–2235.

Møhl, B. 1999. Sperm whale clicks and the Norris/Harvey theory of click generation. *Journal of the Acoustical Society of America* 105(3):1262.

Møhl, B., E. Larsen, and M. Amudin. 1976. Sperm whale size determination: outlines of an acoustic approach. *FAO Scientific Consultation on Marine Mammals*. ACMRR/MM/SC/84.

Møhl, B., M. Wahlberg, P. T. Madsen, L. A. Miller, and A. Surlykke. 2000. Sperm whale clicks: directionality and source level revisited. *Journal of the Acoustical Society of America* 117(1):638–648.

Møhl, B., P. T. Madsen, M. Wahlberg, W. W. L. Au, P. L. Nachtigall, and S. H. Ridgway. 2003a. Sound transmission in the spermaceti complex of a recently expired sperm whale calf. *Acoustics Research Letters Online*. doi:10.1121/11538390.

Møhl, B., M. Wahlberg, P. T. Madsen, A. Heerfordt, and A. Lund. 2003b. The monopulsed nature of sperm whale clicks. *Journal of the Acoustical Society of America* 114(2):1143–1154.

Monastersky, R. 1999. The whale's tale: searching for the landlubbing ancestors of marine mammals. *Science* 156:296.

Moore, M. J., and G. A. Early. 2004. Cumulative sperm whale bone damage and the bends. *Science* 306:2215.

Morgane, P. 1974. The whale brain: the anatomical basis of intelligence. Pp. 84–93 in J. McIntyre, ed., *Mind in the Waters*. Scribner's/Sierra Club.

———. 1978. Whale brains and their meaning for intelligence. Pp. 2:199–217 in S. Frost, ed., *Whales and Whaling*. Australian Government Publishing Service.

Morris, R. J. 1975. Further studies into the lipid structure of the spermaceti organ of the sperm whale (*Physeter catodon*). *Deep-Sea Research* 22:483–489.

Mörzer Bruyns, W. F. J. 1971. *Field Guide of Whales and Dolphins*. Amsterdam.

Mowat, F. 1972. *A Whale for the Killing*. Little, Brown.

Murphy, R. C. 1933. Floating gold: the romance of ambergris. *Natural History* 33(2):117–130; 33(3):303–310.

———. 1947. *Logbook for Grace*. Macmillan.

Murray, J., and J. Hjort. 1912. *The Depths of the Ocean*. Macmillan. Reprint, Weldon & Wesley, Stechert-Hafner, 1965.

Naslund, S. J. 1999. *Ahab's Wife, or, The Star-Gazer*. HarperCollins.

Nasu, K. 1958. Deformed lower jaw of sperm whale. *Scientific Reports of the Whales Research Institute* 17:211–212.

Nemoto, T., and K. Nasu. 1963. Stones and other aliens in the stomachs of sperm whales in the Bering Sea. *Scientific Reports of the Whales Research Institute* 17:83–91.

Nesis, K. N. 1982. *Cephalopods of the World*. English translation, T. F. H. Publications, 1987.

Nickerson, T., O. Chase, and Others. 2000. *The Loss of the Ship Essex, Sunk by a Whale. First Person Accounts*. Penguin.

Nielsen, B. K., and B. Møhl. 2006. Hull-mounted hydrophones for passive acoustic detection and tracking of sperm whales (*Physeter macrocephalus*). *Applied Acoustics* 67(11–12):1175–1186.

Nilsson, G. E. 1996. Brain and body oxygen requirements of *Gnathonemus petersi*, a fish with an exceptionally large brain. *Journal of Experimental Biology* 199:603–607.

Nishiwaki, M. 1962. Aerial photographs show sperm whales' interesting habits. *Norsk Hvalfangst-tidende* 51:395–398.

———. 1966. Distribution and migration of the larger cetaceans caught in Japanese waters. Pp. 171–191 in K. Norris, ed., *Whales, Dolphins, and Porpoises*. University of California Press.

———. 1967. Distribution and migration of marine mammals in the North Pacific area. *Bulletin of the Ocean Research Institute*, University of Tokyo 1:1–64.

———. 1972. General biology. Pp. 3–204 in S. H. Ridgway, ed., *Mammals of the Sea: Biology and Medicine*. Thomas.

Nishiwaki, M., and Y. Maeda. 1963. Change of form in the sperm whale accompanied with growth. *Scientific Reports of the Whales Research Institute* 17:10–11.

Nishiwaki, M., T. Hibaya, and S. Kimura. 1958. Age study of sperm whale based on reading of tooth laminations. *Scientific Reports of the Whales Research Institute* 13:135–153.

Nishiwaki, M., S. Ohsumi, and T. Kasuya. 1961. Age characteristics of the sperm whale mandible. *Norsk Hvalfangst-tidende* 50(12):499–507.

Nordhoff, C. 1856. *Whaling and Fishing*. Moore, Wilsatch, Keys & Co.

Norman, J. R., and F. C. Fraser. 1938. *Giant Fishes, Whales and Dolphins*. W. W. Norton.

Norman, S. A., C. E. Bowlby, M. S. Brancato, J. Calambokidis, D. Duffield, P. J. Gearin, T. A. Gornall, M. E. Gosho, B. Hanson, J. Hodder, S. J. Jeffries, B. Lagerquist, D. M. Lambourn, B. Mate, B. Norberg, R. W. Osborne, J. A. Rash, S. Reimer, and J. Scordino. 2004. Cetacean strandings in Oregon and Washington between 1930 and 2002. *Journal of Cetacean Research and Management* 6(1):87–99.

Normile, D. 2000. Japan's whaling program carries heavy baggage. *Science* 289:2264–2265.

Norris, K. S. 1969. The echolocation of marine mammals. Pp. 991–423 in H. T. Anderson, ed., *The Biology of Marine Mammals*. Academic Press.

———. 1974. *The Porpoise Watcher*. W. W. Norton.

———. 1979. Peripheral sound processing in odontocetes. Pp. 495–509 in R. G. Busnel and J. G. Fish, eds., *Animal Sonar Systems*. Plenum.

Norris, K. S., and T. P. Dohl. 1980. The structure and function of cetacean schools. Pp. 211–261 in L. M. Herman, ed., *Cetacean Behavior: Mechanism and Functions*. Wiley.

Norris, K. S., and G. W. Harvey. 1972. A theory for the function of the spermaceti organ in the sperm whale (*Physeter catodon* L.). Pp. 397–419 in S. R. Galler, K. Schmidt-Koenig, G. J. Jacobs, and R. E. Belleville, eds., *Animal Orientation and Navigation*. NASA.

Norris, K. S., and B. Møhl. 1983. Can odontocetes debilitate prey with sound? *American Naturalist* 122(1):85–104.

Norris, K. S., and J. H. Prescott. 1961. Observations on Pacific cetaceans of Californian and Mexican waters. *University of California Publications in Zoology* 63(4):291–402.

Novacek, M. J. 1993. Genes tell a new whale tale. *Nature* 361:298–299.

———. 1994. Whales leave the beach. *Nature* 368:807.

Nummela, S. 2009. Hearing. Pp. 553–562 in W. F. Perrin, B. Würsig, and J. G. M. Thewissen, eds., *Encyclopedia of Marine Mammals*. Academic Press.

Nummela, S., J. G. M. Thewissen, S. Bajpal, T. Hussain, and K. Kumar. 2004. Eocene evolution of whale hearing. *Nature* 430:776–778.

———. 2007. Sound transmission in archaic and modern whales: anatomical adaptations for underwater hearing. *Anatomical Record* 290:716–733.

Ogawa, T. 1953. On the presence and disappearance of the hind limb in the cetacean embryos. *Scientific Reports of the Whales Research Institute* 8:127–132.

Ogawa, T., and T. Kamiya. 1957. A case of the cachalot with protruded rudimentary hind limbs. *Scientific Reports of the Whales Research Institute* 12:197–208.

Ohsumi, S. 1958. A descendant of Moby Dick, or a white sperm whale. *Scientific Reports of the Whales Research Institute* 15:207–209.

———. 1965. A dolphin (*Stenella caeruleoalba*) with protruded rudimentary hind limbs. *Scientific Reports of the Whales Research Institute* 19:135–136.

———. 1971. Some investigations of the school structure of sperm whale. *Scientific Reports of the Whales Research Institute* 23:1–25.

———. 1980. Catches of sperm whales by modern whaling in the North Pacific. *Report of the International Whaling Commission*, Special Issue 2: Sperm Whales, 11–19.

Okutani, T., and T. Nemoto. 1964. Squids as food of sperm whales in the Bering Sea and Alaskan Gulf. *Scientific Reports of the Whales Research Institute* 18:111–122.

O'Leary, M. A. 1998. Phylogenetic and morphometric reassessment of the dental evidence for a Mesonychian and Cetacean clade. Pp. 133–161 in J. G. M. Thewissen, ed., *The Emergence of Whales*. Plenum.

———. 1999a. Parsimony analysis of total evidence from extinct and extant taxa and the Cetacean-Artiodactyl question (Mammalia, Ungulata). *Cladistics* 15:315–330.

———. 1999b. Whale origins. *Science* 283:1641–1642.

O'Leary, M. A., and J. H. Geisler. 1999. The position of Cetacea within Mammalia: phylogenetic analysis of morphological data from extinct and extant taxa. *Systematic Biology* 48(3):455–490.

O'Leary, M. A., and K. D. Rose. 1995. Postcranial skeleton of the early Eocene mesonychid (Mammalia, Mesonychia). *Journal of Vertebrate Paleontology* 15:401–430.

O'Leary, M. A., and M. D. Uhen. 1999. The time of the origin of whales and the role of behavioral changes in the terrestrial-aquatic transition. *Paleobiology* 24(4):534–556.

Olmstead, F. A. 1841. *Incidents of a Whaling Voyage*. Appleton. Reprint, Charles E. Tuttle, 1936.

Ommanney, F. D. 1971. *Lost Leviathan*. Dodd, Mead.

Omura, H. 1950. On the body weight of sperm and sei whales located in the adjacent waters of Japan. *Scientific Reports of the Whales Research Institute* 4:1–13.

Osborn, H. F. 1905. *Tyrannosaurus* and other Cretaceous carnivorous dinosaurs. *Bulletin of the American Museum of Natural History* 32:91–92.

———. 1924. *Andrewsarchus*, giant mesonychid of Mongolia. *American Museum Novitates* 146:1–5.

Osborn, K. J., S. H. D. Haddock, F. Plcijcl, L. P. Madin, and G. W. Rouse. 2009. Deep-sea, swimming worms with luminescent "bombs." *Science* 325:964.

Packard, A., H. E. Karlsen, and O. Sand. 1990. Low frequency hearing in cephalopods. *Journal of Comparative Physiology* 166:501–505.

Palacios, D. M., and B. R. Mate. 1996. Attack by false killer whales (*Pseudorca crassidens*) on sperm whales (*Physeter macrocephalus*) in the Galápagos Islands. *Marine Mammal Science* 12(4):582–587.

Parry, D. A. 1948. The anatomical basis of swimming in whales. *Proceedings of the Zoological Society of London* 119:49–60.

Pennisi, E. 2010. Hear that? Bats and whales share sonar protein. *ScienceNOW Daily News*, January 24.

Perkins, P. J., M. P. Fish, and W. H. Mowbray. 1966. Underwater communication sounds of the sperm whale. *Norsk Hvalfangst-tidende* 55(12):225–228.

Pilleri, G. 1979. Sonar field patterns in cetaceans, feeding behavior and the functional significance of the pterygoschisis. *Investigations on Cetacea* 10:147–155.

Pitman, R. L. 2009. Indo-Pacific beaked whale (*Indopacetus pacificus*). Pp. 615–617 in W. F. Perrin, B. Würsig, and J. G. M. Thewissen, eds., *Encyclopedia of Marine Mammals*. Academic Press.

Pitman, R. L., D. M. Palacios, R. L. R. Brennan, B. J. Brennan, K. C. Balcomb, and T. Miyashita. 1999. Sightings and possible identity of a bottlenose whale in the tropical Indopacific: *Indopacetus pacificus? Marine Mammal Science* 15(2):531–549.

Pitman, R. L., L. Ballance, S. L. Mesnick, and S. J. Chivers. 2001. Killer whale predation on sperm whales: observations and implications. *Marine Mammal Science* 17:494–507.

Pliny. N.d. *Naturalis Historia.* Reprint, Loeb Classical Library, Harvard University Press, 1933.

Plotnick, J., F. de Waal, and D. Reiss. 2006. Self-recognition in an Asian elephant. *Proceedings of the National Academy of Sciences of the United States of America* 103(5):17053–17057.

Poppe, G. T., and Y. Goto. 1993. *European Seashells.* Hemmen.

Porter, J. W. 1977. *Pseudorca* strandings: eyewitness account of a beaching on Dry Tortugas. *Oceans* 10(4):8–15.

Proulx, J.-P. 1986. *Whaling in the North Atlantic from Earliest Times to the Mid-19th Century.* Canadian Printing Service.

Purrington, P. F. 1955. A whale and her calf. *Natural History* 64(7):363.

———. 1915. *Whale Fishery of New England.* Reprint, Old Dartmouth Historical Society, 1968. State Street Bank and Trust Company, Boston.

Purves, P. E., and G. Pilleri. 1973. Observations on the ear, nose, throat, and eye of *Platanista indi. Investigations on Cetacea* 5:13–58.

Raven, H. C., and W. K. Gregory. 1933. The spermaceti organ and nasal passages of the sperm whale (*Physeter catodon*) and other odontocetes. *American Museum Novitates* 677:1–17.

Ray, C. 1961. A question in whale behavior. *Natural History* 70:46–53.

Read, B. E. 1934. Chinese materia medica: dragon and snake drugs. *Peking Natural History Bulletin* 8(4):297–357.

Rees, W. J., and G. E. Maul. 1956. The Cephalopoda of Madeira. *Bulletin of the British Museum of Natural History* 3:257–281.

Reeves, R. R., and H. Whitehead. 1997. Status of the sperm whale, *Physeter macrocephalus*, in Canada. *Canadian Field Naturalist* 111(2):293–307.

Reidenberg, J. S. 2007. Anatomical adaptations of aquatic mammals. *Anatomical Record* 90:507–513.

Reiss, D., and L. Marino. 2001. Mirror self-recognition in the bottlenose dolphin: a case of cognitive convergence. *Proceedings of the National Academy of Sciences of the United States of America* 98(10):5937–5942.

Rendell, L. E., and H. Whitehead. 2001. Culture in whales and dolphins. *Behavioral and Brain Sciences* 24:309–382.

———. 2003. Vocal clans in sperm whales. *Proceedings of the Royal Society of London B, Biological Sciences* 270:225–231.

Reynolds, J. E., III, and S. A. Rommel, eds. 1999. *Biology of Marine Mammals*. Smithsonian Institution Press.

Reynolds, J. N. 1839. Mocha Dick, or the white whale of the Pacific. *Knickerbocker, New York Monthly Magazine* 13(5):377–392.

Reysenbach de Haan, F. W. 1966. Listening underwater: thoughts on sound and cetacean hearing. Pp. 583–596 in K. S. Norris, ed., *Whales, Dolphins, and Porpoises*. University of California Press.

Rhinelander, M. Q., and S. M. Dawson. 2004. Measuring sperm whales from their clicks: stability of interpulse intervals and validation that they indicate whale length. *Journal of the Acoustical Society of America* 115(4):1826–1831.

Rice, D. W. 1978. Sperm whales. Pp. 82–87 in D. W. Haley, ed., *Marine Mammals of the Eastern North Pacific and Arctic Waters*. Pacific Search Press.

———. 1989. Sperm whale, *Physeter macrocephalus*. Pp. 177–233 in S. H. Ridgway and R. Harrison, eds., *Handbook of Marine Mammals*, vol. 4: *River Dolphins and the Larger Toothed Whales*. Academic Press.

———. 1998. *Marine Mammals of the World: Systematics and Distribution*. Special Publication 4. Society for Marine Mammalogy.

Rice, D. W. and A. A. Wolman. 1971. *The Life History and Ecology of the California Gray Whale (Eschrichtius robustus)*. Special Publication No. 3, American Society of Mammalogists.

Rice, D. W., A. A. Wolman, B. R. Mate, and J. T. Harvey. 1986. A mass stranding of sperm whales in Oregon: sex and age composition of the school. *Marine Mammal Science* 2(1):64–69.

Richard, K., H. Whitehead, and J. M. Wright. 1996a. Polymorphic microsatellites from sperm whales and their use in the genetic identification of individuals from naturally sloughed pieces of skin. *Molecular Ecology* 5:313–315.

Richard, K., M. C. Dillon, H. Whitehead, and J. M. Wright. 1996b. Patterns in kinship in groups of free-living sperm whales (*Physeter macrocephalus*) revealed by multiple genetic analyses. *Proceedings of the National Academy of Sciences of the United States of America* 93:8792–8795.

Richter, C., S. Dawson, and E. Slooten. 2006. Impacts of commercial whale watching on male sperm whales at Kaikoura, New Zealand. *Marine Mammal Science* 22(1):46–63.

Ridgway, S. H. 1966. Homeostasis in the aquatic environment. Pp. 590–747 in S. H. Ridgway, ed., *Mammals of the Sea: Biology and Medicine*. Charles C. Thomas.

———. 1971. Buoyancy regulation in deep diving whales. *Nature* 232:133–134.

———. 1980. Anatomical and physiological measures that might relate to cetacean intelligence [mimeograph]. IWC Conference on Cetacean Behavior, Intelligence, and the Ethics of Killing Whales. Washington, D.C. 2p.

Robertson, R. B. 1954. *Of Whales and Men*. Knopf.

Robson, G. C. 1925. On *Mesonychoteuthis*, a new genus of Oegopsid Cephalopoda. *Annals and Magazine of Natural History*, ser. 9, 16:272–277.

Roe, L. J., J. G. M. Thewissen, J. Quade, J. R. O'Neill, S. Bajpai, A. Sahni, and S. T. Hussain. 1998. Isotropic approaches to understanding the terrestrial-to-marine transition of the earliest cetaceans. Pp. 399–422 in J. G. M. Thewissen, ed., *The Emergence of Whales*. Plenum.

Rojas-Burke, J. 2010. Sperm whales form groups to hunt squid, an OSU scientist says. *Oregonian*, February 22.

Roman, J., and S. R. Palumbi. 2003. Whales before whaling in the North Atlantic. *Science* 301:508–510.

Roper, C. F. E., and K. J. Boss. 1982. The giant squid. *Scientific American* 246(4):96–105.

Roper, C. F. E., and M. Vecchione. 1993. A geographical and taxonomic review of *Taningia danae* Joubin, 1931 (Cephalopoda: Octopoteuthidae) with new records and observations of bioluminescence. Pp. 441–456 in T. K. Okutani, R. K. O'Dor, and T. Kubodera, eds., *Recent Advances in Fisheries Biology*. Tokai University Press.

Roper, C. F. E., M. J. Sweeney, and C. F. Nauen. 1984. FAO Species Catalogue. Vol. 3: Cephalopods of the World: An Annotated of Species of Interest to Fisheries. *FAO Fisheries Synopsis* 125(3):277p.

Rosa, R., and B. A. Seibel. 2010. Slow pace of life of the Antarctic colossal squid. *Journal of the Marine Biological Association of the U.K.* doi:10.1017/S0025315409991494.

Sanderson, I. 1956. *Follow the Whale*. Little, Brown.

Scammon, C. M. 1874. *The Marine Mammals of the Northwestern Coast of North America, Together with an Account of the American Whale Fishery*. Carmany and G. P. Putnam's.

Scheffer, V. B. 1969. *The Year of the Whale*. Scribner's.

———. 1976a. The status of whales. *Pacific Discovery* 29(1):2–8.

———. 1976b. Exploring the lives of whales. *National Geographic* 150(6):752–767.

Schevill, W. E. 1964. Underwater sounds of cetaceans. Pp. 307–316 in W. N. Tavolga, ed., *Marine Bio-acoustics*. Oxford: Pergamon.

Schevill, W. E., and B. Lawrence. 1949. Underwater listening to the white porpoise (*Delphinaptrerus leucas*). *Science* 109:143–144.

Schultz, E. A. 1995. *Unpainted to the Last: "Moby-Dick" and Twentieth-century American Art*. University Press of Kansas.

Schultz, H., and D. J. Yang. 1997. *Pour Your Heart into It: How Starbucks Built a Company One Cup at a Time*. Hyperion.

Scoresby, W. 1820. *An Account of the Arctic Regions with a History and a Description of the Northern Whale-Fishery*. Archibald Constable, Edinburgh. Reprint, David & Charles, 1969.

Scott, M. D., and J. G. Cordaro. 1987. Behavioral observations of the dwarf sperm whale, *Kogia simus*. *Marine Mammal Science* 3(4):353–354.

Scott, T. M., and S. S. Sadove. 1997. Sperm whale, *Physeter macrocephalus*, sightings in the shallow shelf waters off Long Island, New York. *Marine Mammal Science* 13(2):317–321.

Sergeant, D. E. 1969. Feeding rates of Cetacea. *Fiskeridirectoratets Skrifter Serie Havunderskoleser* 15:246–258.

———. 1982. Mass strandings of toothed whales as a population phenomenon. *Scientific Reports of the Whales Research Institute* 34:1–47.

Sherman, S. 1965. *The Voice of the Whaleman*. Providence Public Library.

Siebert, C. 2009. What are the whales trying to tell us? *New York Times Magazine*, July 11, 26–35, 44–45.

Sigler, M. F., C. R. Lunsford, J. M. Straley, and J. B. Liddle. 2008. Sperm whale depredation of sablefish longline gear in the northeast Pacific Ocean. *Marine Mammal Science* 24(1):16–27.

Simmonds, M., and S. Fisher. 2010. Oh no, not again. *New Scientist* 206(2755):22–23.

Simpson, G. G. 1945. The principles of classification and a classification of mammals. *Bulletin of the American Museum of Natural History* 85:1–350.

Slijper, E. J. 1962. *Whales*. Cornell University Press.

Smith, S. C., and H. Whitehead. 2000. The diet of Galápagos sperm whales *Physeter macrocephalus* as indicated by fecal sample analysis. *Marine Mammal Science* 16(2):315–325.

Smith, T. D. 1976. The adequacy of the scientific basis for for the management of sperm whales. *Scientific Consultation on Marine Mammals, Bergen, Norway.* ACMRR/MM/121:1–15.

Spaul, E. A. 1964. Deformity in the lower jaw of the sperm whale (*Physeter catodon*). *Proceedings of the Zoological Society of London* 142(3):391–395.

Stackpole, E. A. 1953. *The Sea-Hunters: The New England Whalers during Two Centuries, 1635–1835.* Lippincott.

——. 1977. *The Loss of the "Essex": Sunk by a Whale.* Kendall.

Starbuck, A. 1878. *History of the American Whale Fishery from Its Earliest Inception to the Year 1876.* Part 4: Report of the U.S. Commission on Fish and Fisheries, Washington, D.C. Reprint, Argosy-Antiquarian, 1964.

Surmon, L. C., and M. F. Ovendon. 1962. The chemistry of whale products. *South African Industrial Chemist* April:62–72.

Taylor, B., J. Barlow, R. Pitman, L. Ballance, T. Klinger, D. DeMaster, J. Hildebrand, J. Urban, D. Placios, and J. Mead. 2004. A call for reasearch to assess risk of acoustic impact on beaked whale populations. *Scientific Report of the International Whaling Commission* SC/56/E36.

Teloni, V., M. P. Johnson, P. J. O. Miller, and P. T. Madsen. 2008. Shallow food for deep divers: dynamic foraging behavior of male sperm whales in high latitude habitat. *Journal of Experimental Marine Biology and Ecology* 354(1):119–131.

Thewissen, J. G. M. 1994. Phylogenetic aspects of cetacean origins: a morphological perspective. *Journal of Mammalian Evolution* 2:157–184.

——. 1998. Cetacean origins: evolutionary turmoil during the invasion of the oceans. Pp. 451–464 in J. G. M. Thewissen, ed., *The Emergence of Whales.* Plenum.

Thewissen, J. G. M., and D. P. Domning. 1992. The role of phenacodontids in the origin of the modern orders of ungulate mammals. *Journal of Vertebrate Paleontology* 12(4):494–504.

Thewissen, J. G. M., and F. E. Fish. 1997. Locomotor evolution in the earliest cetaceans: functional model, modern analogues, and paleontological evidence. *Paleobiology* 23:482–490.

Thewissen, J. G. M., and S. T. Hussain. 1998. Systematic review of the Pakicetidae, early and middle Eocene cetacea (Mammalia) from Pakistan and India. *Bulletin of the Carnegie Museum of Natural History* 34:220–238.

Thewissen, J. G. M., S. T. Hussain, and M. Arif. 1994. Fossil evidence for the origin of aquatic locomotion in archaeocete whales. *Science* 263:210–212.

Thewissen, J. G. M., S. I. Madar, and S. T. Hussain. 1996. *Ambulocetus natans,* an Eocene cetacean (Mammalia) from Pakistan. *Courier Forschungsinstitut Senckenberg* 191:1–86.

Tønnessen, J. N., and A. O. Johnsen. 1982. *The History of Modern Whaling.* C. Hurst & Co. and Australian National University Press.

Townsend, C. H. 1930. Twentieth century whaling. *Bulletin of the New York Zoological Society* 33(1):3–31.

——. 1931. Where the nineteenth century whaler made his catch. *Bulletin of the New York Zoological Society* 34(6):173–178.

——. 1935. The distribution of certain whales as shown by logbook records of American whaleships. *Zoologica* 29(1):1–50.

Uhen, M. D. 1997. *Dorudon atrox:* a first for the Exhibit Museum of Natural History. *LSAmagazine,* University of Michigan, 20(2):9.

———. 1998. Middle to Late Eocene Basilosaurines and Dorudontines. Pp. 29–62 in J. G. M. Thewissen, ed., *The Emergence of Whales.* Plenum.

———. 2007. Evolution of marine mammals: back to the sea after 300 million years. *Anatomical Record* 290:514–522.

Uhen, M. D., and N. D. Pyenson. 2007. Diversity estimates, biases, and historiographic effects: resolving cetacean diversity in the Tertiary. *Paleontologia Electronica* 10.2.10A.

Van Valen, L. 1968. Monophyly or diphyly in the origin of whales. *Evolution* 22:37–41.

Verne, J. 1870. *Twenty Thousand Leagues under the Sea.* Translated by William Butcher. Reprint, Oxford University Press, 1998.

Verrill, A. H. 1926. *The Real Story of the Whaler.* Appleton.

Voss, G. L. 1959. Hunting sea monsters. *Sea Frontiers* 5(3):134–146.

———. 1967. Squids: jet-powered torpedoes of the deep. *National Geographic* 131(3):386–411.

Waters, S., and H. Whitehead. 1990. Population and growth parameters of Galápagos sperm whales estimated from length distributions. *Report of the International Whaling Commission* 40:225–235.

Watkins, W. A. 1977. Acoustic behavior of sperm whales. *Oceanus* 20(2):50–58.

———. 1980. Acoustics and the behavior of sperm whales. Pp. 283–290 in R.-G. Busnel and J. F. Fish, eds., *Animal Sonar Systems.* Plenum.

Watkins, W. A., and W. E. Schevill. 1977. Sperm whale codas. *Journal of the Acoustical Society of America* 62(6):1485–1490.

Watkins, W. A., and D. Wartzok. 1985. Sensory biophysics of marine mammals. *Marine Mammal Science* 1(3):219–260.

Watkins, W. A., M. A. Daher, N. A. DiMarzio, A. Samueles, D. Wartzok, F. M. Fristrup, P. W. Howey, and R. R. Maiefski. 2002. Sperm whale dives tracked by radio tag telemetry. *Marine Mammal Science* 18(1):55–78.

Watson, L. 1981. *Sea Guide to the Whales of the World.* Hutchinson.

Watwood, S. L., P. J. O. Miller, M. Johnson, P. T. Madsen, and P. L. Tyack. 2006. Deep-diving foraging behavior of sperm whales (*Physeter macrocepahus*). *Journal of Animal Ecology* 75:814–825.

Weilgart, L., and H. Whitehead. 1986. Observations of a sperm whale (*Physeter catodon*) birth. *Journal of Mammalogy* 67(2):399–401.

———. 1987. Distinctive vocalizations from mature male sperm whales (*Physeter macrocephalus*). *Canadian Journal of Zoology* 66:1931–1937.

———. 1992. Coda communication by sperm whales (*Physeter macrocephalus*) off the Galápagos Islands. *Canadian Journal of Zoology* 71:744–752.

———. 1997. Group-specific dialects and geographical variation in coda repertoire in South Pacific sperm whales. *Behavioral and Ecological Sociobiology* 40:277–285.

Weilgart, L., H. Whitehead, and K. Payne. 1996. A colossal convergence. *American Scientist* 84:278–287.

Weller, D. W., B. Würsig, H. Whitehead, J. C. Norris, S. K. Lynn, R. W. Davis, N. Clauss, and P. Brown. 1996. Observations of an interaction between sperm whales and short-finned pilot whales in the Gulf of Mexico. *Marine Mammal Science* 12(4):588–594.

Werth, A. J. 2000. A kinematic study of suction feeding and associated behaviors in the long-finned pilot whale *Globicephala melas* (Traill). *Marine Mammal Science* 16(2):299–314.

———. 2006. Mandibular and dental variation and the evolution of suction feeding in Odontoceti. *Journal of Mammalogy* 87(3):579–588.

Wheeler, J. F. G. 1933. Notes on a young sperm whale from the Bermuda Islands. *Proceedings of the Zoological Society of London* 1933:407–410.

Whitaker, I. 1984. Whaling in classical Iceland. *Polar Record* 22(134):249–261.

———. 1985. The King's Mirror (*Konnungs skuggsjá*). *Polar Record* 22(141):615–627.

———. 1986. North Atlantic sea creatures in the King's Mirror (*Konnungs skuggsjá*). *Polar Record* 22(142):3–13.

White, T. 2007. *In Defense of Dolphins: The New Moral Frontier.* Wiley-Blackwell.

Whitehead, H. 1986. Call me gentle. *Natural History* 95(6):4–11.

———. 1990a. *Voyage to the Whales.* Chelsea Green.

———. 1990b. Rules for roving males. *Journal of Theoretical Biology.* 145:355–368.

———. 1993. The behaviour of mature male sperm whales in the Galápagos Islands breeding grounds. *Canadian Journal of Zoology* 71:689–699.

———. 1994. Delayed competitive breeding in roving males. *Journal of Theoretical Biology* 166:127–133.

———. 1995a. Status of Pacific sperm whale stocks before modern whaling. *Report of the International Whaling Commission* 45:407–412.

———. 1995b. The realm of the elusive sperm whale. *National Geographic* 188(5):57–73.

———. 1996a. Babysitting, dive synchrony, and indications of alloparental care in sperm whales. *Behavioral and Ecological Sociobiology* 38:237–244.

———. 1996b. Variation in the feeding success of sperm whales: temporal scale, spatial scale and relationship to migrations. *Journal of Animal Ecology* 65:429–438.

———. 1997. Sea surface temperature and abundance of sperm whale calves off the Galápagos Islands: implications for the effects of global warming. *Report of the International Whaling Commission* 47:941–944.

———. 1998a. Cultural selection and genetic diversity in matrilineal whales. *Science* 282:1708–1711.

———. 1998b. Formations of foraging sperm whales, *Physeter macrocephalus,* off the Galápagos Islands. *Canadian Journal of Zoology* 67:2131–2139.

———. 1998c. Male mating strategies: models of roving and residence. *Ecological Modeling* 111:297–298.

———. 2002. Estimates of the current global population size and historical trajectory for sperm whales. *Marine Ecology Progress Series* 242:249–304.

———. 2003. *Sperm Whales: Social Evolution in the Ocean.* University of Chicago Press.

———. 2009. Sperm whale *Physeter macrocephalus.* Pp. 1091–1098 in W. F. Perrin, B. Würsig, and J. G. M. Thewissen, eds., *Encyclopedia of Marine Mammals.* Academic Press.

Whitehead, H., and P. L. Hope. 1991. Sperm whalers off the Galápagos Islands and the western North Pacific, 1830–1850: ideal free whalers? *Ethology and Sociobiology* 12:147–161.

Whitehead, H., and N. Jaquet. 1996. Are the charts of Maury and Townsend good indicators of sperm whale distribution and seasonality? *Report of the International Whaling Commission* 46:643–647.

Whitehead, H., and B. Kahn. 1992. Temporal and geographic variation in the social structure of sperm whales. *Canadian Journal of Zoology* 70:2145–2149.

Whitehead, H., and S. Waters. 1990. Social organization and population structure of sperm whales off the Galápagos Islands, Ecuador (1985 and 1987). *Report of the International Whaling Commission* Special Issue 12:249–257.

Whitehead, H., and L. Weilgart. 1990. Click rates from sperm whales. *Journal of the Acoustical Society of America* 87(4):1798–1808.

———. 1991. Patterns in visually observable behaviour and vocalizations in groups of female sperm whales. *Behaviour* 118(3–4):276–296.

———. 2000. The sperm whale: social females and roving males. Pp. 154–172 in J. Mann, R. C. Connor, P. L. Tyack, and H. Whitehead, eds. *Cetacean Societies: Field Studies of Dolphins and Whales.* University of Chicago Press.

Whitehead, H., J. Christal, and S. Dufault. 1977. Past and distant whaling and the rapid decline of sperm whales off the Galápagos Islands. *Conservation Biology* 11(6):1387–1396.

Whitehead, H., L. Weilgart, and S. Waters. 1989. Seasonality of sperm whales off the Galápagos Islands, Ecuador. *Report of the International Whaling Commission* 39:207–210.

Whitehead, H., J. Gordon, E. A. Matthews, and K. R. Richard. 1990. Obtaining skin samples from living sperm whales. *Marine Mammal Science* 6(4):316–326.

Whitehead, H., S. Waters, and T. Lyrholm. 1991. Social organization of female sperm whales and their offspring: constant companions and casual aquaintances. *Behavioral Ecology and Sociobiology* 29:385–389.

———. 1992a. Population structure of female and immature sperm whales (*Physeter macrocephalus*) off the Galápagos Islands. *Candian Journal of Fisheries and Aquatic Sciences* 49:78–84.

Whitehead, H., S. Brennan, and D. Grover. 1992b. Distribution and behaviour of male sperm whales on the Scotian Shelf, Canada. *Canadian Journal of Zoology* 70:912–918.

Whitehead, H., M. Dillon, S. Dufault, L. Weilgart, and J. Wright. 1998. Non-geographically based population structure of South Pacific sperm whales: dialects, fluke markings and genetics. *Journal of Animal Ecology* 67:253–262.

Williams, E. M. 1998. Synopsis of the earliest cetaceans: Pakicetidae, Ambulocetidae, Remingtonocetidae, and Protocetidae. Pp. 1–28 in J. G. M. Thewissen, ed., *The Emergence of Whales.* Plenum.

Williams, H. 1964. *One Whaling Family.* Houghton Mifflin.

Williams, T. M., R. W. Davis, L. A. Fuiman, J. Francis, B. J. Le Boeuf, M. Horning, J. Calambokidis, and D. A. Croll. 2000. Sink or swim: strategies for cost-efficient diving by marine mammals. *Science* 288:133–136.

Wilson, E. O. 1975. *Sociobiology: The New Synthesis.* Harvard University Press.

Winge, B. 1921. A review of the interrelationships of the cetacea. *Smithsonian Miscellaneous Collection* 72(8):1–97.

Winter, A. 2010. Gulf oil spill creates "giant experiment" in marine toxicology. *New York Times,* May 20.

Wood, F. G. 1978. The cetacean stranding phenomenon: an hypothesis. In J. R. Geraci and D. St. Aubin, eds., *Analysis of Marine Mammal Strandings and Recommendations for a Nationwide Stranding-Salvage Program.* Final Report to U.S. Marine Mammal Commission in Fulfillment of Contract MM7ACC020. National Technical Information Service.

Wood, F. G., and W. E. Evans. 1980. Adaptiveness and ecology of echolocation in toothed whales. Pp. 381–423 in R.-G. Busnel and J. F. Fish, eds., *Animal Sonar Systems.* Plenum.

Wood, G. L. 1982. *The Guinness Book of Animal Facts and Feats.* Guinness Superlatives.

Worthington, L. V., and W. E. Schevill. 1957. Underwater sounds heard from sperm whales. *Nature* 180:291.

Wray, J. W. 1939. *South Sea Vagabond.* A. H. and A. W. Reed.

Yablokov, A. V. 1994. Validity of whaling data. *Nature* 367:108.

Yamada, M. 1953. Contribution to the anatomy of the organ of hearing of whales. *Scientific Reports of the Whales Research Institute* 8:1–79.

Yano, K., and M. E. Dahlheim. 1995. Orca, *Orcinus orca,* depredation on longline catches of bottom fish in the southeastern Bering Sea and adjacent waters. *Fishery Bulletin* 93:355–372.

Zeidler, W. 1981. A giant deep-sea squid, *Taningia* sp. from South Australian waters. *Transactions of the Royal Society of South Australia* 105:218.

Zenkovich, B. A. 1962. Sea mammals as observed by the Round the World Expedition of the Academy of Sciences of the USSR in 1957/1958. *Norsk Hvalfangst-tidende* 51(5):198–210.

Zenkovich, B. A., and V. A. Arsen'ev. 1955. Short history of whaling and modern conditions at USSR. Pp. 5–29 in S. E. Kleinenberg and T. I. Marakova, eds., *Whaling at USSR (Rybnoe Khozayistvo).* Moscow.

Zemsky, V. A., A. A. Berzin, Y. A. Mikhaliev, and D. D. Tormosov. 1995. Soviet Antarctic pelagic whaling after WWII: review of actual catch data. *Report of the International Whaling Commission* 45(appendix 3):131–135.

Zemsky, V. A., Y. A. Mikhaliev, and A. A. Berzin. 1996. Suplementary information about Soviet whaling in the Southern Hemisphere. *Report of the International Whaling Commission* 46:131–137.

Zhou, X., W. J. Sanders, and P. D. Gingerich. 1992. Functional and behavioral implication of vertebral structure in *Pachyaena ossifraga* (Mammalia, Mesonychia). *Contributions to the Museum of Paleontology of the University of Michigan* 28:289–313.

Zimmer, C. 1995. Back to the sea. *Discover* January:82–84.

Zimmer, W. M. X., P. L. Tyack, M. P. Johnson, and P. T. Madsen. 2005a. Three-dimensional beam pattern of regular sperm whale clicks confirms bent-horn hypothesis. *Journal of the Acoustical Society of America* 117(3):1473–1485.

Zimmer, W. M. X., P. T. Madsen, V. Teloni, M. P. Johnson, and P. L. Tyack. 2005b. Off-axis effects on the multipulse structure of sperm whale usual clicks with implications for sound production. *Journal of the Acoustical Society of America* 118(5):3337–3345.

Index

Acrophyseter deinodon, 84 (fig.), 84, 86, 87

Acushnet (whaling ship), 8, 9, 11, 256

Addison (whaling ship), 241, 249

Aetiocetus, 75, 78

African elephant, 148, 153

Agulhas Current, 273

Ahab

    in *Ahab's Wife*, 36–37

    in film, 9, 43–50, 45 (fig.), 46 (fig.), 50 (fig.), 53

    in *Moby-Dick*, 8, 26, 30–31, 32, 38, 39, 102, 242–243, 268

    in *White as the Waves*, 35–36

Ahab (band), 34

Ahab's Wife (Naslund), 36–37

Albany, Western Australia, whale fishery, 102 (fig.), 187, 272, 277, 278 (fig.), 279

Amazon (whaling ship), 273

ambergris, 37–38, 137–142, 139 (fig.), 141 (fig.), 233, 297

Ambulocetus natans, 68–72, 68 (fig.), 164

American Museum of Natural History, 190 (fig.), 319, 323, 323 (fig.)

American whaling industry

    18th century, 228–231, 237, 273

    19th century, 9–10, 232–234, 256, 270

    20th century, 10, 270–271

    decline of, 235–236

    and Marine Mammal Protection Act, 286, 300, 307

    petroleum discovery and, 235, 237 (fig.), 306

    whale populations and, 237–238

    See also New England whaling

Anderson, Laurie, 33

Andrews, Harry, 49

Andrewsarchus mongoliensis, 60–61, 61 (fig.), 65–66

Angier, Natalie, 310–311

Ann Alexander (whaling ship), 20–21

Antarctic grounds, 271, 280–281, 283, 293–295, 300, 301, 304

antiwhaling movement, 284–285, 290–291, 301 (fig.), 317–319, 325

archaeocetes (extinct whales), 56, 59, 62, 65–66, 69–70, 75, 76–77, 165

Architeuthis (giant squid), 190 (fig.)

    attempts to film, 188–189

    characteristics of, 188 (fig.), 189–190, 206, 211–212

    Cousteau on, 184–186

    Melville on, 174–175

    sperm whale clashes with, 176–178, 176 (fig.), 182–184, 186–187, 193 (fig.), 199–200

    as sperm whale prey, 175–176, 179–180, 189–190, 199, 200, 203, 207, 209

    suckers of, 129, 130 (fig.)

Aristotle, 11, 130, 132, 229

Ashley, Clifford

    on ambergris, 140

    on scrimshanders, 39

    on sperm whale behaviors, 113, 114, 255–256

    on sperm whale size, 98, 101, 261, 262, 263

Audubon magazine, 317, 319, 322, 324, 328

Australian whaling, 277–280, 288, 294, 307

bachelor schools, 135, 145, 153

Bahamonde's beaked whale (*Mesoplodon bahamondi*), 87–88

Baird, Alison, 35–36

Baird's beaked whale (*Berardius bairdi*), 87

Baker, Scott, 300–301, 303

baleen, 78, 234, 263

baleen plate, 57, 78, 96 (fig.), 100, 230

baleen whales

    brain of, 154, 155, 164, 172–173

    echolocation and, 191

    evolution and, 56, 57, 75, 77–78, 79–80, 86–87

    hunting of, 228, 230, 271–272, 280–281, 282, 285–286, 292, 294, 298–299, 301, 303–304, 307

    migration of, 269

    physical characteristics of, 54, 77, 81, 86–87, 96, 100, 104–105

    pollution and, 219

    population decline of, 10, 235, 269, 294, 298, 307, 321

    in *Twenty Thousand Leagues Under the Sea*, 265

    See also blue whale; bowhead whale; Bryde's whale; fin whale; gray whale; humpback whale; minke whale; right whale; sei whale

Barnes, Lawrence, 75, 78, 90, 92
Barrymore, John, 43–45, 45 (fig.), 46 (fig.), 50
Basehart, Richard, 49
Basilosaurus, 62–64, 63 (fig.), 67, 72, 73, 77
Baskin, Leonard, 40, 41
beaching. *See* stranding behavior
beaked whales
    bottlenose as largest of, 83
    feeding habits of, 87
    fossil record of, 74, 79–80
    Japanese hunting of, 303
    physical characteristics of, 73, 87, 89
    and plastic bags, 218–219
    species of, 87–89, 89 (fig.), 93
Beale, Thomas, 106
    on ambergris, 140–141
    as Melville source, 11, 12–13, 23, 24, 28, 29, 104,
        147
    on sperm whale prey, 193–194, 195, 196, 201, 203
    on sperm whales, 98, 102, 113–114, 115, 120, 252,
        282
    on whaling danger, 255
Bel'kovich, Vladimir, 123–124, 195
Bell, John, 279–280
beluga whale, 108, 165
Bennett, Frederick Debell
    as Melville source, 11, 12, 23, 24, 28, 29–30, 104,
        107
    on sperm whales, 114, 125, 128, 150
Bennett, Joan, 44, 46 (fig.), 48
Bernardini, Craig, 34
Berta, Annalisa, 64, 69–70, 75, 77–78
Berzin, A. A., 124, 126, 149, 193–194, 195–197, 295, 297
Bianucci, Giovanni, 83, 84–86
bioluminescence, 56, 189, 191, 203–204
blowholes, 69, 75, 76, 104–105, 107–108, 110, 120,
    131–132, 226 (fig.)
blubber, 113, 116–117, 126, 161, 178, 219, 228–229, 253,
    257–259, 260 (fig.), 263–264, 296, 296 (fig.)
bluefin tuna, 86, 95, 279, 309, 313
blue whale, 94 (fig.), 95 (fig.), 160 (fig.), 320
    hunting of, 10–11, 238, 271, 280–281, 292
    Melville on, 27–28
    physical characteristics of, 10, 54, 77, 80, 81, 86, 96,
        100, 104, 164
    population decline of, 10, 269, 294, 298
    protection for, 284
Bolstad, Kat, 208
Bonham, John, 34
bottlenose dolphin, 83, 115, 154 (fig.)
    brain of, 154, 155, 164
    in captivity, 94, 160–161, 164, 171
    echolocation of, 119–120, 121, 160–161, 317

hearing capability of, 117, 119
intelligence of, 160, 162–163, 164
play behavior of, 171–172, 171 (fig.)
U.S. fisheries for, 274
bottlenose whales (*Hyperoodon*), 83, 88, 90, 106
bowhead whale, 10–11, 23
    depletion of, 235, 307
    habitat of, 321
    hunting of, 234, 236, 270, 288
    physical characteristics of, 77, 96, 100
BP. *See* Deepwater Horizon explosion
Bradbury, Ray, 49, 50
Brancheau, Dawn, 101
Brazil Banks, 267, 273
British whaling, 10, 19–20, 22, 232, 234, 293, 294,
    321
Brokaw, Tom, 5 (fig.)
Browne, J. Ross, 11–12, 14, 246, 247, 248–249, 259,
    261
Bryde, Johan, 273–274, 285
Bryde's whale, 78, 273–274, 320
    IWC quota on, 307–308
    Japanese whaling and, 285–287, 304
    physical characteristics of, 77, 96, 100
    and plastics, 219
Budker, Paul, 224, 225
Bullen, Frank T., 12, 22, 98, 127, 138, 149, 177–180,
    199–200
Butcher, William, 267

Cachalot (fictional whaling ship), 22, 138, 177, 179
California whaling, 236, 286, 294, 299–300
Calypso, 51, 185
Canton (whaling ship), 251 (fig.)
Cape Hatteras, North Carolina, 161, 268, 274
Carcharodon megalodon (giant shark), 85, 86–87
Carroll, Robert, 61–63
Cawthorn, Martin, 201
Ceres (whaling ship), 233
cetaceans
    aquatic lifestyle of, 55, 59
    behaviors of, 96, 168–172
    brains and intelligence of, 54, 108, 125, 154–165,
        172–173, 188, 318
    and decompression sickness, 209–210
    and echolocation, 57, 109–110, 112, 121–122, 125–
        126, 158, 160–161, 192
    elephant parallels with, 147–148, 153, 163
    evolution of, 55–59, 61–72, 73, 75–78, 81–82, 90–
        92, 142, 163–164
    hearing of, 115–116, 118–119, 158
    humans compared to, 172–173, 310–311
    Melville on, 27–28

sleep behavior of, 130–132
stranding by, 132–137, 133 (fig.), 135 (fig.),
    136 (fig.), 138 (fig.), 227–228, 227 (fig.)
Charles W. Morgan (whaling ship), 172, 250
Chase, Owen, 15, 16–17, 18, 29
Chase, William, 15–16
Cheever, Henry T., 11, 12–14, 108, 138–139, 176
Chinese river dolphin, 116, 325
Christson, Robert, 253
Church, Albert Cook, 251–252
Clapham, Phillip, 292, 299
Clark, Christopher, 222–223
Clark, Genie, 51
Clarke, Malcolm R.
    and spermaceti organ function, 111–112, 121–122
    on sperm whale diving, 216
    on sperm whale and squid interaction, 186–187,
        197–198
    sperm whale population estimate by, 291
    and sperm whale squid consumption, 187, 189,
        197–198, 205, 206–207, 211, 214
Clarke, Robert, 141, 180, 199–200
clipper ships, 250, 251
Coast of Arabia Ground, 268
Coast of Japan Ground, 268
Coast of New Holland Ground, 268
colossal squid *(Mesonychoteuthis hamiltoni)*, 129–130,
    207–209, 207 (fig.), 209 (fig.), 214
Commodore Preble (whaling ship), 13–14, 176
Connor, Edric, 49
convergence, 147–148, 159, 163–164, 191–192
Cope, Edward Drinker, 59–60
Costello, Dolores, 43, 44, 45 (fig.)
Cousteau, Jacques, 22, 51–53, 96 (fig.), 184–186, 320
Cranford, T. W., 111
Cretaceous period, 91
Curtsinger, William, 97
cuttlefish *(Sepia officinalis)*, 139, 178, 195, 203

Dagoo, 8, 9, 53, 174, 178. See also *Moby-Dick*
    (Melville)
Dana, Richard Henry, 12, 31, 246
Dannenfeldt, Karl, 139
Darwin, Charles, 63
Davis, Edgerton V., 224–225
Davis, William, 247, 249–250, 263
Deepwater Horizon explosion, 221, 222–223
Delagoa Bay Ground, 267, 273
Diolé, Philippe, 22, 51, 96 (fig.), 184, 185
DNA analysis, 78, 89, 191–192, 220
dolphins
    brain of, 157, 158, 165–166
    echolocation of, 68, 109–110, 116, 192

and evolution, 55–56
play behavior of, 171–172
sleep behavior of, 130–131
sound transmission in, 70
speed of, 96
See also cetaceans
dorsal fin, 82, 96–97
Dorudon, 72, 73
Dorudon atrox, 72–73
Dosidicus gigas (Humboldt squid), 152, 211–212,
    212 (fig.), 213 (fig.), 215 (fig.)
Drake, Edwin, 306
Duffield, Debbie, 137
dugongs, 55, 69, 72, 158
Dutch whaling, 10, 227–228, 234–235, 293,
    294, 321
dwarf sperm whale *(Kogia simus)*, 82, 87, 90

echolocation
    in baleen whales, 191
    in bats, 158, 159, 165, 191
    in bottlenose dolphins, 119, 121, 160–161,
        317
    and convergence in mammals, 191–193
    in dolphins, 68, 109, 116, 192
    in odontocetes, 57, 77, 156–159, 165, 192–193
    in sperm whales, 109, 112, 118–119, 121, 122, 125, 217
    and stranding, 132–133
Egeland, Jacob, 274
Ellis, Timothy, 154 (fig.)
Emlong, Douglas, 78
encephalization, 155–156, 159, 164–165
Eocene period, 59, 65–67, 75
Essex (U.S. frigate), 232
Essex (whaling ship), 15–18, 20, 29, 37, 112, 127, 232
evolution
    and baleen whales, 56, 57, 75, 77–78, 79–80, 86–87
    cetacean, 53–54, 55–59, 61–72, 73, 75–78, 81–82,
        87, 90–92, 142, 163–164
    convergence and, 159, 163, 191–192
    extinction as part of, 56, 87
    mammalian, 55, 69, 192
extinction, 56, 79, 81–82, 87, 116, 306–307, 311–313,
    325. *See also* archaeocetes (extinct whales)
extinct whales. *See* archaeocetes (extinct whales)
Exxon Valdez, 221

factory-ship whaling
    in Antarctic, 271, 280–281, 283, 293–295, 306–307
    banning of, 301
    and IWC quotas, 271–272, 293
    in North Pacific, 27, 213–214, 238, 293, 296 (fig.)
    See also Japanese whaling; Soviet whaling

false killer whale (*Pseudorca crassidens*), 128–129, 132, 134

Ferber, Dan, 220

Ferecetotherium, 81, 83, 91

fin whale, 51, 320
  hunting of, 271, 284, 287, 294, 304
  physical characteristics of, 54, 77, 80–81, 96, 100
  population decline of, 269, 298

Fish, F. E., 70–71, 72

Fitzgerald, Erich, 79

Flask, 8, 9, 49, 53. See also *Moby-Dick* (Melville)

Flower, William, 75

flukes, 58, 97, 105 (fig.), 136 (fig.), 218 (fig.), 297
  in cetacean evolution, 62, 65, 69, 70–72, 73
  "lob tailing" of, 114, 170
  as propellants, 75, 103–104
  as weapons, 254–255

Ford, John, 165–166

Foyn, Svend, 273, 306

Frank, Stuart, 14–15

French Rock Ground, 268

Fristrup, Kirk, 118, 202–203

Galápagos Islands, 128, 153, 167–168, 263, 268, 269

Gambell, Ray, 302

Garrett, Tom, 303 (fig.)

Gaskin, David, 75, 114, 200–203, 255

Genn, Leo, 49

giant squid. See *Architeuthis* (giant squid)

Gilmore, R., 134, 135 (fig.), 269

Gimbel, Peter, 277

Gingerich, Philip, 62, 64–65, 66–68, 72–74, 76–77

Goltzius, Hendrick, 227 (fig.), 228

Gould, Stephen J., 71–72

Grampus griseus, 27, 156

gray whale, 320, 321
  in cetacean evolution, 77, 78
  Cousteau and, 51, 53
  physical characteristics of, 96, 100
  population of, 307
  for scientific research, 286

Great Britain. *See* British whaling

Greenland right whale, 23–24, 38, 227, 228

Greenland whaling, 230, 234–235, 308

Gulf of Aden, 88

Gulf of Alaska, 213, 217, 293

Gulf of California, 135 (fig.), 152, 212–213

Gulf of Mexico, 91, 129, 221–222, 268

Hall, Richard Melville, 34

Haplodectes, 60

Harbison, Kirk, 118, 202–203

Harlan, Richard, 63–64

Harmer, Sidney, 23, 234

harpoon, whaler's
  grenade cannon, 26–27, 238, 271, 273, 276 (fig.), 286 (fig.), 288, 306
  hand-thrown, 253, 271, 288
  and prussic acid, 253–254
  Temple toggle design of, 253, 327

Hart, Joseph Coleman, 11

Harvey, G. W., 112, 121, 122, 123, 124–125

Heezen, Bruce, 196, 216

Heggie, Jake, 34

Hemingway, Ernest, 32–33

Henry IV, 107

Herrmann, Bernard, 35

Heteroteuthis dispar, 204

Hill, David, 317, 318, 322–323

Himalayacetus, 65

Hirohito (Japanese Emperor), 284

Hirota, Kiyoharu, 90, 92

Histioteuthis, 189, 204, 204 (fig.)

Hjort, Johann, 183–184

Hof, Patrick, 172, 173

Hogarth, William, 308

Hohman, E. P., 243

Holthuis, L. B., 93–94

Hopkins, W. J., 98–99, 129–130

Hudnall, Jim, 320–321

Humboldt Current, 211, 212

Humboldt squid (*Dosidicus gigas*), 152, 211–212, 212 (fig.), 213 (fig.), 215 (fig.)

humpback whale
  in cetacean evolution, 64, 77
  communication of, 170, 219, 317
  Cousteau and, 51, 53
  hunting of, 230, 270, 274, 277–278, 280–281, 287–288, 292, 304, 308
  in Melville, 27
  Migaloo, 26
  migration of, 26, 274, 279
  physical characteristics of, 96, 100, 172–173
  population decline of, 269, 277, 294, 298, 307, 321
  protection for, 278, 284
  stranding behavior of, 170

Hussain, T., 66, 68

Hussey, Christopher, 228, 230, 305

Husson, A. M., 93–94

Huston, John, 9, 35, 49, 50–51, 50 (fig.)

Iceland whaling, 299, 308, 309–310

Ichthyolestes, 66

Idiophorus patagonicus, 80

Idiophyseter merriami, 80

Indian Ocean, 51, 52, 88–89, 185, 273, 301

Indocetus ramani, 66–67
Indopacetus pacificus (Longman's beaked whale), 88–89, 89 (fig.)
Inge, M. Thomas, 39, 50–51
intelligence and brain size, 154–156, 164
International Conference on Whaling, 58, 271. *See also* International Whaling Commission (IWC)
International Convention for the Regulation of Whaling, 285–286
International Whaling Commission (IWC), 154, 291 (fig.)
    ban on humpback whaling, 277, 278
    and factory-ship whaling, 271–272, 293
    Japanese whaling and, 281, 285–286, 287, 293, 304, 308, 309–310
    quotas, 238, 280, 285, 288–290, 294, 298–302, 302 (fig.), 307–308
    weak enforcement by, 285, 307, 308
    whaling moratorium, 276, 287, 300, 302–304, 303 (fig.), 307, 308–310
Ishmael, 8, 9, 18, 36, 45–46, 48, 49, 53. See also *Moby-Dick* (Melville)
IWC. *See* International Whaling Commission (IWC)

Jacob A. Howland (whaling ship), 262 (fig.)
Janjucetus hunderi, 79, 79 (fig.)
Japanese whaling, 286 (fig.)
    in the Antarctic, 280–281, 283, 300, 301, 304
    and consumer products, 283 (fig.)
    IWC and, 281, 285–286, 287, 293, 308–310
    and "mis-reporting" of whales, 292
    in the North Pacific, 27, 213, 238, 278, 281–284, 288, 290, 304, 318
    opposition to, 284–285
    "scientific" whaling and, 285–287, 303–304, 307
    in South America, 294
    sperm whales and, 23, 27, 239, 281, 283–284, 285 (fig.), 304
    and whale meat, 282–283, 286–287
    See also factory-ship whaling
Japan Grounds, 232, 263, 306
Jaques, Francis Lee, 212 (fig.)
Jardine, Robert, 29
"jawing back," 255–256
Jenkins, Thomas, 21–22
Jerison, H. J., 154, 155–156
Johnsen, A. O., 273, 282
Johnson, Noble, 46
Jonah, 25, 104, 140, 224
Josselyn, John, 140
jumbo squid *(Dosidicus gigas),* 135

Kelley, Seamus, 49
Kellogg, Remington, 58–59, 67, 80, 81, 153–154

Kent, Rockwell, 40–41, 320
killer whale *(Orcinus orca),* 83, 85 (fig.), 95, 100–101, 117, 128, 147, 154–155, 165–166
Koefoed, Einar, 182
Kondo, Isao, 292
Kubodera, Tsunemi, 110, 189

Lacépède, Comte Bernard-Germain de, 313–314
Lacy Ann (whaling ship), 233
Lagoda (whaling ship), 250, 262–263, 327–328
Lambert, Olivier, 84–86
lance, whaler's, 253–254
Lavery, Trish, 305
Lawrence, Barbara, 321
Lawrence, D. H., 31–32, 143
Lawrence, Mary Chipman, 249
Ledebur, Friederich, 49
Led Zeppelin, 34
Leidy, Joseph, 80, 83–84
Lerner, Michael, 215 (fig.)
Leviathan melvillei, 85–86, 87
Ley, Willy, 184
Lilly, John Cunningham, 161, 317
Lindberg, David, 57
Linnaeus, 93, 224
Llanocetus denticrenatus, 78
Loates, Glen, 181 (fig.), 193 (fig.)
Logan (whaling ship), 234–235
Loligo pealei, 7, 118
Longman's beaked whale *(Indopacetus pacificus),* 88–89, 89 (fig.)
Lowe, Tristin, 13 (fig.), 20
Lowenstein, Jerold, 66

MacArthur, Douglas, 280
Mackintosh, N. A., 281, 289
Madsen, Alex, 50
Mahé Banks, 268, 273
Maiacetus innuus, 74
Malloy, Mary, 256–257
manatees, 55, 69, 72, 80, 106, 158, 165
marine mammal fossils, 58–59. *See also* whale fossil record
Marine Mammal Protection Act (1972), 286, 300, 307
marine reptiles, 64, 79, 81, 86, 91
Marino, Lori, 159, 162, 163–164, 165
Mastodon (band), 34
Mate, Bruce, 128–129, 136, 152, 222–223
Matthews, L. Harrison, 97, 126, 149, 180, 187, 234
McCann, Charles, 130, 182–183
McShea, Daniel, 165
McVay, Scott, 317

mechanized whaling, 9–10, 26–27, 238–239, 271, 295–299, 307. *See also* factory-ship whaling

Melville, Herman
  books by, 8, 11–12, 32, 248
  cetological expertise of, 10–11, 27–29
  descriptive powers of, 41–42
  on French whalers, 234
  on Japan fishing grounds, 306
  on Nantucket, 34
  on New Bedford, 313
  portrait of, 10 (fig.)
  reputation of, 30–32, 38
  sources of, 11, 12–13, 15–18, 23, 24, 28, 29, 104, 147, 194
  sperm whale glorification by, 23–24, 230
  on sperm whales, 9, 102–103, 104, 111–113, 127–128, 147, 169, 319–320
  on sperm whaling, 107, 140, 248, 253, 254, 257–258
  on squid, 174–175
  See also *Moby-Dick* (Melville)

Melville Society, 32, 34

mesonychids, 59–62, 61 (fig.), 65–66, 68, 70, 74. *See also* whale fossil record

Mesonychoteuthis hamiltoni, 129–130, 207–209, 207 (fig.), 209 (fig.), 214

Mesonyx, 60–61

Mesoplodon perrini, 89

mesoplodonts, 83, 87. *See also* beaked whales

Migaloo, 26

migration
  of baleen whales, 269
  of California gray whale, 321
  of humpback whales, 26, 274, 279
  and 19th-century whaling, 256
  of sperm whales, 20, 26, 94, 145–146, 148–149, 153, 269, 279, 281–282, 299

Miller, Patrick, 168–169

mini-sperm whale, 80, 82, 82 (fig.), 87, 90–91, 222

minke whale, 77, 78, 96, 100, 127, 294, 301, 304, 320

Mitchell, E. D., 288–289

mixed schools, 145, 210

Moby-Dick (Melville)
  Ahab in, 8, 26, 30–31, 32, 38, 39, 102, 242–243, 268
  editions of, 40–43, 44, 42 (fig.), 320
  in film, 9, 43–51, 45 (fig.), 46 (fig.), 50 (fig.), 53, 321
  Ishmael in, 8, 9, 18, 36, 45–46, 48, 49, 53
  Pequod in, 8, 22, 26, 112, 117–118, 127, 234, 268, 329
  in popular culture, 38–39
  Queequeg in, 8, 9, 36, 46, 47, 48, 49, 53, 111
  Starbuck in, 8, 9, 49, 53

Stubb in, 8, 9, 49, 53
  works inspired by, 32–35
  See also Melville, Herman

Mocha Dick (sculpture), 13 (fig.), 20

Mocha Dick (sperm whale), 18–20, 30, 126

Møhl, Bertel, 124–125, 196

"monkey's muzzle." See *museau du singe*

Monstro (*Pinocchio*), 225, 225 (fig.), 321

mormyrids, 155

Moroteuthis, 202 (fig.), 207, 213

Moroteuthis ingens, 200–201

Moroteuthis knipovitchi, 214

Moroteuthis robustus, 200, 201 (fig.), 206, 213–214

Moser, Barry, 10 (fig.), 40, 41, 42 (fig.)

Mowat, Farley, 133–134, 320

MSR (mirror self recognition), 162–163

Muizon, Christian de, 84–86

Mumford, Lewis, 31

Murphy, Robert Cushman, 141–142, 199

Murray, J., 183, 184 (fig.)

museau du singe, 4, 94, 110, 111, 120, 123

mysticetes
  evolution of, 56–57, 64–65, 69, 75–79
  modern, 56–57, 69, 75, 80
  See also baleen whales; cetaceans; odontocetes

Mystic Seaport (Conn.), 230, 232, 250, 322, 326

Nalacetus, 66

Nantucket (Mass.)
  Hemingway in, 32–33
  Melville on, 34
  in *Miriam Coffin,* 11
  and whaling industry, 13, 107, 230–232, 237, 250, 273, 305–306, 318
  See also New England whaling

"Nantucket sleigh ride," 36, 114

Nasu, Keiji, 196

Nemoto, Takahisa, 196, 213–214

Netherlands. *See* Dutch whaling

New Bedford (Mass.)
  and Melville, 8
  Pequod from, 329
  textile factories, 327
  whaling industry, 13, 107, 230–232, 233 (fig.), 236, 242 (fig.), 244, 250, 313, 326–327
  See also New Bedford Whaling Museum; New England whaling

New Bedford Whaling Museum, 43, 160 (fig.), 250, 262, 322, 327–333, 328 (fig.), 329 (fig.), 331 (fig.), 332 (fig.), 333 (fig.)

New England whaling, 9–10, 232–233, 233 (fig.)
  accidents and injuries, 249–250, 249 (fig.), 254–255

364 INDEX

catching whales, 252–256, 255 (fig.), 257 (fig.)
Cheever on, 13–14
decline of, 272
in film, 258 (fig.)
heyday of, 250, 318
life on ships, 240–242, 241 (fig.), 244–249, 256, 270
logbooks from, 245–246, 270, 292
right whales in, 228, 230
ship design, 250–252, 251 (fig.)
sperm whales in, 230–231, 237
wages and profits, 242–244, 262–263
women on ships, 244–245, 249
See also American whaling industry; Nantucket
(Mass.); New Bedford (Mass.)
New York Aquarium, 7, 161, 162
New York Times, 20–21, 310–311, 322
Niobrarateuthis, 91
Nishiwaki, M., 141, 216, 281–282
Nordhoff, Charles, 175, 200, 248
Norris, Ken, 120 (fig.)
acoustic theory of, 111–112
on dolphins, 171–172, 317
on odontocete hearing, 115–117
in *The Porpoise Watcher*, 300
on sleep behavior of pilot whales, 131–132
as sperm whale authority, 3
on sperm whale sound production, 120, 121, 122–
125, 196
northern fishery, 234
North Pacific grounds
bowhead whaling in, 234
factory ships in, 271–272, 280
IWC quotas in, 299–300, 302 (fig.)
Japanese whaling in, 27, 213, 238, 278, 280–282,
283–284, 285 (fig.), 288, 290, 293, 295, 304,
318
Soviet whaling in, 238, 278, 280, 284, 288, 295,
299–300, 318
sperm whales in, 268, 269
squid in, 213–214
Norwegian whaling, 10, 273–274, 277–278, 284, 294,
299, 303, 308–310
Nummela, Shippa, 115, 117

Obama, Barack, 310
Octopoteuthidae, 214
odontocetes
evolution of, 58–59, 65, 75–78, 85, 90
hearing apparatus in, 115–119
modern, 56–57, 69, 76, 77, 85, 115
sound production in, 165–166
sperm whale largest of, 80–81
See also cetaceans; mysticetes; toothed whales

Odyssey (floating laboratory), 219–220
Ogawa, Tezio, 173
Ohsumi, Seiji, 97, 98, 145, 146, 238, 281, 281 (fig.), 282,
284
oil spills, 221–223
O'Leary, Maureen, 60, 66
Oligocene period, 59, 65, 75, 78, 79–81, 82–83, 90
Olmstead, Francis, 11, 14, 245–246, 247
Ommanney, F. D., 200, 274, 275–276
Onassis, Aristotle, 281
Ontocetus oxymycterus, 80
onychoteuthids, 214
Orcinus orca (killer whale), 83, 85 (fig.), 95, 100–101,
117, 128, 154–155, 156, 165–166
Orycterocetus quadratidens, 83–84, 87, 91
Osborn, Henry Fairfield, 60–61
O'Shea, Steve, 188–189, 208
Owen, Richard, 63

Pachyaena, 59–60
Pacific giant squid *(Moroteuthis robustus)*, 200,
201 (fig.), 206, 213–214
Packwood-Magnuson Amendment, 285
Page, William, 255 (fig.)
Pakicetus, 62, 65–67, 68, 65 (fig.), 70, 76, 164
Palacios, Daniel, 128–129
Paleocene era, 59
Payne, Katy, 148, 317
Payne, Roger, 148, 219–220, 310, 317
Peck, Gregory, 49–50, 50 (fig.)
Pelly Amendment, 285
Pequod
in *Ahab's Wife*, 37
in film, 49–50
in *Moby-Dick*, 8, 22, 26, 112, 117–118, 127, 234, 268,
329
in *White as the Waves*, 35–36
Perrin, William, 89
persistent organic pollutants (POPs), 219–220
Peruvian beaked whale *(Mesoplodon peruvianus)*,
87–88
Peruvian whaling, 270–271, 303, 307–308
petroleum industry, 233, 235–236, 289, 306, 327
Philbrick, Nathaniel, 11, 18
photophores, 204–205, 205 (fig.), 208
Physeter catodon, 93
physeterids, 81–82, 90
Physeter macrocephalus. See sperm whale
Physty, 1–7, 2 (fig.), 3 (fig.), 5 (fig.), 6 (fig.), 131, 133,
150, 163
pilot whales *(Globicephala melas)*, 85, 129, 131–132,
155, 185–186
pinnipeds, 58, 95

Pitman, Robert, 88–89, 128
plastic. *See* pollution
platanistids (freshwater dolphins), 67–68, 116
politics and whale conservation, 298, 324, 328
Pollock, Jackson, 40
pollution, 218–223
porpoises, 115–117
Prescott, J. H., 131–132, 172
Protocetus atavus, 59, 70
Pyenson, Nick, 57
pygmy killer whale *(Feresa attenuate)*, 128
pygmy sperm whale *(Kogia breviceps)*, 80, 82, 82 (fig.),
    87, 90–91, 222

Queequeg, 8, 9, 36, 46, 47, 48, 49, 53, 111. See also
    Moby-Dick (Melville)

RARE (Rare Animal Relief Effort), 317–319
Ray, Carleton, 134
Read, Bernard, 139–140
Reeves, Randall R., 221
Reiss, Diana, 162–163
Remingtonocetus harudiensis, 67–68
Reynolds, Jeremiah N., 18, 19, 126
Rice, Dale, 62, 94, 136, 200, 206, 214, 286, 291
right whale, 10–11, 96 (fig.), 320
    in cetacean evolution, 78
    depletion of, 235, 298, 307, 321
    hunting of, 228, 230, 256, 292
    in North Carolina fishery, 233
    physical characteristics of, 54, 96, 100
Rodhocetus, 70, 74
Rondelet, Guillaume, 224
Roper, Clyde, 188–189, 205–206, 207
rorquals. *See* blue whale; Bryde's whale; fin whale;
    minke whale; sei whale
Rosa, Rui, 208–209
rough-toothed dolphin *(Steno bredanensis)*, 129, 160

sablefish *(Anoplopoma fimbria)*, 217, 218, 218 (fig.)
Sadove, Sam, 6, 7
Sanderson, Ivan, 197
Sandlofer, Michael, 4
Scaldicetus shigensis, 89, 90, 91
Scammon, C. M., 12, 106
    on New England whaling, 231, 270
    on pilot whales, 31
    on sperm whales, 97, 98, 114, 151, 194–195
    on whaling ships, 252
Scheer, Gene, 34
Scheffer, Victor, 12, 99, 100, 127, 180–181, 225, 291
Schevill, William E., 108, 116, 119, 122, 166, 321–322
schoolmasters, 145, 147

Schultz, Elizabeth, 41, 42–43
Schultz, Howard, 53
"scientific" whaling, 285–287, 289, 303–304, 307
Scoresby, William, 11, 12, 23–24, 28, 233
Scott, Peter, 303 (fig.)
Scott, Timothy, 7
scrimshaw, 39–40, 40 (fig.), 100–101
Sea of Japan, 304
Seibel, Brad, 208–209
Seifert, Douglas, 110
sei whale, 77, 78, 96, 100, 127, 269, 284, 298, 304, 320
Sepia officinalis (cuttlefish), 139, 178, 195, 203
Shakespeare, William, 32, 107
sharks
    and evolution, 56
    and extinction, 87
    as predators, 128, 151, 162
    as prey, 167, 214, 216
    as scavengers, 74, 258, 259, 277
Sherman, Stuart, 245
shore-based whaling, 272, 276–277, 288
Siebert, Charles, 167
Simpson, George Gaylord, 55
sirenians, 56, 69, 72, 158
Slater, P. J. B., 169
Slijper, E. J., 106, 195
Smith, Tim, 290
Smithsonian Institution, 58, 188, 280
South African whaling, 272, 273–277, 275 (fig.),
    276 (fig.), 288
southern fishery, 234
Southern Ocean, 213, 218, 234, 269, 280, 286, 288,
    302 (fig.), 304, 320
Soviet whaling
    in the Antarctic, 293–294, 294–295
    end of, 303, 307
    factory ships, 141 (fig.), 293–297, 295 (fig.),
    296 (fig.), 301
    illegal whaling, 291–292
    IWC and, 290, 294, 298–300, 301
    in the North Pacific, 238, 278, 280, 284, 288, 295,
    299–300, 318
    right whales and, 292
    sperm whales and, 295–298
    See also factory-ship whaling
spermaceti candle, 27, 108, 228, 229, 276
spermaceti oil, 14, 108, 120, 121–122, 229, 229 (fig.),
    263–264, 276, 297. *See also* tryworks
spermaceti organ
    composition of, 111, 120
    of female sperm whale, 113
    in fossil record, 80, 81, 83, 86, 91
    function of, 111–112, 120–122

Melville on, 24
of mini-sperm whales, 82
and sound production, 57, 120–121, 125–126
Soviet processing of, 296–297
See also spermaceti oil
spermaceti whale. *See* sperm whale
sperm whale, 16 (fig.), 79 (fig.), 91 (fig.), 96 (fig.), 285 (fig.)
    African elephant compared to, 147–148, 153, 163
    aggression toward, 128–130
    ambergris, 37–38, 137–142, 139 (fig.), 141 (fig.), 233, 297
    blowhole, 76, 104–105, 107–108, 110, 120, 131, 132, 226 (fig.)
    blubber, 113, 116–117, 126, 161, 178, 219, 228–229, 253, 257–259, 260 (fig.), 263–264, 296, 296 (fig.)
    as bottom feeder, 216–217
    brain, 54, 108, 125, 158–159, 164–165, 188
    breathing mechanism, 106, 110
    breeding behavior, 145, 146, 147–148, 148–153, 167–168
    in captivity, 1, 147, 162, 163
    catch statistics, 27, 271, 272 (fig.), 278, 280–281, 284, 287, 291–292, 294–295, 300
    and deep-sea cables, 126, 196, 216
    diet of, 90–91, 139, 141, 199, 207, 214–217
    dimorphism in, 146 (fig.), 148, 152
    distribution of, 268–269
    diving behavior, 114–115, 210, 216
    echolocation and, 57, 109–110, 112, 121, 122, 125–126
    evolution and, 53–54, 56–57, 78, 87, 90–92, 142
    exploitation of, 312–313
    feeding habits, 85–86, 90, 91, 125–126, 152, 179, 192–204, 210–211
    flukes, 75, 103–104, 105 (fig.), 114, 136 (fig.), 170, 218 (fig.), 254–255
    in fossil record, 59, 81–86, 85 (fig.)
    head, 103 (fig.), 107, 107 (fig.), 112–113, 120
    hearing, 116, 118–119
    hunting of, 23, 27, 98, 107, 108, 178, 239, 281, 283–284, 285 (fig.), 295–298, 304, 318
    jaws, 100–101, 117, 181, 202, 255–256, 270
    life span, 7, 153–154
    in literature, 25, 53, 265–267, 267 (fig.)
    meat of, 106, 282
    migration of, 20, 26, 94, 145–146, 148–149, 153, 269, 279, 281–282, 299
    nose, 54, 90–91, 94, 110–111, 120, 132
    osteonecrosis (bone death) in, 209–210
    play behavior, 170–172
    pollution and, 218–219, 226
    population of, 239, 288–290, 291

predators to, 128, 151, 167
prey, 124–126, 175–182, 176 (fig.), 181 (fig.), 189, 196–200, 204 (fig.), 206–207, 210–211, 213–214
reproductive organs, 111–112
sexual maturity of, 152–153
size of, 54, 80–81, 97–98, 99–100, 99 (fig.), 101, 116, 147–148, 261, 262, 263, 270
skull, 61, 125
sleep behavior, 131
social organization, 134, 145–147, 152–153, 167–170
sound production of, 54, 57, 94, 106–107, 108–109, 119–120, 121–125, 166–167, 188, 196
speed of, 113–114
stranding and, 133, 133 (fig.), 134–137, 135 (fig.), 136 (fig.), 138 (fig.), 227 (fig.), 227–228
See also Melville, Herman; *Moby-Dick* (Melville); spermaceti oil; spermaceti organ; sperm whale teeth
sperm whale teeth, 101 (fig.)
    to estimate age, 154
    and feeding, 57, 86, 101–102, 124, 152, 193, 199
    in fossil record, 82–86, 84 (fig.), 85 (fig.)
    in lower jaw, 81, 94, 100–102, 102 (fig.), 179
    and upper jaw, 80, 81, 91, 101
    See also scrimshaw
square-rigged whalers, 10, 26, 36, 233 (fig.), 235, 236–237, 250, 271, 293, 318
squid
    bioluminescence and, 189, 202, 203, 204
    hunting behavior of, 203
    limited knowledge of, 207
    as sperm whale prey, 124–126, 175–182, 176 (fig.), 181 (fig.), 196–200, 206–207, 204 (fig.), 210–211, 213–214
    See also *Architeuthis* (giant squid); *Dosidicus gigas* (Humboldt squid)
Starbuck, 8, 9, 49, 53. See also *Moby-Dick* (Melville)
Starbuck, A., 114, 138, 231, 256, 261, 270
Starbuck's Coffee, 53
steam whalers, 235, 236
Stella, Frank, 40
Straley, Jan, 217–218, 218 (fig.)
stranding behavior, 132–137, 135 (fig.)
Strombus gigas (queen conch), 316 (fig.)
Stubb, 8, 9, 49, 53, 234, 254. See also *Moby-Dick* (Melville)
Styx (whaling ship), 247, 259
suction feeding, 85–86
Sumich, J. L., 77–78
swordfish, 95, 98, 109, 127, 220, 265

Taningia danae, 204–206, 205 (fig.)
Tashtego, 8, 9, 45, 53, 111, 263

Temple, Lewis, 253, 327
Thewissen, J. G. M., 58, 66, 68–71, 72, 81–82
Tønnessen, J. N., 273, 282
toothed whales
    brain of, 164–165, 172–173
    echolocation and, 191–192
    evolution of, 56–59, 75–81
    sperm as largest of, 54, 116, 147–148
    squid as prey to, 57, 81, 118–119, 198
    See also cetaceans; odontocetes
Townsend, Charles Haskins, 237–238, 269, 287–288,
    291, 292–293
trying out. *See* tryworks
tryworks, 228–229, 250–251, 256–259, 260 (fig.),
    261–262, 261 (fig.), 263–264
Twenty Thousand Leagues Under the Sea. See Verne,
    Jules
Tyack, Peter, 169, 170
Typee (Melville), 8, 11, 32

Uhen, Mark, 66, 73, 165
Union Whaling Company (South Africa), 214–215,
    274, 275 (fig.), 276–277
United Nations, 276, 300, 307, 317–318

Van der Gucht, Estel, 172, 173
Van Valen, Leigh, 59, 62, 75
Vecchione, M., 205–206
Verne, Jules, 265–268, 267 (fig.)
Verrill, A. Hyatt, 138
Villiers, Alan, 49
Vincent, Howard, 9, 12, 19, 20, 23, 26, 30, 32

Wanderer (whaling ship), 233 (fig.), 236, 293
Watkins, W. A., 115, 119, 166
Watson, Lyall, 291
Weaver, Raymond, 21
Wehle, Richard, 325–326, 326 (fig.)

Welles, Orson, 35, 49
Werth, A. J., 85
Westergaard, Peter, 34
Western Interior Seaway, 91
whalebone whales, 59, 269, 270
whale fossil record, 58–59, 62–71, 65 (fig.), 68 (fig.),
    73–75, 77–86, 79 (fig.), 85 (fig.), 164
whale meat, 282–283, 286–287
whale phylogeny. *See* evolution
whale shark *(Rhincodon typus)*, 86
Whale World Museum, 101 (fig.), 280
whaling grounds, 149, 231–232, 234, 263, 268–269,
    271–273, 281–282, 287, 293, 306
White, Thomas, 164
Whitehead, Hal, 144 (fig.)
    on cetacean culture, 166–170, 172
    on cetacean order, 311
    on the spermaceti organ, 112, 113
    sperm whale population estimate of, 239
    sperm whale research of, 98, 128, 131, 143–145,
        147–148, 151, 153, 211
    in *Voyage to the Whales*, 143–144
Williams, Eliza Azelia, 244–245
Wolman, A. A., 286
Wood, F. G., 133, 317
Woolf, Leonard, 31
World Wildlife Fund, 143, 303 (fig.), 317, 318
Wray, J. W., 182
Wu, Tony, 110

Yablokov, Alexei, 123–124, 195, 291–292

Zanzibar Grounds, 232, 263, 268, 273
Zeidler, Wolfgang, 206
zeuglodonts, 58–59, 63
ziphiids, 73, 83, 87, 89–90
Zygophyseter varolai, 83, 91
Zygorhiza, 73